Information Systems Engineering

Springer
London
Berlin
Heidelberg
New York
Barcelona
Hong Kong
Milan
Paris
Singapore
Tokyo

Sjaak Brinkkemper, Eva Lindencrona
and Arne Sølvberg (Eds)

Information Systems Engineering

State of the Art and Research Themes

Springer

Sjaak Brinkkemper
Department of Computer Science, The Norwegian University of Science and Technology, Trondheim, Norway

Eva Lindencrona
Swedish Institute for Systems Development, Kista, Sweden

Arne Sølvberg
Baan Company R&D, Barneveld, The Netherlands

ISBN 1-85233-317-0 Springer-Verlag London Berlin Heidelberg

British Library Cataloguing in Publication Data
Information systems engineering : state of the art and
research themes
1.Systems engineering 2.System design
I.Brinkkemper, Sjaak II.Lindencrona, Evan III.Solvberg,
Arne
004.2'1
ISBN 1852333170

Library of Congress Cataloging-in-Publication Data
A catalog record for this book is available from the Library of Congress

Printed in Great Britain

Typesetting: Camera ready by contributors
Printed and bound at the Athenæum Press Ltd., Gateshead, Tyne and Wear
34/3830-543210 Printed on acid-free paper SPIN 10766886

Preface

This book is a collection of 24 State-of-the-Art contributions in Information Systems Engineering. It was compiled on the occasion of the retirement in February 2000 of Prof. Dr. Janis A. Bubenko jr. from his Professorship at Kungl Tekniska Högskolan (The Royal Institute of Technology) in Stockholm, and from the University of Stockholm. The papers were presented at the Information Systems Engineering Symposium in Stockholm on 5 and 6 of June 2000. The symposium was organised as a pre-conference event to the CAiSE'00 conference.

The contributors were invited among the many friends and colleagues of Janis, in Sweden and abroad. During the last 40 years Janis has made many, many friends in the international research community of his chosen fields of study, in databases and information systems engineering. To select whom to invite to contribute has not been easy. People have been more than eager to contribute. The friends are so numerous that it is not possible to avoid offending many that rightfully may feel that they should also have been invited.

Janis did not know about this initiative. The initiators and editors of this book are Sjaak Brinkkemper of Baan Company R&D, Eva Lindencrona of the Swedish Institute for Systems Development (SISU), and Arne Sølvberg of the Norwegian University of Science and Technology (NTNU). The selection of contributors is ours. Given the limited size of a book we have selected as best we could. We have invited friends from different realms of Janis' scientific life, from the domestic scene, from the Nordic scene and from the wider international scene. Some of the invitees have co-operated in writing their contributions, some have contributed together with their students, and some have done it alone. We hope that the blend is satisfactory.

When planning the book we decided that we would go for a "free-for-all" attitude in deciding on their own themes, instead of asking for contributions of particular kinds. Nevertheless the papers have come out fairly coherent, to be naturally divided in three parts which we have called (I) Information society, (II) Approaches to Information Systems Engineering, and (III) Concepts for Information Systems. Knowing the character of Janis' scientific interests this division should come as a surprise to nobody

We are very pleased with the totality of the book that we have had the good fortune of editing. We wish to thank Rebecca Mowat of Springer Verlag, London, for her efficient support in the publication of this book. Benkt Wangler and Lars Bergman provided generously the hospitality for the symposium during CAiSE'00.

It is an honour to be allowed to present this gift to Janis on the occasion of his retirement, as a token of gratefulness for his contributions to IT research.

Sjaak Brinkkemper Eva Lindencrona Arne Sølvberg (editors)

June 2000

Biography of Janis A. Bubenko jr

Janis A. Bubenko was born in 1935 in Riga, Latvia. His family immigrated to Sweden in 1945, where Janis went to school and university. He received a MSc in civil engineering (Chalmers Univ. of Technology, 1958), a Licentiate of Technology in Structural Mechanics (Chalmers, 1965), and a Ph.D. in Information Systems (Royal Inst. of Technology, 1973). He was a professor of Computer and Systems Sciences at University of Gothenburg and Chalmers Univ. of Technology 1977-81, and at the Royal Institute of Technology and University of Stockholm (KTH/SU) 1982 - 2000. He retired in February 2000. During his years as professor Janis Bubenko guided and promoted a large number of Ph.D. and licentiate students to their degrees. He developed and taught numerous University courses and inspired generations of students.

After four years as manager of systems and programming at Univac Scandinavia in the early 1960s, Janis returned to research in 1965. In 1969 he founded the CADIS (Computer Aided Design of Information Systems) research group at the department of Computer and Systems Science, KTH/SU in Stockholm. In CADIS he designed an associative database management system CS1, which later was developed into a commercial Swedish product DREAM/CS5. In 1980, Janis founded the SYSLAB research laboratory which is still active. SYSLAB conducts research in information systems, requirements engineering, business engineering, databases, and in software engineering. At SYSLAB Janis was, among other things, in charge of the development of an initial version of a meta-case tool, which was later developed and deployed in a number of projects for CASE tool development. In 1984 he initiated the establishment of the Swedish Institute for Systems Development, SISU, and was its Managing Director from 1985-92. SISU was supported by the public research board STU/NUTEK, and by a number of Swedish industries and organizations. In the late 1980s, Janis initiated SISU's participation in a number of European Community projects such as KIWIS, TEMPORA, MILORD, NATURE, and F3.

Janis is the author/co-author or editor of eight influential textbooks in the areas of Information Systems Development Methods, Performance Analysis of Data Processing Systems, Operating Systems, Databases, and Conceptual Modeling Methods. He is the author/co-author of more than 140 research reports and published articles. Current research includes Requirements Engineering and Enterprise Modeling methods, as well as Information System modeling and design issues. He has participated in leading positions in the ESPRIT projects KIWIS, TEMPORA, F3, NATURE, and ELEKTRA.

Janis Bubenko is one of the internationally leading scientists in the field of Information Systems. He was the general chairperson of IFIP TFAIS (Theoretical and Formal Aspects of Information Systems) working conference 1985, Sitges, CAISE'90 Stockholm, EDBT'94 in Cambridge, U.K., and IFIP 8.1/13.2 Joint Working Conference on Domain Knowledge for Interactive System Design, Geneva, 1996. He was program co-chair of the VLDB (Very Large Data Bases) Conference in 1978 in West Berlin, of CAISE'91 (Conference on Advanced Information System Engineering) in Trondheim, the Baltic DB'94 workshop on National Infrastructure Databases in Vilnius, ADBIS'95 in Moscow, and WITS'98 in Helsinki. He has been Programme Committee member of more than 50 international conferences and is a member of the Editorial Advisory Boards of the Journal of Information Systems, the Information Systems Journal, Data and Knowledge Engineering, the International Journal of Intelligent and Co-operative Information Systems, and the Journal of Information Technology and

Management. He has been employed a reviewer for the National Science Foundation of the United States, the European Commission, and other funding organizations in UK, Canada, Australia, Switzerland, Latvia, and Austria. Bubenko was vice president of the VLDB Endowment from 1985 - 1989, and was its president 1990 - 1993. He is a member of ACM, IEEE Computer Society, and IFIP Working Groups 8.1 and 2.9. He is also member of the board of the EDBT Foundation (Extending Data Base Technology).

Janis Bubenko is married to Karin Lindquist, who is retired from her job as a Medical Doctor working for an insurance company and a Stockholm hospital. Janis and Karin have two children: Mikael, married to Caroline Varnauskas, and Anna, married to Per Engström, and four grandchildren, named Marcus, Clara, Sara, and Thomas. They enjoy living in Danderyd, a small suburb in the north of Stockholm, while spending their time with listening to music, especially chamber music, and playing golf. However, through writing, presenting and debate, Janis will most certainly enjoy being involved in advancing the state-of-the-art in Information Systems Engineering for years to come.

Table of Contents

I. Information Society

II. Approaches to Information Systems Engineering

III. Concepts for Information Systems

I.

Information society

Imperfection and the Human Component: Adding Robustness to Global Information Systems

Peter C. Lockemann, Gergely Lukács
Fakultät für Informatik, Universität Karlsruhe
Karlsruhe, Germany

1 The Uncertainty Principle

A – or perhaps the – dominant influence on modern society is the convergence of computing and communication technologies. It fosters communication between people, and access to computing and information resources, no matter where, when and by whom. A dense network of fixed-wire communication links, the enormous spread of mobile phones and the all-pervading services of the Internet produce a mixture that has the potential of keeping everyone on an equal level of knowledge, and of making every piece of publicly available information equally accessible. Sure, there are some problems still to be overcome, like being able to withstand the flooding with information, or protecting one's privacy. But a suitable infrastructure seems just over the horizon, one that allows a more discriminating access to information according to one's present needs, and that provides better transparency of what can be obtained from what source at what price. And protection measures ranging from smart cards to encryption to firewalls are on their way to ensure privacy. So all in all, we are on our way to what is called the "information society" or, even better, the "knowledge society".

We have to recognize, though, that due to large-scale distribution global information is inherently imperfect. An insider within the system will never be able to observe the global state all at once, but will have to piece together any state from a sequence of observations spread over time. Hence, there is something to distributed information systems what we may call the "uncertainty principle in information systems". The same is also true for any changes administered to the global information base: It takes time to spread across the network. Hence, neither will one ever achieve a globally consistent state of the information base, nor could one observe one if there ever was one.

2 A first hypothesis: imperfection breeds robustness

The first impulse is to view imperfection as something inherently harmful, to be avoided at all cost. Indeed, this is true if we think of banking transactions. But think instead of a travel advisory announcing a traffic congestion a mile ahead together with the official detour. If everyone takes the detour all that will have been achieved is inducing a second congestion. The example already gives a notion of what we refer as the *robustness of a system*: It is the capability of a system to spread a local disturbance across the system such that disturbances elsewhere remain negligible. We call an information system *robust* if it provides the information to its users such that disturbances in the real system are kept to a minimum.

Apparently, we achieve robustness if not every user takes the same action. Being able to take different actions implies a certain amount of latitude for opportunistic decision making by the individual user. In turn, latitude is provided by a lack of certainty, i.e., by an imperfect state of knowledge. The challenge, then, is to provide the user with exactly the dose of imperfection such that, by optimizing his own actions, he contributes to the collective robustness of the system. It seems that this is what mankind has always done, to draw on experience, expertise and the advice of others to judge the situation and thus obtain the leverage to adapt to changing environments.

Consequently, robust information systems should incorporate a good deal of human judgement. We draw two conclusions. One is that if the system can only provide imperfect information it should only take limited, well-circumscribed attempts to reduce the imperfection. Second, if the system is sure of the perfection of the information, it may very well add spurious imperfection ("disinformation") to each individual information passed along.

Our hypothesis, then, is that global information systems should evolve in a way that they assist human decision makers rather than replace them by algorithms that only work in a consistent and perfect environment. Future information systems must become an essential link to the robustness and adaptability of the information society.

3 A scenario

To support our hypothesis we consider a traffic scenario taken from an interdisciplinary project in which the authors currently participate. The project is motivated by the observation that the current centralized monitoring and advisories to highway traffic – drivers of private and commercial vehicles – have reached their limits which, as everyone will readily attest to, have done little to relieve congestion. There are two major reasons: To the extent that the control has indeed improved the situation, it just has induced new traffic, and where the system has given advisories it has given the same advisory to all drivers with the result of just shifting the congestion to new locations. The basic premise of the project is to replace centralized decision making with local decision making by drivers, by giving them relevant information but leaving much of the response to their own experience, knowledge and observations of local circumstances, communication with others nearby or individual preferences.

This requires vehicles to be equipped with electronic means such as sensors for, e.g., road conditions, distance and relative location of nearby vehicles, a GPS system for determining the current location, robust mobile radio links to other vehicles or stationary communication devices such as satellites or cellular senders and receivers. Vehicles also carry processors for analyzing sensor data, and for executing transmission protocols. In order to provide assistance functions they carry more powerful PC-like processors, often referred to as "personal travel assistants" (PTA). These PTAs are capable of interpreting observed situations and communicate them to other partners in the information system. Equally important, they process data received from a variety of sources in the information system to prepare them for presentation to the driver or to execute some sort of automated control for, e.g., automatic convoys. Together with the processing capabilities go storage facilities that allow to collect data for subsequent analysis, to keep histories of the past or to maintain material whose repeated transmission would strain the system, such as roadmaps among which suitable excerpts can be selected on the basis of the current location.

Given all these facilities, where does inherent imperfection come into play? A major source is lack of timeliness. For example, vehicles may temporarily be disconnected from the communication links so that they cannot provide information, or even worse, there will be many vehicles on their way that are not suitably equipped at all. Drivers may not immediately realize the importance of an observation, or not be in a situation where they can communicate it right away. Centralized systems may not recognize the slow evolution of congestion, and lack the means to predict where there is a latent danger of congestion, what length it will have over time and how long it will take to dissolve. Information on road construction, temporary traffic lights or detours seems notoriously out of date. For maps, drivers will rely on their local CDs or on geographic information systems, and these maps have a tendency to be updated only over longer periods of time. A second source of imperfection is non-availability of information that may prove relevant only in certain situations. Given a map so that a driver may choose his own detour, there may be obstacles not shown such as poor road conditions, speed limits, weight limits, environmental regulations, that the driver will only detect when he runs into the traffic signs. A third source is lack of predictability of the behavior of other drivers. Given a choice of detours, is there any way to predict how traffic will evolve on any of them, i.e., what choice other drivers will take? A fourth source is lack of knowledge of one's preferences by the system, and the lack of predictability of one's own behavior. For example, when the traffic situation becomes known, the driver may decide to take a break or modify his destination. In the commercial world of freight hauling, the route may have to be rescheduled by the company's freight loading disposition on the basis of an incoming stream of new orders.

Now take a driver under these conditions of imperfection. His PTA may flash a warning that some 100 kilometers ahead on his way traffic density is such that congestion is a real possibility. Should he take a different route? He may inspect a map for that purpose. Which route among several that are possible? Can one predict how conditions will evolve on each of them? Depending on the answers that the driver receives from a central source that does appropriate simulations, he will inquire with a local road maintenance database whether there are additional conditions that may have an effect. After settling on a first choice, the driver may become curious whether other vehicles within a radius of, say, 10 kilometers, received the same warning and are in a similar quandary. Can these drivers somehow coordinate their activities so that they ultimately spread across several of the alternative routes? Or the system may itself come up with a recommendation, but it may consist of several alternatives with different probabilities with regard to expected minimum, average and maximum travel time.

Robustness under these circumstances means leaving the drivers "somewhat in the dark". That is, they should be aware of, or deceived into believing, the imperfection of the information. However, robustness also means that there should be some limit to the perceived imperfection in order to avoid that the system slides into a chaotic state.

4 A second hypothesis: imperfection reduces vulnerability

As the information society comes more and more to rely on the all-pervading information network, it also becomes more dependent on the proper functioning of this network and more vulnerable to its failures. Certainly, the first line of defense is making the system technically as reliable as possible. But not all technical failures can be corrected right away. Rather than shut down the system altogether at this point, it makes better sense to continue wherever possible and treat the consequences of the failure as some kind of imperfection. Now, take the uncertainty principle in information systems as explained in Section 1, and it becomes clear that imperfections due to failures or mistakes and inherent imperfections become indistinguishable to the system user. In other words, by getting users accustomed to the idea of imperfection we broaden the scope of situations covered by a robust information system.

It seems just natural to extend the principle towards immunity against malign manipulations and intrusions. Altogether then, if one accepts that global information is imperfect one will also have to accept that there is no way one can distinguish whether imperfection is due to inherent reasons alone or also to corruption by mistakes, attacks and manipulations. Hence, the same robustness measures that allow people to live with inherent imperfection should also allow them to cope with misinformation.

To return to our scenario of Section 3, assume that the traffic warning has been flashed by an intruder with the purpose of ridding a certain stretch he wishes to have all for himself from all vehicles. While the other drivers will probably respond in the intended way, less damage is done because no congestion builds up elsewhere. Or instead, drivers may simply not believe the warning because it goes against their own experience.

To summarize, information systems built on the premise of imperfection should exhibit overall robustness. And since dealing with the content of data rather than just their forms is the domain of information systems, it falls to information systems research to devise clever means for protecting against unreliable content.

5 Imperfection

Imperfection is a very general term. As noted in the previous sections, it may have to do with the time it takes for changes to spread across the system, and with local

system failures. It may also have to do with a user's perception of the system state. Consequently, we should take more care to distinguish various ways in which imperfection manifests itself. By doing so, we also hope to gain a solid scientific basis so that there is some objectivity and rigor to whatever advice the global information system gives to the human decision makers. Unfortunately, there are neither generally accepted, established categories, nor a corresponding vocabulary for the different types of imperfections. However, in a recent survey Bosc and Prade give a classification of imperfection in data management for information systems [1] which we regard as very useful.

- *Uncertainty*. Uncertainty refers to the lack of sufficient information about the state of the world, i.e., a statement is not known to be true or false. One example for an uncertain statement is "I will travel 40 minutes to work today", because travel-time is influenced by many unforeseeable factors. Another is "The next highway interchange is congested" where factors like traffic density on the intersecting highway have an influence. Even though full information is not available on the truth of the statement, we may have some information about the tendency of the statement to be true.
- *Imprecision*. Imprecision refers to the content of the statement with respect to the granularity of the language used. E.g., the statement "I traveled 35 minutes today " is precise, while the statement "I traveled between 40 and 60 minutes today " is imprecise. Imprecise information includes disjunctive information (e.g., "He took either A5 or A61"), negative information (e.g., "He did not travel by car"), range information (e.g., "Travel time is between 40 and 50 minutes") and error margins (e.g., "Travel time is 45 plus/minus 5 minutes"). Imprecision may have dire consequences, for example if a driver knows that the traffic jam ahead has a length of between 3 and 4 kilometers and, hence, the optimal decision, whether to take the next exit or stay on for the following exit, is uncertain.
- *Vagueness*. Vagueness refers to gradual predicates that evaluate to a certain degree of truth instead of only true or false. As an example, "I am close to a highway exit" is a vague statement, when the predicate "close" is gradual, as it is meant most of the time. Vague predicates may have a context dependent meaning. The word "close" will have a different meaning in the statements "My job is close to my home" and "My weekend house is close to my home". A statement that is not vague is often referred to as a "crisp" statement.

To distinguish between types of imperfections is more than an academic exercise. First, they may differ as to their best use. For example, vague predicates are close to human thinking, and are therefore especially adequate for man-machine communication. Uncertain and imprecise predicates are candidates for system internal representations although they may also occur in man-machine communication.

As one consequence, the three types of imperfections may occur in combination. Take as an example the statement "The loading docks of my customers in town X are close to my position". It is both imprecise and vague. First, suppliers and their loading docks are located at distinct places in town X, and consequently, they do not have the same distance, second, "close" is a gradual predicate with a vague meaning. As another example, the statement "There is a high chance that there will be a traffic jam today." is uncertain, as we are not sure if there will actually be a traffic jam. The level of uncertainty in turn is expressed vaguely, "high chance" does not have a crisp meaning.

As a second consequence, processing imperfections may have to combine several mathematical models. Mapping between the models may be possible, though. For example, uncertainty and imprecision are closely related. An imprecise statement can typically be converted into a set of uncertain statements. E.g., the statement "I traveled between 30 and 35 minutes" can be converted into a disjunction of five precise statements: "I traveled 30 minutes.", "I traveled 31 minutes." etc. Each of the statements is uncertain, but it is known that one of the five statements is true.

6 Representing imperfection

Presumably, the success in processing imperfections rests on the availability of adequate mathematical models. There are a number of them, which differ along criteria such as types of imperfections which can be expressed, the operations for combining and updating knowledge, the consistency checking of available knowledge, or the computational complexity of deriving conclusions [3].

6.1 Missing and disjunctive values

When the value of a data item is not perfectly known, the simplest approach is to ignore all partial information that may be available. The value of the data item is then declared to be missing. Another approach is to handle a set of possible values for the data item. The value of the data item is known to be in this set, but there is no information available to favor any value over any other. This kind of information is called *disjunctive values*. A number of types for missing and disjunctive values have evolved over time to naturally describe uncertain and imprecise data.

6.2 Probability theory

A probability is a number expressing the likelihood that a proposition is true or an event occurred. The probability of a proposition or an event is in the interval [0,1], where 0 means almost certainly false and 1 means almost certainly true. Probabilistic variables, i.e. variables the value of which can be expressed by probabilities, are usually described by probability distributions, e.g.,

$$P(t<30\ min) = 0.3$$
$$P(30 \leq t < 40) = 0.4$$
$$P(40 \leq t) = 0.3$$

A typical interpretation of this probability distribution is that in 30 % of the time the travel time will be less than 30 minutes, in 40 % of the time between 30 and 40 minutes and in the remaining 30 % of the time over 40 minutes.

Probabilities can be viewed in three major different ways. In the *propensity view*, probability is a physical property of something, e.g., the tendency of a particular dice to come up 6, knowing that the dice is symmetric. In our scenario, the propensity view will have no place. In the *frequency view*, probability is a property of a collection of similar events and, hence, a matter of neutral observation independent from the human observer. In our scenario, it would be the result of the system's collection of data by electronic sensors, e.g., the traffic load distribution on stretch of highway over the 24 hours of a weekday. In the *subjective view*, probability expresses a person's confidence based on previous experiences and current knowledge, that a proposition is true or an event occurred. This view may enter the information system to the extent that it collects messages and observations from drivers. And it is certainly the view that drivers will apply when deciding on their next actions.

Probability theory fits naturally for data with imprecision and uncertainty type imperfections, if one wishes to associate numeric measures with them. Note that uncertainty and imprecision can often be traded against each other. Take a probability density function of travel time. On the basis of this, several imprecise statements – which are also uncertain – can be made. E.g., "Travel time is between 45 and 55 minutes with 0.6 probability", or "Travel time is between 52 and 55 minutes with 0.1 probability".

There exist several representations that have their basis in probability theory. One of the best known is Bayesian networks [4]. They give a description of probabilistic variables and contain a qualitative and a quantitative part. The qualitative part describes which variables are (in-)dependent in a probabilistic sense. The quantitative part consists of conditional probability distributions of the variables. If the values of some of the probabilistic variables are known, the network allows one to draw conclusions on the values of the other variables.

Another, the Dempster-Shafer theory of evidence [5], is suitable for describing a special type of uncertain and imprecise information, one that results from multivalued mappings. A typical example for the application of the theory is the "unreliable witness": Our evidence of a particular fact stems from an unreliable witness, a witness who is known to make mistakes with a certain probability. Dempster's rule of combination allows combining pieces of such information, e.g., in the above example the pieces of information stemming from several unreliable witnesses observing the same fact.

6.3 Fuzzy set theory

The basic concept of fuzzy set theory [6] is the fuzzy set. A fuzzy set F is a set of elements, where each element has an associated grade of membership, meaning the degree to which the element belongs to the set. Grades of memberships are taken from the [0,1] interval, where 0 means that the element is not a member of the set and 1 means that the element is fully a member of the set. An example of a fuzzy set is (using a common notation):

$$near = \{10/1.0, 15/1.0, 20/1.0, 25/0.8, 30/0.6, 35/0.4, 40/0.2, 45/0, 50/0\}$$

This means that, e.g., 10 (minutes) belongs to the set "near" with 1.0 grade of membership, while, e.g., 30 (minutes) with only 0.6 grade of membership

Operations on sets, e.g., union, intersection, have been extended for fuzzy sets. Note, however, that the number of possible correct fuzzy extensions of these operations is infinite.

Fuzzy set theory suits naturally for vague data. It is typically used in connection with humans, either to communicate information to human users in an easily interpretable way, or to acquire information from human users. As an example of the latter, take a driver who reports to other drivers that he has run into a traffic jam that appears “rather long”. Such vague information has nonetheless its value for information systems, since human knowledge for controlling complex, nonlinear systems – and traffic systems are of this kind - is often collected by means of fuzzy statements, which have an intuitive meaning, and are easier to tune than complex mathematical models.

7 Limiting imperfection

Imperfection, particularly uncertainty and imprecision, can be regarded as a somewhat constrained n-dimensional space within which we know one point to be true. An earlier example for such a constraint was a numerical interval. In general, though, constraints need not just be numeric but may take the form of a proposition such as a first-order predicate logic formula. One could then use deductive means to determine whether a given situation or inquiry satisfies the constraints, i.e., falls within the space circumscribed by the constraints.

Constraints could thus play the role of a plausibility check. On data input constraints could act as some kind of filter. In its assistance to the decision maker, the system could check whether computed data satisfy the constraints. If the decision maker is to be informed of the result, a bit more is needed than just a note that constraints are being met or not. Therefore, classical logic should be extended to multi-valued, modal or probabilistic logic.

Of particular importance to our scenario seem statistical constraints. They detect data items that are potentially incorrect, provided we can somehow describe what we mean by correct or incorrect in a statistical sense [7]. If the probability of such an item falls below a certain threshold, the system should treat it as a so-called “outlier”, interpret it as a “suspiciously looking” item, and present it to a human decision maker for further evaluation. Dealing with the outlier problem holds the best promise for detecting manipulations.

Like traditional constraints, one should be able to describe statistical constraints in a declarative way. Examples are linear and nonlinear regression models for numerical data and decision trees for categorical data [8].

8 Managing imperfection

8.1 Querying imperfect data

We expect that queries contain a heavy dose of vague query conditions representing vague user preferences, and similarly, answers do. In the database itself, imprecision and uncertainty will be the dominant forms of imperfection. Nonetheless, all three may occur in combination in all, queries, answers and stored data. We claim that the type, the size and eventually the source of the imperfection should be handled explicitly such that the answer to a query contains these aspects, and the user has a clear perception of how and where the answer is imperfect. Only in this way will the user truly participate in the decision. Answers should be given either on a high level, where imperfections are aggregated appropriately to allow a quick overview, or on a low level explaining details. A set of tools has to be available to the user to help him understand the nature and size of imperfections.

We gained first experience from a construction database for architects [9]. An object-oriented model was extended by fuzzy data to describe possible design alternatives and actual design decisions. The user – the designer – can express his preferences by means of imperfect constraints. The system keeps track of the actual level of consistence of the design with the design constraints and gives appropriate support to control the level of consistency.

A geographic information retrieval system is the case study of the ongoing Ph.D. research work of the second author of this paper [10]. The goal is to query digital libraries with respect to travel and transportation possibilities. Here, the information system has only an imprecise and uncertain model of the world because, e.g., the geographic references, the addresses of the data objects in the digital libraries are incomplete, or because the geographic databases used to interpret the addresses are of limited resolution. Also, vague user preferences are supported to improve user friendliness.

8.2 Modifying imperfect data

Two types of modifications are defined in [11]: update and revision. Update, or change-recording update, refers to changing the data, the model of the world, in response to a changing world. Revision, or knowledge-adding update, refers to a modification of the data that reflects new knowledge on a world, which itself is static. Both modifications may play a role in our context. E.g., the user's preferences, expressed by vague query conditions, may change, or our knowledge of the world may become more certain and precise.

The design database in [9] supports creating new worlds, so the modification operations differ slightly from both of the above two concepts. There are design constraints and imperfectly described designs. The level of consistency of a design describes the level of satisfaction of the design constraints. The user has direct control over the acceptable level of inconsistency. Two modification operations are supported: expansion and revision. Expansion means adding new design constraints to the existing ones. This may result in a decreasing level of consistency. Revision means adding a new design constraint to the previous ones and removing other design constraints so that the acceptable level of consistency is maintained.

In the ongoing Ph.D. work, revision of available data in the sense of [11] is supported. Statistical data is used to extend known data, and a framework is built to support gathering new pieces of data on the world. This framework consists of two parts: informing the user and allowing him to enter new data. The first consists of identifying and prioritizing important attributes with unknown values, and giving information on what improvement is expected if some of the values become known. The second allows the user to set the values of the identified attributes. Once a new attribute value becomes known, the system updates the query result.

9 The scenario revisited

Our scenario impresses by its sheer size in data volume and distribution. Data are stored across a distributed network of sources. There are a few interconnected centralized databases such as geographic databases, road condition databases, road maintenance databases, traffic pattern databases, and a number of other centralized systems for weather prediction, traffic prediction, which use comprehensive computation and simulation models. In addition, there are a huge number of local databases, notably on the automobiles but also for local traffic control. They will periodically send statistical aggregates to the centralized databases, and they will also send individual items if these seem to be of more than local interest. It makes sense to associate metadata with them that characterize the kind and degree of their imperfection. If a driver queries the system, or the system communicates on its own with a number of vehicles, data may have to be drawn from several of the distributed repositories. The system would then have to judge what further imperfections arise with the transmission and collection.

Which of the aforementioned techniques would be useful? Road capacities, determined by traffic safety and environmental criteria, can be best described by constraints. However, these constraints are vague, ensuring some flexibility, rather than strict: there is some optimal limit, but slightly higher values may also be acceptable. User communication can be improved by supporting vague query conditions and answers. Geographic databases may be on different scales and may thus vary in their degree of precision. Traffic patterns are best expressed in terms of probabilities. Traffic and weather predictions are more in the direction of imprecision and uncertainty, which could again be translated into probabilities. Clearly then, even when using the same mechanisms, e.g., probabilities, there may be different interpretations among which to choose in the context of a given query.

10 Challenges

How well is present information systems technology prepared to support human decision making in such a way that there is a beneficial division of responsibilities between information system and decision maker? In other words, how well does the technology deal with imperfect data in a systematic and theoretically sound way? We claim that it still has a long way to go and that there are numerous challenges left to the scientific community.

Since our scenario draws on large repositories, one set of challenges can be expected for database technology. Indeed, there has been considerable interest in the recent past to extend the relational and the object-oriented data models by various representations for imperfect data, foremost fuzzy theory and probability theory [12,13]. The emphasis is still on storage and direct retrieval, so that extensions hardly go beyond simple operations such as selection and join. Decision support in general and our scenario specifically require more ambitious processing, though, e.g., functions such as aggregation or grouping. Support for combined representations of imperfect data is also a neglected issue so far. Also, performance has attracted little attention even though processing of imperfect data is known to be highly compute-intensive. First discussions for fuzzy databases are found in [9]. Performance remains a difficult issue for the processing of constraints even though (simple) consistency constraints have been known in databases for a long time. Query planning and processing in a distributed environment with high communication cost and unreliable connections remain open topics as well.

Another set of challenges has to do with the integration of several types of imperfections and representations. Because replying to queries may involve combining data from numerous sources, one needs to develop a comprehensive framework for handling imperfect data with several types of representations.

A further central issue is characterization of the kind and degree of imperfection with the data sources. Obviously this is a problem of scalability: With the huge number of data sources one has to find the means to derive the characterizations without too much human intervention, i.e. largely by automatic means. A first attempt at concepts that model the reliability of information sources can be found in [14] as the Information Source Tracking Method (IST-Method). One may argue that, after collecting data over a longer period of time, much of the needed knowledge is hidden in the amassed data. Hence, one should be able to derive probability distributions from the databases, and by applying data mining techniques one should also be able to determine constraints, fuzzy rules and dependency rules. Some of these rules could then be employed while the system generates answers or messages, for example by reducing the imperfection that is encountered with the raw data.

A mechanism to guarantee robustness of database systems is transactions. They guarantee robustness by preserving the integrity of databases. However, their definition of robustness relies on the strict agreement (consistency) between the database and the real world, and they try to isolate system disturbances rather than thin them out across the system. What is needed are transaction models that permit a more flexible definition of what is meant by robustness. Modern database systems already offer a small degree of relaxation in the form of "dirty reads". For the purpose of improved concurrency new operations have been added, which define the consistency of so-called "hot spots" (frequently accessed data items) in terms of numerical intervals. A more recent approach to controlled inconsistency is epsilon-serializability [15], which requires a metric space for the relevant attributes.

As a final challenge, our hypothesis requires the introduction of controlled imperfection even in situations of precision. In general, then, how much imperfection is to be added to or removed from an answer in order to provide the right level of latitude for the decision maker, and what kind of mechanisms are to be used?

In summary, if our hypothesis is true that imperfection holds the clue to the robustness of information systems and thus to the proper functioning of the information society, then information systems technology still has a long way to go to supply the technical means. Our conjecture is that robustness of information systems offers a whole new area of scientific endeavor in informatics. What we hope for is that enough researchers are willing to meet the challenges head on.

Acknowledgement: Gergely Lukacs is partially supported by the project OTKA T 030586 and the KAAD. We would like to thank Gerd Hillebrand for his careful reading and thoughtful comments and suggestions.

References

[1] Bosc P, Prade H. An introduction to the fuzzy set and possibility theory-based treatment of flexible queries and uncertain or imprecise databases. In [2], pp 285-324

[2] Motro A, Smets P. editors. Uncertainty management in information systems. Kluwer, 1997

[3] Walley P. Measures of uncertainty in expert systems. Artif Intell 1996; 83: 1-58

[4] Perl J. Probabilistic reasoning in intelligent systems: networks of plausible inference. Morgen Kaufmann, 1988

[5] Shafer G. A mathematical theory of evidence. Princeton Univ. Pr., Princeton, New Jersey, 1976

[6] Klir G.J, Yuan B. Fuzzy sets and fuzzy logic; theory and applications. Prentice Hall, 1995

[7] BarnettV, Lewis T. Outliers in statistical data. John Wiley, 1994

[8] Hou W.C. Extraction and applications of statistical relationships in relational databases. IEEE Trans Knowl Data Eng 1996; 8:939-945

[9] Boss B. Fuzzy-Techniken in objektorientierten Datenbanksystemen zur Unterstützung von Entwurfsprozessen. Infix, Sankt Augustin, 1996

[10] Lukács G. Geographic information retrieval with loosely integrated information systems. In Databases and Information Systems, Proc. Of the Third International Baltic Workshop, April 15-17, Riga, Latvia. Institute of Mathematics and Informatics, University of Latvia, 1998

[11] Keller A.M, Wilkins M.W. On the use of an extended relational model to handle changing incomplete information. IEEE Trans Software Eng 1985; 11(7):620-633

[12] Dyreson C.E. A bibliography on uncertainty management in information systems. In [2], pp 413-458

[13] Chen G. Fuzzy logic in data modeling; semantics, constraints, and database design. Kluwer, 1998

[14] Sadri F. Integrity constraints in the information source tracking method. IEEE Trans Knowl Data Eng 1995; 7:106-119

[15] Pu C, Leff A. Autonomous transaction execution with epsilon serializability. In Philip S. Yu, editor, RIDE-TQP'92, Second International Workshop on Research Issues on Data Engineering: Transaction and Query Processing, 1992, pp 2-11

The B2B E-commerce Revolution: Convergence, Chaos, and Holistic Computing

Michael L. Brodie
GTE Laboratories Incorporated
Waltham, Massachusetts, USA

Abstract

The Internet has triggered the Fourth Information Revolution[1], which is leading to fundamental and irreversible changes in the way we do business and in the requirements for the supporting core technologies. Business-to-Business (B2B) e-commerce, the Information Revolution driver, is estimated to be between a $2.7 and a $7.3 trillion market in 2004. This pot of gold is resulting in massive competition in every business sector. Millions of Internet experiments (i.e., dot-coms) are testing hypotheses for more efficient business processes and technology solutions. No one knows which experiments will succeed or which technology requirements or solutions will emerge.

At the heart of the revolution are two diametrically opposed forces—convergence and chaos, in both the evolving business processes and the enabling technologies. Technology and business convergence is redefining the entire problem domain, while Internet experiments are appearing like the chaos in a gold rush. For technologists, chaos manifests in myriad noninteroperable existing and emerging technologies. B2B chaos will persist as long as the business processes are in flux. The technology chaos may persist longer.

How should technologists deal with the revolution and chart their map to the gold mine? What really matters in this revolution are the business processes that will come to dominate individual businesses. For technologists, what matters are the resulting technical requirements. These should be derived, from the top down, from the fundamental economic model, from the economic-chain, and then from specific business models and processes.

This chapter offers ideas from the now inseparable domains of economics, business, and technology. It examines the revolution and its challenges, and it proposes a holistic orientation for information technology (IT). The holistic view attempts to provide guidance through the chaos for requirements for core technologies that will underlie the future Internet-connected world.

---

[1] The previous three Information Revolutions were based on writing 5,000 years ago in Mesopotamia; on the written book in 1,300 BC in China; and on Gutenberg's printing press in 1455 [1].

1 The Irreversible Information Revolution

1.1 The Fourth Information Revolution

Peter Drucker [1, 3] argues that we may be able to predict aspects of the Fourth Information Revolution based on previous technology-based revolutions. Just as the steam engine was the trigger for and symbol of the Industrial Revolution, the computer and the Web triggered the current revolution. As with the Industrial Revolution, the fundamental and irreversible changes of the current revolution will be in the ways business will be conducted. Whereas train transportation reduced the costs of manufacturing and distribution in 1820, B2B e-commerce reduces the costs (e.g., time) of most business processes. In the Industrial Revolution, it took over 40 years for new opportunities to be understood and incorporated into new business processes. It will take five to ten years to understand and incorporate the emerging opportunities into a new generation of business processes.

The revolutionary period will be marked by massive attempts to innovate and create better processes. A chief characteristic of this period of reinvention will be chaos in the affected businesses and in the supporting technologies. Existing processes will be improved directly by simple cost reduction. Drucker calls these improvements "routinization," to which he attributes the associated economic boom. Other processes will emerge, unanticipated, due to new possibilities. He observes that, so far, the train boom is very similar in size and impact to the Internet boom. The economic boom and the related chaos will continue until the experimental business processes stabilize and new business processes emerge. Hence, there may be waves of chaos and economic boom as models and processes in each business sector evolve and mature. According to the theory of growth economics [2], these waves will continue as long as innovation leads to value creation.

Some lessons are already clear. Connectivity and information access are leading to a shift in the balance of power from vendors to consumers. B2B leaders are using the Internet to increase customer value and build deeper relationships with partners and customers. This is a key characteristic of B2B business models.

Drucker argues that the most profound change will arise beyond routinization in unanticipated areas. His candidate for the Fourth Information Revolution is e-commerce "... the explosive emergence of the Internet as a major, perhaps eventually the major, worldwide distribution channel for goods, for services, and, surprisingly, for managerial and professional jobs. This is profoundly changing economies, markets, and industry structures; products and services and their flow; consumer segmentation, consumer values, and consumer behavior; jobs and labour markets. But the impact may be even greater on societies and politics and, above all, on the way we see the world and ourselves in it" [3]. E-commerce may be the basis of a class of new economic models.

In short, the economic boom and the associated chaos will continue for some time. More significantly, the chaos essential to innovation means that the ultimate business processes, business models, and perhaps even economic models are currently unpredictable. Entirely new industries may emerge. The long-term

beneficiaries of the technology boom will be ordinary companies that sell ordinary products and services to ordinary people—in short, real life.

1.2 B2B Is Fundamental and Irreversible

Technology is rife with predictions of world-shattering "paradigm" shifts. Yet none of the recent shifts (e.g., expert systems, client/server, 4GL, distributed object computing, business rules, ERP) has materialized as advertised. The Internet may look like a technology of the same ilk, but it is not. Predictions based on the Web cannot be overstated. Every business sector is on its way to the Web [4].

Like the Industrial Revolution, it is not about technology. It is about economics and business. Once you reduce the costs of a business process, there is no going back. It has already changed the way we do business in many domains, e.g., 50% of travel and 50% of car purchases are initiated on the Web. In 2004, the worldwide B2B e-commerce market in estimated to be $2.7 trillion by Forrester [5] and $7.29 trillion or 7% of the forecast total global economy by Gartner [25]. The world's largest corporations have committed to move to e-commerce models. General Electric (GE) has made a commitment to reinvent itself as an e-commerce company as has General Motors Corp. (GM) with its e-commerce supply-chain and Web-enabled design, manufacture, and distribution plan. Ford Motor Co. (Ford) is neck-and-neck with GM. Wal-Mart, Warner-Lambert, and Colgate-Palmolive have already moved substantial parts of their operations to the Web. These commitments have been confirmed by real profits. GM credits a significant part of its 1999 record sales and earnings to its e-commerce ventures. In 1998, Wal-Mart projected a $1 billion annual savings due to its e-commerce supply-chain.

A Forrester survey [5] estimates that 66% of all businesses will be on the Web to some degree in 2002. By 2005, e-commerce will include 25% of the auto supply-chain business; 17% of all energy trading; 20% of all shipping and warehousing; and 14% of all pharmaceutical and medical product trading. The related stock market boom that in January 2000 valued AOL at more than GM, Ford, and the entire American steel industry combined reflects investors' belief in the significance of e-commerce. In January 2000, the U.S. Department of Labor attributed the U.S. annual 5% productivity increase, in large part, to the Web.

1.3 Role and Significance of Information

Whereas the Information Revolution is about economics and business and not about computing, it is fueled by information. Everything on the Web is digitized information. Prior to the Web, 5%–10% of routine business was automated and represented in digital information. Current estimates suggest this will rise above 20% within four years, more than a doubling in four years of the digital information it has taken 50 years to accumulate. This takes into consideration neither the digitization required to use the Web nor the related increase in transactions. In the personal medical domain alone, a significant percentage of the annual 30 billion medical transactions is expected to move to the Web.

Current information technology is inadequate to meet the requirements of the Information Revolution, as discussed below. A significant challenge will be the development of new information technologies to capture, generate, store, search, analyze, and disseminate information in volumes previously unimagined. Yet, information is more than part of the technology of the Web and an enabler of the Information Revolution. Information is its fuel or base currency (see 2.3), an integral part of the revolution. Information is required not only to conduct business but also to bring about and realize business change, the Information Revolution driver. As Claude Shannon, founder of information theory said, "Information causes change. If it doesn't, it's not information." The full role of data and information has yet to be appreciated [6]. This chapter examines the B2B-driven revolution with a focus on the role and significance of information technology.

2 B2B E-commerce: Business Convergence

Convergence is the synergistic coming together of people, ideas, or processes in novel ways. Traditional business processes, e.g., retail sales and manufacturing, developed over decades or centuries. The U.S. home-building supply-chain was established in the World War II era. Computing helped to reduce costs by automating internal functions and, with considerable cost and difficulty, some established relationships (e.g., EDI). Business homeostasis and technical challenges impeded change in business relationships, processes, and business models. The Internet changed all that. Its ability to connect everyone led to experimentation in new business processes that disintermediated existing partners and included new partners. Some experiments (e.g., GE, GM, Amazon) were dramatically synergistic (i.e., vast cost/benefit ratio). Clayton Christiansen [7] calls such businesses disruptive because they alter the conventional business entirely and irreversibly. Jack Welch, GE CEO (and reputedly the world's best CEO), coined the phrase "DestroyYourBusiness.com" to warn that if CEOs do not experiment with their businesses to increase efficiency, then others will and will put them out of business. This applies to all businesses, including large established ones like Toys "R" Us, GE, and GM. This is an enormous incentive to experiment with traditional business processes not only by the established companies but also by anyone with a novel, synergistic idea and a business plan.

The Internet has spawned the B2B revolution that is changing the way we do business in every business domain. The novelty, diversity, and richness of current B2B experiments (i.e., dot-coms) are astounding. The business processes details will determine future technology and information requirements; chief among them are integration and interoperability. This section examines B2B applications that require enterprise integration and B2B processes that require integration with partners.

2.1 Enterprise Integration Applications

The first phase of the current Information Revolution was based on the premise that enterprise success required understanding and managing a business in its entirety. Organizations with finance and employee information distributed in many business units or over a large number of heterogeneous disconnected systems were hard to manage effectively. Starting in the early 1990s, enterprise integration was enabled by a succession of enterprise applications, the most visible and successful of which are the 40 or more Enterprise Resource Planning (ERP) suites. ERP systems, like SAP R/3 and PeopleSoft, provided a complete set of functional components that could run the back-office of the world's largest companies (e.g., Exxon Mobil Corp.). Back-office functions include the complete financial operations (e.g., general ledger, asset management, financial and business analysis and reporting), complete human resource operations (e.g., hiring, benefits, payroll, tax reporting), and other functions. By 1998, 60% of the Fortune 1000 firms had implemented ERP, which in less than 10 years became a $10 billion/year business, as large as the DBMS market.

ERP success had little to do with technology. Existing technologies were used to provide homogeneous solutions. ERP made significant technical contributions to making commercial off-the-shelf (COTS) applications a dominant means of software development and delivery and to making business processes the focus of enterprise operations; hence, process-orientation is now the focus of enterprise computing. ERP success was due to the competitive requirement, real or perceived, for enterprise integration, supported by massive cost reductions in back-office operations.

A key ERP requirement and benefit, business and technology integration, led to new enterprise integration applications such as Business Intelligence and Knowledge Management. Each requires integrated access to information already in ERP systems plus considerable additional information. Business Intelligence and the related Enterprise Performance Management (EPM) provide senior management with "what if" tools to plan, execute, and measure the effectiveness of their long-term business goals. For example, they help estimate how much it costs to make a potential product or service, the profitability, and the potential customers. Knowledge Management requires access to all enterprise "knowledge," whatever that means. Success requires access to all relevant information that is both precise and up to date. These requirements pose significant technology integration challenges.

B2B has overwhelmed enterprise integration since it addresses only part of the picture and inhibits enterprise interoperation beyond ERP [8]. B2B focuses outside the enterprise on customers and partners. This section concludes with the hottest B2B-driven application, Customer Relationship Management (CRM), which underlies all B2B processes, which are examined in the next section.

The shift of power from vendors to customers plus the ability to connect to all customers forces companies to increase customer value and provides the opportunity to build deeper connections with partners and customers. CRM involves the synchronization of all customer information systems, both data and functions. It started with the market-leading Siebel Systems, Inc., with sales force automation, which largely failed. It rocketed to acceptance when CRM was "sold" as addressing every customer-related activity, including customer profiles, sales and marketing, campaign management, order entry, and customer service. Like ERP, the first CRM wave was based on an integrated COTS application internal to the company. Siebel, Vantive, and Clarify offered the leading CRM applications. CRM was then touted as assisting in building a complete, real-time understanding of every customer, based on all employee-customer interactions, called eRelationship Management. This is leading to the rapid demise of internal CRM applications in the face of Web-based CRM, such as offered by BroadVision and SilkNet [9]. The CRM story illustrates the rapid evolution of business processes and the rapid demise of even new, apparently successful, technology solutions.

2.2 B2B Business Processes

B2B is changing the fundamentals of every business process and model. Fifty percent of all U.S. car purchases and of all air travel initiates on the Web. Five percent of all news is obtained on the Web. Egreetings, one of many such vendors, send five million e-greeting cards per month. Sixty percent of U.S. farmers are now on the Internet along with their dealers, distributors, and manufacturers, who will spend an estimated $20 billion buying and selling chemicals, seed, and other agricultural products in 2002. One of the first agriculture successes was a manure portal created by a single woman farmer. The Internet and the music format MP3 are changing most music industry business processes. With financial instruments trivial to digitize, the financial and investment industry has seen the most dramatic changes, with 30% of its processes predominantly Web-based. U.S. trucking is becoming Web-based. Independent truckers are using the Internet to find loads and avoid making empty return trips. Few truckers travel without a laptop. U.S.-wide Yellow Freight Inc., which invested hundreds of millions of dollars to reinvent itself, manages every order, shipment, and activity through its Web-based central dispatch [10]. Yellow's real-time access to its entire network provided a gold mine of information that led to a 20% productivity gain. The core processes of U.S. presidential elections, campaigning, organizing, communicating, fund raising, polling, and platform development, have been reinvented through the Web. Every corner of the economy is affected [4]. To see the devil is in the details, we now look at several basic business processes.

In early 2000, GM, one of the world's largest manufacturers, said that it would reinvent itself by moving its core design, manufacturing, selling, and shipping of cars substantially to the Web [11]. GM wants to do for cars what Dell Computer Corp. has done for computers—take orders online, custom produce products, and deliver them. This is a conversion of one of the world's largest companies from a "build to stock" business model to a "build to order" business model. The move includes reinventing the way cars are manufactured. Rather than producing a car from 3,000 parts, they want to produce a car from approximately 30 modules or major subassemblies. In comparison to the current eight-week period, GM wants to reduce the total elapsed time from order to delivery to between one and four days. Consumer satisfaction should rise when buyers get a car they design rather than one they choose from the lot by closest match.

Amazon.com is the best known retail revolutionary. With "One Click," customers purchase books 24 hours a day from anywhere with delivery to anywhere. Amazon's business processes are simple but profound. The corner bookstore is an endangered species. The Amazon story is a good example of rate of changes in retail and possibly all other domains. As Amazon says [www.Amazon.com], "Today, Amazon.com is the place to find and discover anything you want to buy online. We're very proud that 13 million people in more than 160 countries have made us the leading online shopping site. We have Earth's Biggest Selection™ of products, including free electronic greeting cards, online auctions, and millions of books, CDs, videos, DVDs, toys and games, and electronics." Many details distinguish Amazon's e-business, including building and maintaining profiles on millions of customers, recommending purchases based on the behaviour of similar customers, services available over cellular phones, and partnering with myriad vendors (e.g., drugstore.com, Gear.com, HomeGrocer.com, and Pets.com). Could we have anticipated the bookseller to omni-seller transition, the books, pets, and grocery connections, or other details? E-retailers have threatened brick and mortar retailers to the point that U.S. malls and stores are being redesigned to better meet customer needs [12].

BuildNet.com intends to revolutionize the 55-year-old home-building supply-chain by providing electronic communications between builders and suppliers, enabling, for example, Internet searches that take seconds to locate items. The current chain, which includes tens of thousands of material providers and far more builders and subcontractors, is vastly inefficient. It often takes months to find materials dispersed over many warehouses, which must be physically searched. The Internet-based home-building supply-chain is in its infancy. It is impossible to predict the processes that will emerge from the current divergent experiments: top-down, bottom-up, and middle-out. USBuild.com hopes to completely replace subcontractors with electronic bidding and distribution and to offer discounts on aggregated volumes. Equalfooting.com is working from the bottom up, attempting to build the supply-chain from the small builders and suppliers to assist them with the communications and volume discounts obtained through cooperation. Finally, ImproveNet is working on the supply-chain on behalf of the consumer, providing connectivity between buyers, contractors, and suppliers. It will take time for it to shake out. In that time, each candidate process will require very fast means of creating and separating from potential networks of consumers, builders, and suppliers. The technical requirements are interoperability, flexibility, and speed.

With 273 million customers, 550,000 doctors, and 30 billion yearly health-care transactions, the enormous U.S. health-care industry has resisted many attempts at reformation. Healtheon/WebMD Corp., the largest E-health-care company, is about to change that with its network of 450,000 doctors. E-health-care is attempting to reformulate the currently paper-dependent health-care processes that include the major players—doctors, patients, hospitals, HMOs, pharmacies, insurance companies, and labs. Potential savings are estimated at $280 billion due to unnecessary administrative and clinical expenses. Business processes currently being changed to reduce cost and dramatically increase patient satisfaction include appointment scheduling, insurance claims, referrals, prescriptions, lab test data transmission, patient monitoring, and the exchange, integration, and transmission of medical records. Substantial benefits beyond cost reduction include time savings for all players; improved precision; automated checking; access to a wide range of providers; error reduction due to reductions in duplicated data or human errors; and accurate medical records that arrive at any medical appointment before the patient. The size of this change both technically and socially is mind boggling.

A final domain to consider is e-procurement and supply-chain, which are required by almost every e-commerce activity. E-procurement has a buy side, a sell side, and the connection of the two. On the buy side, a customer such as a company purchasing agent needs to access information on all relevant products, including product specifications, comparisons with all competitive products, pricing including discounts, delivery arrangements, and promises. The seller must have all relevant information on the buyer, including company, finance, credit, contact, logistics, preferences, and legal. On the sell side, the vendor must provide all relevant, up-to-date catalogue information from hundreds or thousands of suppliers together with real-time inventories and pricing. For a sale, transaction details must be irrefutably committed on both sides, and reflected in the inventory and financial systems. But that is just the purchase. The purchased goods must traverse the supply-chain from the manufacturers through the stages of assembly into the final products to form the customer order to be delivered according to agreed-upon terms. Supply-chain management tools let companies configure products to order, confirm availability, and track orders and delivery schedules in real time. This is critical to e-businesses as well as to traditional businesses in increasingly competitive situations. Market leaders are Ariba, Commerce One, and Oracle [13]. Forrester claims that Oracle's lead is based on its understanding and support of interoperability.

Although supply-chain is a big story for all e-commerce, Ford and GM made the biggest supply-chain move in late 1999. Collectively, Ford and GM have 50,000 suppliers for their annual $300 billion in purchases. Originally, these plans required GM suppliers to use GM's TradeXchange open online marketplace and Ford suppliers to use Ford's Oracle-based AutoExchange. In February 2000, Ford, GM, and DaimlerChrysler AG announced a single automotive parts exchange that combines AutoExchange and TradeXchange and avoids a third from DaimlerChrysler. Most automakers worldwide are negotiating to join the exchange. As it is, the new exchange will form the world's largest Internet company and trade exchange. Similar exchanges are being established in other industries, including aerospace (MyAircraft.com), paper (PaperExchange.com), agriculture (ASAg.com), chemicals (ChemConnect), and steel (E-steel).

These exchanges are similar in concept to those already implemented, e.g., by Wal-Mart. Some include the ability to alter their supply-chains by altering a vendor's shop floor and delivery schedules and to analyze supply-chains for such things as supplier effectiveness. In addition to optimizing existing processes, Drucker's routinization, there are unanticipated results, such as GM's entry into the supply-chain business and Ford's entry into the Web design and hosting business.

2.3 B2B's Fundamental Characteristics

The B2B Information Revolution is all about business and not about technology. To understand it and to build the best technical support, we must understand the bigger picture. For example, we could focus first on the detailed business requirements in specific businesses and generalize where possible. The dominant characteristic of B2B business models is efficient interaction with all relevant partners, established on demand, sometimes at a very intimate level and in real time. Internal enterprise integration is necessary but far from sufficient, as illustrated by ERP. Every business model and process is being reinvented in a race where speed-to-market is king. The result is a chaotic range of diverse, rich business models that destroy both established and new models. Amazon, the world's leading online retailer, exemplifies both rapid evolution and unanticipated convergence. The current scope and scale of the B2B revolution is phenomenal, but the potential, illustrated by e-health-care, is vastly greater. Whereas some characteristics are common, the current diversity, chaos, and unanticipated events make the ultimate characteristics unpredictable.

The above B2B characteristics are details within something more fundamental, the economic model. Consider this in an analogy to database or programming language type systems. A data model defines a class of possible schemas that in turn define a class of instances. Similarly, an economic model defines a class of business models that in turn define a class of business processes, which have operational instances.

Twentieth century economics was based on wealth and value creation from assets such as land, labour, and capital[2]. Fifteenth century accounting practices for measuring value were augmented with twentieth century cost accounting that provided means for setting and measuring business objectives, namely profits. Twentieth century accounting focused on measuring assets and activities within an enterprise with the objective of maximizing profits and minimizing expenses by eliminating waste. To do this, it assigned values to assets and transactions or operations within a business. Traditional accounting was focused internally within the enterprise [1].

A new model of accounting, economic-chain, with origins in 1908, is now coming to the fore. Economic-chain focuses not on the internal operations of an organization but on external opportunities (and threats) and on the end-to-end processes (i.e., the chain of partners) required to realize them. A chain could start with raw materials and end with the consumer. The external context sets the parameters for efficient internal operations.

Economic-chain was developed in the United States by William C. Durant in the 1920s [1] and was applied in Britain in the 1950s and by Sam Walton as the basis of Wal-Mart's success in 1975. Economic-chain accounting addresses costs throughout the entire economic-chain, from the supplier to the ultimate customer. It is now being applied under concepts called value-chains in manufacturing, retail, services, and many other industries. It has influenced the focus on business processes and is leading to a more sophisticated notion of value webs. Economic-chain is the economic model of which B2B business models are instances. To understand what business models and processes are possible, one must understand economic-chain. The primary characteristics of economic-chain will also be those of B2B business models, which in turn define the requirements of next-generation technologies. Economic-chain has three main characteristics relevant to B2B. First, business is conducted as a process through a chain of partners. This requires the ability to do business with any potential partner, competitor, or customer. Second, business must address the entire end-to-end business process required to achieve the business objective, including planning, implementing, monitoring, and

[2] The objective of an economic model is wealth and value creation. In the information economy, twenty-first century economics has shifted from land, labour, and capital as base assets to hardware, software, and wet ware [19]. These assets permit value to be transferred at almost no cost. Twentieth century economics had major barriers to entry and exit (e.g., the factories, machines, and land) that are hard to change. Twenty-first century economics has a similar problem in software [23] and less so in hardware. This indicates the significance of technology, "knowledge" (e.g., wet ware), and information, which is far beyond its technical value. They are an integral part of the current chaos and economic growth.

management. Third, flexibility is required to alter the business process and partners, to maximize opportunities and minimize threats.

To make the above arguments more concrete, consider SAP, AG's path to its recent mySAP.com product offering [www.mysap.com]. SAP's origins in the late 1980s were in ERP, with the objectives of cost reduction and business integration. The objective of their second-generation suite was "Inter-Enterprise Co-operation," with a focus on the supply-chain. SAP's current focus, mySAP.com, attempts to achieve for its customers "e-Community Collaboration" within their business communities, called e-marketplaces. This vision under construction is pure economic-chain.

3 B2B Technical Requirements

B2B has a vast number of detailed technical requirements, with far more to evolve as B2B matures. This section addresses four critical technical requirement areas.

3.1 Integration and Interoperation

The fundamental economic-chain requirement of B2B translates to technical requirements for interoperation and integration with both internal and external partners. Interoperation concerns processes, workflows, or systems interacting to achieve business objectives. Integration concerns the required combination of information resources to achieve business objectives. B2B's predecessor and prerequisite, enterprise integration, requires integrated access to information resources. B2B requires potentially intimate integration with partners' information resources.

Consider internal integration. Most organizations have many information repositories (e.g., GTE has more than 1,500), which have been developed over a long period of time, each designed to meet the requirements of a particular application. For example, there are often as many independent customer information repositories as applications that require customer data. CRM involves the synchronization of all customer information systems, both data and functions. Integrated customer information across the organization is currently considered to be mission critical for many enterprises. The objectives include providing a complete, real-time understanding of all customers, expanding customer contacts to all employee-customer interactions, and synchronizing customer relationships across all communication channels. Hence, significant effort has been invested in identifying and integrating all customer information in all information repositories. Similar arguments could be made for most information subjects (e.g., parts, inventory, finances, and products). Such arguments have been made for knowledge (knowledge management), finance and human resource data (ERP), and parts (MRO). The Global Data Management challenge is to be able to deal with all the information resources of an enterprise as a single information repository in which the same information subjects (e.g., customer profiles) can be accessed as if they were meaningfully integrated. The Global Data Management Challenge has been faced by every enterprise from the time it developed more than one database. The intensity of the need is reflected in the data warehouse craze, which provides read-only access to out-of-date, partly integrated data. Global Data Management is mission critical in B2B for specific topics (e.g., customer and supply-chain data). Codd's goal of data independence is still elusive.

Consider external integration. The current most critical B2B and Business-to-Customer (B2C) (e.g., retail) applications are those that support purchasing, procurement, and delivery (e.g., supply-chain, e-procurement, fulfillment). Success in these domains depends on efficient, smooth partnering that requires integration between specific partner systems at a level similar to internal integration. The purchaser (e.g., Warner-Lambert, Colgate) may want direct access to suppliers' systems to plan, monitor, and alter production and delivery schedules. Each external integration requirement will depend on the details of the business processes being implemented.

The interoperation and integration requirements are augmented by other B2B requirements. Internet speed reduces the time to build interoperable solutions. The potential number of partners makes conventional point-to-point mappings economically infeasible. More significantly, B2B computing and data architectures may differ radically from those of today. Analysts [5] estimate that a significant portion of all business will be conducted on the Web. Speed to market forces most enterprises to connect their legacy data repositories, systems, processes, and workflows to the Web. The resulting architecture, with as many gateways as there are systems, will not support the transaction and data volumes. Nor will legacy systems support the anticipated number of trading partners. A significant portion of current and future computing resources will have to move from internal enterprise architectures to some new architecture on the network, as has long been predicted by Sun Microsystems and others [14]. The network is the computer.

3.2 Scalability, Reliability, Real-Time Access, and Accessibility

Four traditional database requirements, scalability, reliability real-time access, and accessibility are now mission-critical B2B requirements. All three stem from the nature of the Web. B2B business processes can be made available to anyone on the Web. Hence, they must be accessible 24 hours a day and scalable to the access requirements. Stories abound about the failure of a Web-based application to scale to the number of users. A recent survey [15] states that scalability is a serious problem for almost all B2B players and that 99% accessibility was seen by most players as inadequate for nonstop e-commerce. These factors not only limit business transactions, they significantly affect quality of experience, which is seen as the critical success factor in B2C [16].

Real-time access to up-to-date data is critical for a large number of B2B processes. Internet speed and competition increases the need, possibly artificially. For example, retail sales, supply-chain transactions, and transactions involving changing prices (e.g., due to time, inventory, competition, airline reservation seat inventories) all require data at the instant of the transaction. Online retailers and wholesalers may sell products from thousands of suppliers. Pricing and delivery are based on inventory and fulfillment systems. The seller may require real-time read and write access to these systems to be able to make and complete the sale. Old data may cause the transaction or business to fail (e.g., selling goods or services too low or too high, selling already booked seats, selling products that cannot be delivered). Data warehouses that are updated periodically are inadequate. The data must be obtained in real time. This leads to the requirement for real-time access to heterogeneous or federated databases.

3.3 Transactions and Cooperation

The initial Web driver, B2C, produced relatively low volumes of transactions (e.g., sales) directly into the sellers' databases. The technology was there to make them robust and reliable. Challenges will come as B2B generates the projected massive transaction volumes [5] for which there will be legal requirements for robustness, reliability, or at least nonrepudiation. Whereas trust and work will be required to make single operations transactionable between two partners, there is no technology to support transactionable multistep operations. The requirement for reliable, robust, and non-repudiable business processes is a direct result of the fundamental B2B requirement of supporting economic-chains, end-to-end.

Another basic technical requirement flows from fundamental B2B requirements. Economic-chain requires that enterprises continuously find partners, collaborate on a business opportunity, and cooperate to execute on the agreement. Forrester says that by 2003, live collaboration will become as much a part of business as e-mail is now. The characteristic that distinguishes Web-based collaboration and cooperation is time. These activities currently happen without computers but take considerably longer than with computers. Without the Web, enterprises automate partnering agreements using system interfaces that take months or years to build. The Web currently facilitates human-based collaboration and cooperation by dispersed people via Groupware, e-mail, messaging, calendaring, and bulletin boards, augmented by documents, database sharing, workflows, and project management. As these activities result in more process improvements (e.g., more efficient design and development cycles), there will be a requirement to automate collaboration and cooperation. You may want to find a partner over a very short time frame, possibly down to the single transaction level. In the near future, firms must form and disband partnerships at lightning speed: in days, hours, and minutes [17]. This will lead to the requirement to find partners without human intervention, maybe only for a single transaction especially in situations where systems do the work, e.g., trading, negotiation, and project and resource management. Fast partnering requires fast system-to-system integration. But more than that, it will require forms of discovery of services and capabilities, negotiating terms of a partnering agreement, and the planning, execution, monitoring, and termination of the partnership.

3.4 IT Provisioning

An enterprise's information systems have traditionally been the responsibility of a centralized IT organization. Centralized IT can be efficient when all IT resources are under one control that can set standards, leverage IT skills and solutions, and manage the entire IT environment. Central IT has attempted to reduce the problems of interoperation by establishing standards and reducing the number of technologies involved in an IT operation. The centralized IT model must be reinvented to meet the requirements of the B2B revolution. Relevant B2B requirements are business-driven, end-to-end solutions in an economic-chain.

B2B IT activities cannot be managed by a central control which is too slow to act [18] but, more significantly, very hard to create since there is no central control of processes on the Web. There may be 2 to 20 partners participating in implementing and offering a process. Although there may be a dominant player in one process, it is not likely that that player can act as the central IT for the process, since the required resources exist in the partner IT organizations. B2B's chaotic nature, which will persist for years, requires flexibility and speed. From moment to moment, different players, partners, and competitors may join or leave a process. Central IT is seldom known for flexibility or speed. Partner links, which used to be created over months and years, may now exist for seconds (e.g., for the duration of a transaction). Scalable and reliable end-to-end solutions must be constructed from the technology components of the current process partners.

A significant result of the success of ERP is the business-IT partnership for systems development, provisioning, and operation. This is entirely consistent with the fundamental requirements of B2B, and it is changing the face of IT. In the past, IT "owned" the entire information system life cycle. Information systems were considered complex technology that only IT could understand and manage. ERP systems embody complex business models and processes. Managing those business models and processes is the core responsibility of a functional organization. For example, in ERP human resource systems, many policy decisions, e.g., on employee reimbursement, are analyzed and implemented for the relevant employee populations not in paper memos but directly in the systems. Human resource organizations must be directly involved. As a result, massive ERP systems are now "owned," planned, designed, and operated by the functional owners with IT as a partner. The corresponding B2B situation is more complex. Business processes that cross organizational boundaries will involve multiple functional owners and IT organizations.

The context that core technologies must support and in which information systems must be developed and provisioned defines requirements for next-generation infrastructure technologies that are as significant as technical requirements, such as data access and transaction processing. Consider the fundamental B2B requirement for end-to-end solutions. Problem owners (i.e., now functional and not IT staff) face real-world problems with many facets for which they require end-to-end solutions. For example, emerging e-marketplaces require a complete solution, including the sell side, the buy side, and sales fulfillment, for every sales transaction. An end-to-end solution must meet the full set of requirements, including political, business, technical, and operational. Due to the complexity of real-world problems and the novelty of B2B, there are seldom off-the-shelf comprehensive technical solutions. Solutions must be created to meet the requirements from components solutions so that the business requirements are met and the component solutions fit together so as not to inappropriately raise the cost or complexity of the solution or place the overall solution at risk.

Traditionally, technology solution vendors offer solutions to very specific technical challenges, e.g., storage, search, data management, communications, applications. They tend to understand their technology area (and possibly closely allied technologies or products) in great depth. However, they tend to not understand the problem owner's problem. An example of this is data management. Data management solution vendors understand data management in remarkable detail, including the most recent advances in databases, data management, and all the allied data management technologies and products. These are often wonderful solutions for well-understood data management applications. However, problem owners seldom have clear-cut, well-understood data management problems (e.g., the Global Data Management problem, for which data solution vendors have no solution). Functional ownership of information systems is leading to the development of end-to-end solutions. At the turn of the millennium, CRM is viewed as one of the highest priority IT solutions for competitiveness. As a result, a large number of data management solutions have been developed to address the Global Data Management problem restricted to CRM. These solutions do not come from the technology solution vendors, but from system integrators who create solutions from component technologies or products.

In summary, real problems require real solutions that inevitably involve many technical and business domains in conjunction. Historically, the IT community has not readily provided such solutions. Researchers, solution vendors, and IT organizations have focused on narrow technical and product areas. This has increased the technical chaos. Hopefully, next-generation technology will respond to the fundamental B2B requirements of economic-chains and end-to-end solutions.

3.5 Core B2B Technical Requirements

The core technical requirement of the B2B revolution is fast, efficient, and reliable interoperation of processes, workflows, and systems and meaningful integration of information resources, with internal and external partners. Access may be real-time but certainly always-on to support nonstop e-commerce. Multistep business processes will be in volumes and at speeds previously unimagined and must be transactional or at least non-refutable and traceable. Complete and flexible end-to-end solutions must be provided at Internet speeds by next-generation technology or by composing existing component solutions to meet complex and constantly changing business processes requirements. New technology solutions must include network-based architectures that provide reliable, efficient access directly on the network to the vast anticipated and changing set of business partners. Many of these requirements necessitate an ability to deal with the semantics of processes and information far beyond what is possible today. The step beyond these requirements is the automation of cooperation and collaboration.

4 Integration Solutions: Overcoming Chaos

4.1 Significance and Difficulty of Integration

Integration and interoperation involve providing technology solutions to meet end-to-end business requirements from multiple, heterogeneous technical components.

This section reviews the nature and significance of the challenge and briefly reviews the dominant solutions.

A core challenge for more than 50 years, integration solutions tend to be ad hoc, proprietary, and largely labour intensive. In the past decade, the need for much more effective and efficient solutions has increased. It is a core requirement of dominant business trends, including corporate globalization, mergers, and acquisitions. It is a fundamental requirement of B2B. Mercer Consulting reports that Internet companies enter into mergers after 6 months of operations compared to 72 years for the top 50 Fortune 500 firms. The problem is compounded by the chaos of the diversity of emerging and disruptive [7] business processes and technical solutions, the number of trading partners, and the nodes in the network. Wintergreen Research estimates that 50% of the 1999 worldwide IT budget of $900 billion (U.S.) was devoted to the problem. Systems integration products and consulting services have become one of the largest sectors of IT.

Integration and interoperation solutions are mission critical. It is common that such problems dramatically slow the progress and increase the cost of the related business activities. Major mergers and acquisitions have failed or have nearly failed due to the integration challenges posed by their systems. System integration challenges in the Norfolk Southern Corp. merger with Consolidated Rail Corp. resulted in a $40 million loss of business in the first month [20]. Freight cars were misrouted, crews were misscheduled, and customers temporarily lost their products. It cost $29 million a month to repair the system over several months, followed by the cost of regaining the lost market share.

4.2 Integration Solutions

4.2.1 Middleware: Distribution, Databases, and DOC

Data independence, the ability of a database to provide data for many applications, was one of the greatest interoperation solutions in computing history. Unfortunately, the success and widespread use of databases led to noninteroperable solutions and the Global Data Management problem. Distributed and federated databases, conceived in the early 1980s to address these problems, were limited in their functionality and scalability, so were unsuccessful. Recent federated database solutions (e.g., IBM's DB2 Data Joiner and Garlic, Cohera's Data Federation System, and Metagon's Dqbroker and Dqview) resolve earlier limitations but have yet to be widely used. Whereas they facilitate interoperation solutions, they do not address the core semantic problems. Perhaps for similar reasons, distributed object computing (e.g., CORBA, COM+), which is more comprehensive yet more basic, has not been widely adopted [21]. As recent TPC benchmarks indicate (www.tpc.org), COM+ works wonderfully in a homogeneous environment. Workflow, an interoperation solution, has met with conceptual acceptance but limited application in heterogeneous environments. The Internet application need for interoperation is leading to a broad acceptance of XML and JavaBeans. These are more amenable to addressing semantic issues since they support a degree of metadata representation. However, semantic solutions still require the ability to understand (e.g., through standardization), which they do not support.

4.2.2 Enterprise Application Integration: Build It Yourself

Enterprise Application Integration (EAI) are suites of tools and technologies (e.g., messaging, transactions, workflows, and data translations) with which to build interoperation and integration solutions. Market leaders include IBM's MQSeries, New Era Of Networks' E-business and EAI solutions, Tibco Software's TIB/Active Enterprise suite, and Mercator's Enterprise Broker and E-Business Broker. These products provide standard data and transaction integration layers that are placed on top of databases and systems. The solutions must be built with these relatively low-level tools (i.e., each data mapping and transaction must be defined) or tailored from their growing libraries of translators between common DBMSs and applications (e.g., ERP suites). Although EAI is one of the fastest growing markets, it still appears that organizations want to build their own solutions without these tools.

4.2.3 COTS: Integration in Isolation

COTS application suites, such as ERP, provided the greatest interoperation and integration solution in the 1990s. ERP application suites provided a single, homogeneous solution for a significant number of back-office functions. This biggest software trend of the 1990s typically provided integrated finance, human resources, and manufacturing/supply-chain business processes, systems, and infrastructures. ERP products and their partner products provided up to 30 functions covering the majority of back-office processing. An ERP integration and interoperation solution is based on common models for enterprise organization, security, business processes, functions, and data, as well as being based on a single language and architecture. ERP solutions are much deeper than mere technical interoperation. They provide common semantics at every level of the solution, from the enterprise, through the business processes, down to the data elements. Heterogeneity of legacy applications can be overcome by migrating to a homogeneous ERP solution. Migrations are typically massive challenges that are costly and that take considerable time, resources, and expertise. With some notable exceptions, the benefits vastly outweigh the costs. This attests to the significance of effective interoperation and integration solutions.

Yet ERP solutions are strongly limited, especially considering B2B requirements. First, ERP solutions exceed the complexity barrier. The functional and technical complexity of a single system that supports the entire back-office operations of the world's largest corporations is beyond human comprehension. This poses problems of understanding not only for ERP customers but also for ERP vendors. Limitations related to this complexity include the inability to specialize the system to the requirements of entire industry sectors, e.g., telecommunications, let alone an individual company. ERP suites provide powerful interoperation and integration solutions within the scope of the suite. Integration and interoperation outside the suite is as difficult as for non-ERP system, except that ERP systems must interact with hundreds of systems. For some ERP products, integration with external systems is made more difficult by the fact that the technical and functional aspects of the ERP product are designed assuming the required resources will be built within the ERP system. This assumption is seldom true because organizations cannot always eliminate non-ERP systems since ERP systems support only a small portion of the functionality required by an enterprise. ERP homogeneity and inflexibility is counter to core B2B requirements.

4.2.4 B2B Integration Solutions: Portals and E-marketplaces

At the turn of the millennium, an emerging B2B integration and interoperation solution is arising in the form of portals and e-marketplaces. The primary motivation for portals and e-marketplaces is economic. Those who control an e-marketplace attempt to monopolize the Web for a specific service, product, or industry. For example, search engines, such as Yahoo!, attempt to maximize their market share for Web searches. They maintain and build the customer base for their core and related businesses. Control of a portal permits the imposition of solutions that address Web heterogeneity. An e-marketplace is a recent and potentially very sophisticated form of portal. An e-marketplace is an online market for all stakeholders in a given market. For example, it would include all the buyers, sellers, and others (such as banking and fulfillment) required to support the worldwide pharmaceutical industry. Twenty-five percent of those surveyed by Forrester [5] anticipate that most of their online trade will flow through e-marketplaces in 2002. The most dramatic e-marketplaces in terms of their projected size are already in place (e.g., the Ford-GM-DaimlerChrysler auto parts exchange; SAP's e-marketplaces in chemical, pharmaceutical, oil and gas, and health-care equipment; and others named in Section 2.2). Although e-marketplaces are economic entities, they provide the opportunity for those in control to overcome the heterogeneity chaos. The November 1999 plans called for the 30,000 Ford suppliers to use Ford's Oracle-based AutoExchange and a similar number of GM suppliers to use GM's Commerce One-based TradeXchange. The merger of these exchanges was forced not only by economic reasons (e.g., control of the supply chains by auto companies and not the technology companies) but also to overcome heterogeneity. The exchange will establish one standard for auto part exchange. Ford and GM have forced Commerce One and Oracle to work together. It is likely that e-marketplaces will arise for almost all domains in 2000, and standards will be set in 2001. As the GM and Ford examples illustrate, there may be multiple standards per industry.

4.2.5 IT as a Service: Let the Experts Do It

The technical challenges posed by the current incomplete interoperation and integration solutions lead an increasing number of enterprises to outsource the problems. Many forms of outsourcing are emerging, each providing a portion of the interoperation solution. A dominant ERP model is to outsource their development to systems integrators. Since ERP systems are not comprehensive and organizations can have more than one, portfolio assemblers have arisen. They are systems integrators that work with you to select and develop the optimal portfolio of ERP and other systems to support an enterprise. Once developed, ERP systems operations can be outsourced to an Application Service Provider (ASP) that is a world-class expert at ERP operations. An ASP variant is a Business Service Provider (BSP), to whom individual business process are outsourced. There are also ASPs that permit organizations to outsource their entire IT function. Hence, organizations can outsource any part of their information systems, from selection, design, and development to operations, all on the basis of individual business processes, business domains (e.g., human resources), or any subset.

A typical premise of outsourcing is that the outsourced activity is not the core business over which you require complete control and the necessary skill sets. Another premise is that the outsourcing organization is world class in its domain, allegedly far better than central IT in an enterprise. For example, Oracle is reinventing itself as a full-service ASP. Oracle's depth of knowledge in data management, back-office (e.g., ERP), and some front-office (e.g., CRM) business domains may explain their leadership in procurement applications [13]. An Oracle solution is likely to be a homogeneous, thin-client Oracle solution but with market-leading [13] interoperation capabilities.

Just as there are business-process-specific ASPs, there is an emerging ASP for B2B online data provisioning. Online data brokers [22] support the critical B2B requirement for data and are world-class experts at data integration as well as information analysis, usually restricted to specific domains (e.g., credit information). As a domain-specific B2B data portal, the online data broker must move their data resources to the network. They attempt to be masters at making data available online tailored to fit the requirement—the right data designed to optimize business performance (e.g., transactions, real-time, dynamic pricing, and accessibility).

4.3 Integration Solution Status

The current solutions do not address the core B2B technical requirements dominated by the need for end-to-end solutions. Some solutions avoid the problem by outsourcing. Other solutions attempt to reduce heterogeneity with homogeneous solutions (e.g., ERP); still others provide intermediate interoperability layers or gateways (e.g., CORBA, standards) between heterogeneous technologies and augment them with functions needed to complete the particular distributed computing model [23]. Interoperation layers will be necessary as long as there are legacy systems. The trillions of dollars of legacy systems must be included in solutions for some time since there is often neither a target solution to which to migrate nor the resources and tools with which to migrate [23]. Current solutions are labour intensive. There may not be enough labour to build the solutions required by the anticipated growth in interconnection.

5 Holistic Computing

The current generation of infrastructure technologies does not meet the requirements of B2B e-commerce. Although it will continue to satisfy the requirements of conventional applications, it is coming to the end of its useful life for next-generation applications. Just as business is being reinvented, so might technology be reinvented to meet current B2B requirements as well as future requirements that will emerge from the unanticipated opportunities that will arise and come to have a greater impact on the world than the initial results of the revolution [3]. This final section proposes ideas toward the process of technology reinvention.

5.1 The Challenge

Next-generation technologies must directly support future economic models. Current computational models and information technology are based on twentieth century economic models designed to manage the internal resources of an enterprise [1]. Twenty-first century computing, starting with B2B e-commerce, should support advanced economic models such as economic-chain. Hence, a fundamental requirement of twenty-first century computing is cooperation or collaboration with resources inside and outside the enterprise [24].

Seen from today, a future generation of infrastructure technology might be inherently distributed and support interoperation and integration as primitive operations independent of any programming or computational languages, models, or architectures. But it is far too early in the revolution to tell what will emerge. The revolution has just begun, driven by emerging economic and business models and processes. The chaos due to changes introduced by routinzation will continue for five to ten years. Consider automobile manufacturing. GM projects that it will take until 2003 for them to reinvent their manufacturing processes and supply-chain. Will B2B e-commerce-based automobile manufacturing processes be stabilized industry-wide by then? The e-commerce revolution is going through each industry at a different pace. The revolutionary five to ten years will be marked by continuous and dramatic changes in business models and process. These changes may disrupt companies [7], processes, or even entire industries (e.g., retail book sales, the music industry). The changes will lead to new, possibly unanticipated technical requirements.

What is a technologist to do, especially under the pressure of Internet speed and myriad emerging and converging technology opportunities (e.g., network and mobile appliances)? The results of adopting apparent solutions too early and the waves of alleged paradigm-shifting technologies has littered the IT landscape with heterogeneous legacy systems and technologies. This is just the problem we want to avoid. Due to the vast investment in and value of legacy systems and the migration challenges, legacy resources must be part of almost all near-term solutions. Even Amazon must obtain information resources from thousands of partners whose businesses operate on legacy systems.

5.2 A Holistic View

The Information Revolution is driven by economic and business models and is enabled by technology. Technical solutions to B2B e-commerce challenges and opportunities require an understanding of how economics, business, and technology relate. An economic model defines a class of business models that are, in turn, defined in terms of their core business processes. The business requirements of a process determine the requirements of the technology solution that will support it. This big picture provides a holistic framework for the role of technology in the Information Revolution. This framework is the next-generation virtuous cycle beyond that of Intel and Microsoft for processing capacity and need. A depth of understanding of economic models, such as economic-chain, and B2B business models and processes will help immensely in developing technology solutions for B2B e-commerce.

Technology triggered the current Information Revolution by enabling connectivity. But this technical solution would not have succeeded had it not met a critical business requirement. The coincidence was fortuitous and, as they say, the rest is history. With the hope that such a rare coincidence will rise again, some technologists may continue to invent technology *que* technology. With the significance of B2B e-commerce and the urgent need for technical solutions, we should turn our attention to a deeper understanding of economic and business requirements and their relationship to the requirements for the next generation of infrastructure technology. For example, there is some volatility in the evolution of economic models, leading to entire new classes of business models. There is enormous volatility in the evolution of B2B business models. It is almost impossible to develop a full technical solution until the full business requirement is understood. It may be possible to develop long-lasting, general-purpose technical solutions in specific areas that will meet the requirements of many business processes based on an understanding of current requirements.

One might imagine an e-commerce computational model composed of basic operators or services called e-services. Current B2B business processes may lead to a set of e-services such as "order procurement, online trading, customer relationship management, product promotion, or real-time car navigation and traffic information services (www.hpl.hp.com/hosted/tes2000)." Experimentation with potential e-services is exactly the right research direction. However, it will take a much deeper understanding of B2B e-commerce as well as time for its evolution before the full requirements and technical opportunities are known. How might this necessary technical experimentation be conducted?

5.3 Holistic Experimentation

Technical experimentation could be done within a holistic framework such as proposed above. Economic-chain suggests holistic guidelines. First, technical solutions should be for comprehensive end-to-end business processes. Like many vendors [16], SAP, AG is using a scenario-driven approach to develop e-services in areas such as procurement, collaborative bidding, auctioning, collaborative forecasting, CRM, distributor/reseller management, available to promise, online training, online consulting, and Web marketing. Second, as current architectures will be reinvented as resources move to the network, architectural views of systems may not be as helpful as process-oriented views. Technical optimization will involve the execution of operators and the flow of information and control. Processes provide those requirements for next-generation architectures. Technology components have specific dependencies (e.g., interfaces) in current IT architectures (e.g., the levels of the ISO stack). These dependencies may change radically as network-based architectures evolve. Processes may indicate potential next-generation technology components and their dependencies. Finally, it is essential for technologists to partner with the relevant business and economics subject matter experts to ensure the relevance of experiments.

Experimentation can follow the conventional scientific method but should be as close to reality as possible (e.g., in real dot-coms or in real e-marketplaces). Let a real context provide the ultimate evaluation. There is a wonderful, possibly apocryphal story about an Amazon marketing experiment. Allegedly, senior executives were discussing two potential marketing strategies. Should they give airline miles or frequent-book-buyer credit? Jeff Bezos, Amazon's CEO, said that they didn't need to speculate as to which would be the more effective. All they needed to do was to run live with one and then with the other option and evaluate both. The experiment was both decisive and took less time than the anticipated discussion. The world has changed, and new opportunities abound!

6 Summary

Business-to-Business e-commerce is driving the Fourth Information Revolution that was triggered by the computer and the Web. The revolution is leading to fundamental and irreversible changes in the way we do business in every domain. As a result, all business processes are in flux and there appears to be chaos, as in a gold rush, in diverse and creative Internet experiments.

B2B's Web-centric requirements are beyond the capabilities of current infrastructure technologies and legacy systems. Intelligence (i.e., applications, workflows, data) must move from individual resources in internal IT architectures to integrated resources on the network. This re-frames current integration challenges and provides new opportunities. A new generation of infrastructure technologies is required.

The requirements of the new generation will be those of the B2B business models and processes that will emerge from the current apparent experimentation. Following previous technology-based revolutions, the chaotic period will last for some time and have unanticipated results. With incomplete requirements for five to ten years, how do we design the next generation of infrastructure technologies?

While it is premature to determine the next generation, challenges of the current generation remain and are further exacerbated by B2B. Specifically, heterogeneity and interoperability, far from resolved today, are mission critical in the connected world of B2B e-commerce. The flood of new technologies and the Internet speed at which solutions are required further complicate these problems.

This chapter examined the current chaos of B2B e-commerce and the nature of significant business processes. It considered B2B technology requirements and reviewed some current solutions. It concluded by proposing a holistic view for the period of intense experimentation that will lead to a new generation of infrastructure technologies to support an interconnected world.

7 Homily

The most profound changes resulting from technology-based revolutions are the unanticipated ones [1, 3]. The Gutenberg press led to secularization and the Reformation. The steam engine and the Industrial Revolution led to factories, the rise of cities, and the middle class. The current Information Revolution has the potential to change our world as much as have electricity and the telephone.

Technologists may not be able to anticipate or understand the coming changes, nor control them. However, we can work with those who might to more fully understand what we are doing. More fundamentally, we can appeal to basic morality and ethics. We should be aware of the potential for the B2B e-commerce revolution to change the way we conduct business and to impact everyone on the planet. These are powerful forces. Let's develop them for the common good.

References

1. Drucker PF. The Next Information Revolution, Forbes ASAP, Aug. 24, 1998.
2. Romer P. with Evans, G and Honkapojha, Growth Cycles, S. American Economic Review, Vol. 88, No. 3, June 1998, 495-515.
3. Drucker PF. Beyond the Information Revolution, The Atlantic Monthly, Oct. 1999.
4. e-volve: Dot-Com and Beyond, A Comprehensive Study of the Second Stage of the Internet Revolution, Bear Sterns, New York, Feb. 2000.
5. EMarketplaces Boost B2B Trade, Forrester Research Inc., Cambridge, USA, Feb. 2000.
6. Brodie ML. Broadening the Database Field, www2.aucegypt.edu/vldb2000, Dec. 1999.
7. Christensen CM. The Innovator's Dilemma: When New Technologies Cause Great Firms to Fail, Harvard Business School Press, Boston, 1997.
8. Brodie ML. Rethinking ERP: The Second Generation, 1st Int'l Conference on Enterprise Management and Resource Planning, Venice, Nov. 1999 e-proceedings (http://mosaico.iasi.rm.cnr.it/emrps99).
9. The Demise of CRM, Forrester Research Inc., Cambridge, USA, July 1999.
10. High Tech Truckers, National Public Radio (www.npr.org), Feb. 8, 2000.
11. GM Retools to Sell Custom Cars Online, Wall Street Journal, Feb. 22, 2000.
12. Faced With the Online Threat, Developers Race to Improve Their Layouts and Maps, Wall Street Journal, Feb. 8, 2000.
13. Oracle Leads SAP to Procurements Apps, Forrester Research Inc., Cambridge, USA, Aug. 23, 1999.
14. Intelligence Is Moving to the Web, The Economist, November 13, 1999.
15. Is Nonstop Enough? Forrester Research Inc., Cambridge, USA, Feb. 2000.
16. Why Most B2B Sites Fail, Forrester Research Inc., Cambridge, USA, Dec. 1999.
17. Brokered Partner Integration, Forrester Research Inc., Cambridge, USA, Jan. 2000.
18. The Death of IT, Forrester Research Inc., Cambridge, USA, Jan. 2000.
19. Karlgaard R. "Digital Rules" Technology and the New Economy, Forbes, Aug. 10, 1998.
20. Merged Railroads Still Plagued by IT Snafus, Computerworld, Jan. 17, 2000.
21. Brodie ML. *The Emperor's Clothes Are Object Oriented and Distributed,* in Cooperative Information Systems: Trends and Directions. M.P. Papazoglou, G. Schlageter (eds.), Academic Press, 1998.
22. Online Data Brokers, Forrester Research Inc., Cambridge, USA, Feb. 2000.
23. Brodie M.L. and M. Stonebraker. *Legacy Information Systems Migration: The Incremental Strategy*, Morgan Kaufmann Publishers, San Francisco, 1995.
24. Brodie ML. The Cooperative Computing Initiative: A Contribution to the Middleware and Software Technologies Sub-Subcommittee of the Presidential Information Technology Advisory Committee, Jan. 22, 1998.
25. Triggering the B2B Electronic Commerce Explosion, Forecast Analysis, Gartner Group, Stamford, Jan 31, 2000.

Experience-Based Knowledge Management

Matthias Jarke
RWTH Aachen, Informatik V, and GMD-FIT
Aachen, Germany
jarke@informatik.rwth-aachen.de

Abstract

Experience-based knowledge management is the art of capitalizing on failures and missed opportunities. We explain this idea and study three possible approaches within the context of a cooperative information systems framework, building on experiences in a number of interdisciplinary research projects.

1 The Knowledge Management Problem

In the late 1990s, knowledge management has emerged as a hot topic in the management literature as well as in the information systems literature. While information is often considered as interpreted data (i.e. data with semantics), knowledge is considered action-oriented information, i.e. subjectively available information which is able to improve the quality and success ratio of my actions.

However, knowledge creation and management mean different things to different research communities. For example, KM research can be based on social theories of organizational learning [NT95], on knowledge representation research in AI [O'Le98], or on business reference models [Sch95]. To find out what these views have in common, it may be useful to resort to a simple statistical description of the role of knowledge [Dh98].

In figure 1, the cloud in the middle denotes the precision of our knowledge about the success-failure possibilities of some action. Along the horizontal axis, we can divide this cloud by setting the point where we accept or reject the action. Along the vertical line, the division between success and failure is marked. These two lines divide the cloud into four areas, true negatives, true positives, failures of action (type I errors) and missed opportunities for action (type II errors).

Increased knowledge is indicated by narrowing of the cloud. At perfect knowledge, the cloud degenerates into a family of curves with clear-cut decision points and no remaining risks; each such curve corresponds to a precisely characterized situation. Simplistically then, knowledge management (KM) is the effective exploration and narrowing of such clouds in an organizational rather than individual setting. Several important conclusions concerning KM can be drawn from this simple interpretation.

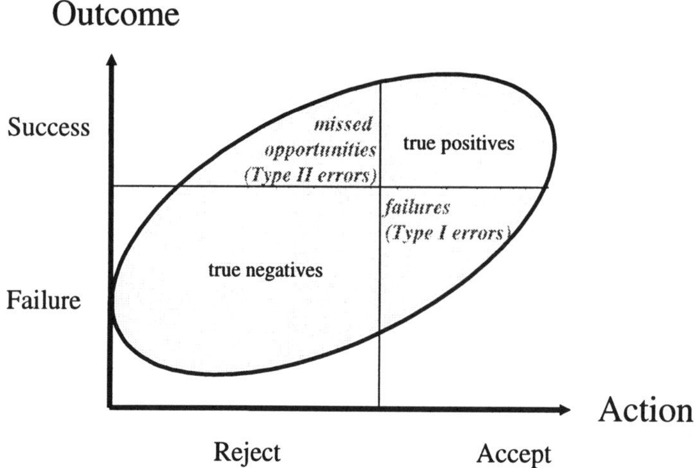

Figure 1: Statistical foundation of knowledge management [Dh98]

In traditional organizations, type I errors are recognized with much higher likelihood that type II errors. Such organizations have a tendency to move the decision line to the right, i.e. they become more and more cautious and bureaucratic, making no errors but missing lots of opportunities.

Learning organizations need an environment in which type I errors can happen without endangering the organization. This implies a certain degree of chaos, which allows to play around with various possible futures. In order to avoid endangering the organization with such experiments, they can be executed as simulations [Dh98] or limited to certain areas, i.e. the organization needs to be fractalized or *virtualized.*

Organizational learning will only occur if these decentralized experiences (positive or negative) are *shared* across the boundaries of organizational units. Sharing increases the communications overhead and has thus traditionally limited the degree of virtualization. Changes in information and communication technology are radically re-defining this trade-off, thus enabling a new generation of knowledge management strategies.

In order to obtain an overview of these new options, this paper interprets knowledge management as a special case of *cooperative information systems* [DD+98]. Cooperative information system support an interplay between three facets: human work practice which corresponds to the social reality in the organization, organizational models as external representations of the organizational structures, processes, and goals; and (information) technology in which an system integration layer provides the glue between hardware and software components.

This framework one of whose predecessors can be found in work by Janis Bubenko during our ESPRIT project NATURE [JB+93, Bub93] is shown in fig. 2.

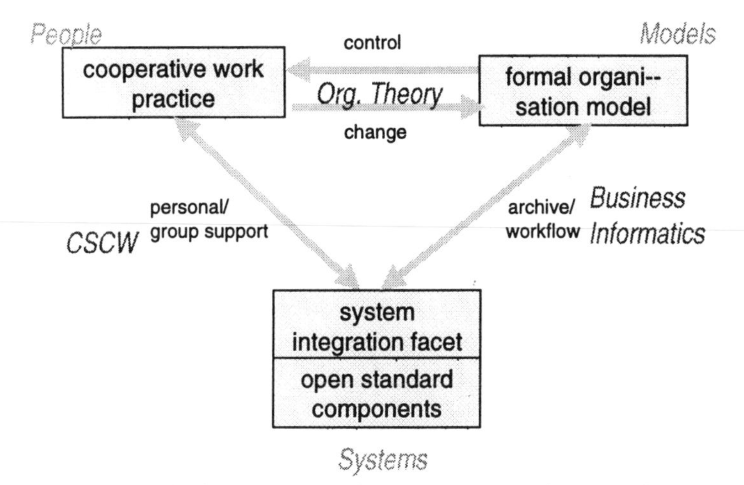

Figure 2: Cooperative information systems framework

The figure shows a fragmentation between scientific disciplines that study interactions between the three facets. Organizational theory has dealt with the interplay of organizations trying to *control* work practice and work practice to *change* organizations. It is interesting to observe that two rather different disciplines, with barely overlapping research communities, study IT support for this.

Computer-supported cooperative work (CSCW) focuses on the empowerment of work practice, whereas business informatics takes a managerial standpoint where IT is supposed to enforce organizational models rather than help individual or group work practice.

Only recently, some tendencies can be observed to bring these perspectives together – a prerequisite for successful knowledge management in a rapidly changing world. [DD+98] argue that this is due to advances in the flexibility of system integration technology, such as the advent of light-weight component software, the liberation of control from application programs into separately specifiable scripts or workflows, the integration of human and computerized communication e.g. via the Internet, and – last not least – reflective process and repository technologies that allow for schema evolution during operation.

Knowledge management can start from different parts of the framework. In the remainder of this paper, we illustrate three such approaches which (along the lines of breakfast definitions in international hotels and not completely seriously) we characterize as the "American" (pragmatic, technological), "Continental" (formal, model-driven) and "Eastern" (collaborative, process-centric) approach. We shall see that each of these approaches delivers valuable contributions which could complement each other in actual knowledge management solutions.

2 The "American" Approach

Pragmatic, technology-centered solutions have taken two paths into knowledge management, one focusing on support for organizational tasks, the other on empowerment of cooperative work (cf. figure 3). We illustrate both approaches with extensions to a workspace system for internet collaboration called BSCW which was initially developed jointly between GMD-FIT, RWTH Aachen, and Janis Bubenko's SISU in the CoopWWW Telematics Engineering Project [AJ98].

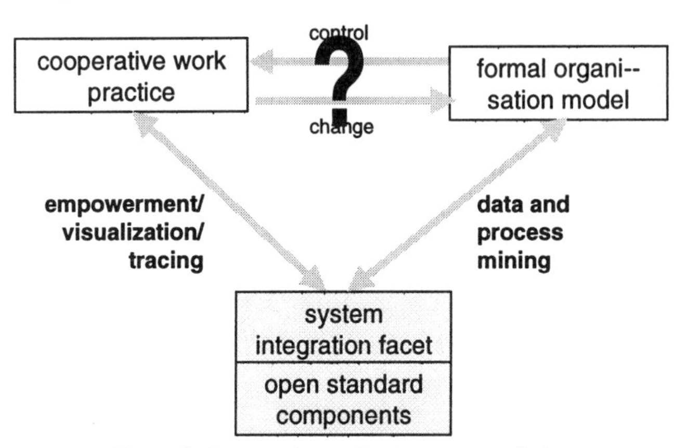

Figure 3: Pragmatics, technology-centered view

2.1 Restructuring a Workspace Systems by Document Mining

In the CAPE-OPEN project, the worldwide chemical industries attempt to standardize a component-oriented approach to the simulation of chemical processes for plant design and operation. The plan was to structure this UML-based standardization process according to a normative architecture of process simulation software, as shown in the hierarchy on the left of figure 4. Accordingly, a folder for each component was set up on a shared workspace on the internet to coordinate the cooperation between the roughly 60 participants from 15 companies.

Unfortunately, even the first step of generating the approximately 160 basic use cases showed that this scheme may not have been fully optimal and that, moreover, several important use cases cut across the pre-defined categories. It turned out to be very difficult for project management to retain an overview of these problems and relationships.

Relying on the fact that a certain degree of terminology coherence was available in the project through the shared chemical engineering background of the use case developers, MIDAS – an exploratory data mining tool developed in our group – was employed to re-classify the use cases according to their full-text similarity [JB+99]. MIDAS offers a map-like interface in which the (dark) mountain ranges separate valleys of similar use cases (shown as dots of different shape and color on the map).

Document similarity is computed using a mix of neural network and information retrieval techniques. The map on the right of figure 4 automatically re-creates much of the pre-specified use case hierarchy but also identifies some additional clusters and cross-relationships whose introduction improved the overall organization. These suggestions proved helpful when linking the use cases to the subsequent UML class and interaction diagrams.

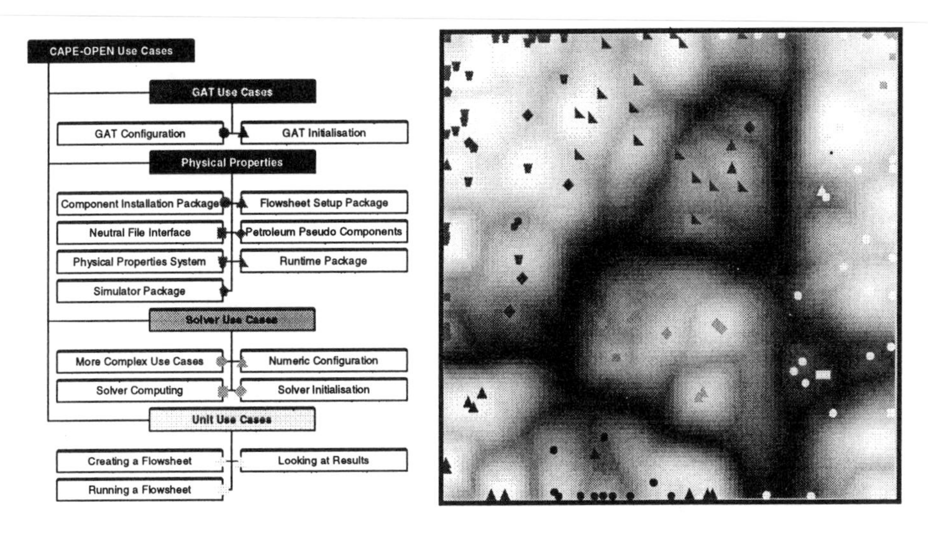

Figure 4: Document mining as an aid to knowledge management

2.2 Ambient Awareness Displays

The traceability of artifacts, technical or organizational, to requirements, goals, and constraints is a key theme in any kind of experience-based knowledge management [RT99]. While traceability in the long term -- an organizational memory of dependencies -- is mainly an organizational issue (cf. section 3), awareness of relevant concurrent activities is a key to successful distributed work practice.

The NESSIE environment developed at GMD-FIT [MFS94] provides a visual mapping of user activities on shared workspaces to a virtual reality (cf. figure 5). BSCW workspaces (left on figure 5) are mapped to rooms, while users are mapped to Lego-like avatars. Operations on folders, such as reading, writing, requesting a video meeting, etc., are mapped to visible moves in the virtual reality. This way, project participants are enabled for chance encounters with co-worker who happen to be available for discussion of issues of mutual interest at a given moment in time. Empirical studies of teleteaching show that this ability to discuss upcoming issues with peers immediately is one of the most important pre-requisites for success. In terms of organizational KM theory (cf. section 4), the important element of socialization is supported. Ongoing research at GMD-FIT aims at ambient awareness displays where physical reality and virtual reality are mixed in cooperative design and awareness features are deeply embedded in the physical environment of users.

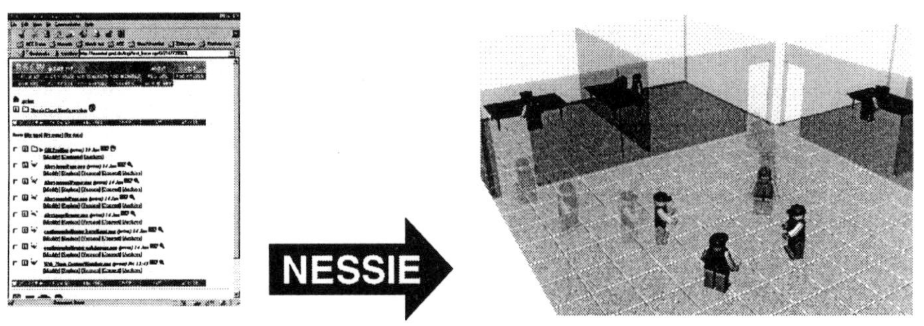

Figure 5: Groupwork awareness using avatars in shared workspaces

3 The "Continental" Approach

Pioneered by SAP, enterprise resource planning (ERP) systems have been a stunning success of conceptual modeling and model-driven information systems development [Sch94]. Starting from continental Europe (hence the section title!), most large organizations worldwide have adopted versions of this approach in order to radically reduce the effort spent on standard administrative software. As figure 6 shows, the approach is centered around *reference models* which are abstracted from practice via case studies. Reference models are intended to standardize work practice but can be *customized* to the needs of specific organizations. The customized reference models are then mapped rather directly to ERP standard software and can thus be used to control work practice in a fairly strict manner.

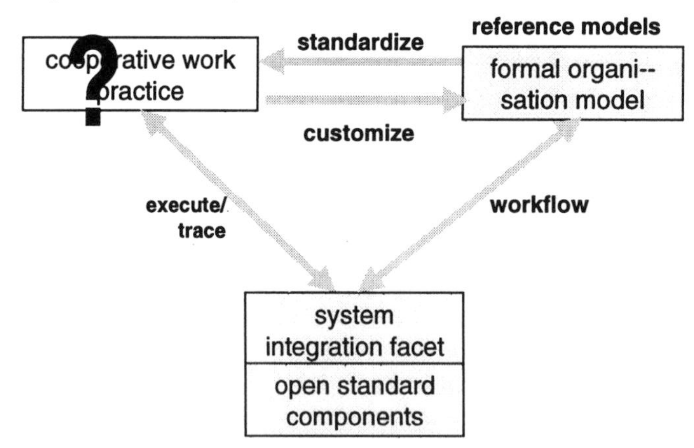

Figure 6: Model-centric approach to KM as in ERP systems

If we look at the use of such systems over longer periods of time, we find a process pattern as shown in figure 7 [JB+93]. A change definition is based on an abstraction of reality created by a combination of as-is analysis in the present system and as-should ideas, then implemented taking into account the social and technical legacy context. As evolution continues for a while, the models tend to become less understandable, sharing of the understanding among stakeholders is lost, and further coherent change becomes more and more difficult. Moreover, a major critique of such models has been that they focus too much on the normal case and are very clumsy when dealing with exceptions.

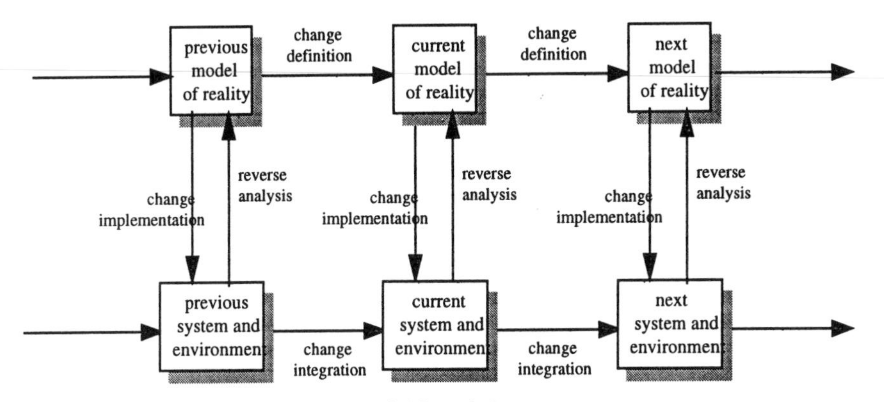

Figure 7: Model-based change process

The European CREWS project [Jar99] has therefore studied the augmentation of this model-based approach by the systematic usage of text-based scenarios and media-based scenes, as shown in figure 8. Scenarios are introduced as middle-ground abstractions between reality and models. In terms of organizational memory they are known to have three major advantages over the pure use of formal models [JBC98]: (1) they focus attention on the *system use* and on the *differences* between old and new system; (2) due to their informal and practical nature, they delay commitment but increase participation of stakeholders, thus leading to more innovative solutions; and (3) due to their closeness to reality, they increase memorization and reuse. An empirical study of the role of scenarios in current practice is reported in [WP+98].

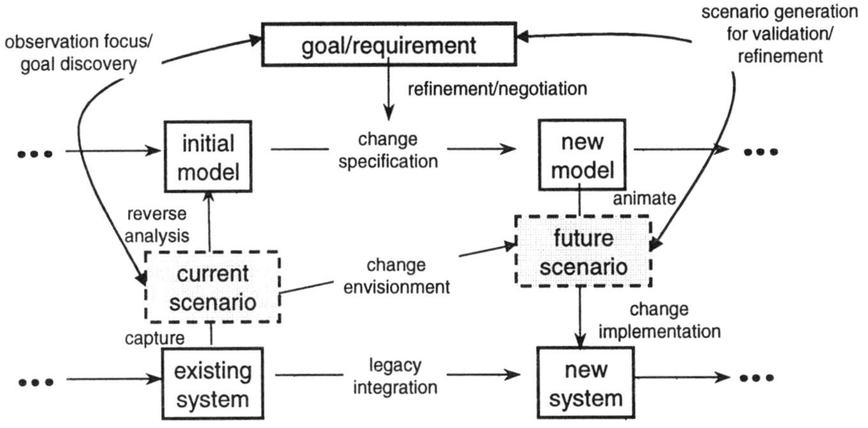

Figure 8: CREWS change process enhanced by scenarios and goals

The introduction of scenarios as a middle-ground abstraction between reality and models has the potential to increase traceability of models and design decisions and thus serve as a better kind of process knowledge repository [RT99]. However, this requires that scenarios are carefully selected and their linkage to the models is preserved in a meaningful manner. The CREWS-EVE prototype shown in figure 9 employs three kinds of links in order to achieve this goal.

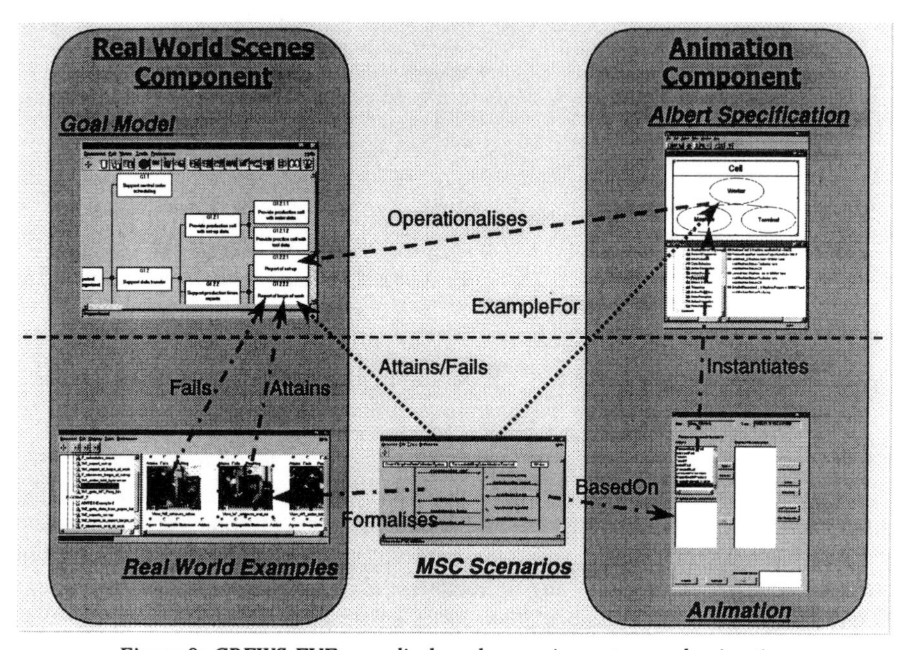

Figure 9: CREWS-EVE -- media-based scenario capture and animation

Firstly, real-world observations from the as-is analysis are captured in multimedia and linked to a goal hierarchy models as positive or negative examples [HPW98]. Secondly, these goals are operationalized to formal as-should specifications using the ALBERT agent language [HD98]. These specifications are then animated in as-should scenarios. Third, using UML message sequence charts to formally describe both the as-is scenes and the as-should scenarios makes them comparable and allows stakeholders to relate the as-should animation to the multimedia representation of current reality, thus giving a better grounding for design reviews.

4 The "Eastern" Approach

In their famous book "The Knowledge-Creating Company", Nonaka and Takeuchi [NT95] describe organizational knowledge creation as moving between tacit work practice and explicit representations. This framework can be mapped into our cooperative information systems framework, as shown in figure 10.

Socialization is the imitation and informal tradition of work practice from one person to another. *Externalization* makes lessons from work practice explicit through documentation or modeling. If such models are sufficiently formal, they can be formally manipulated, analyzed and integrated in what [NT95] call *combination.* However, useful organizational knowledge is only created if it is brought back into work practice by teaching or workflow advice in the *internalization* step.

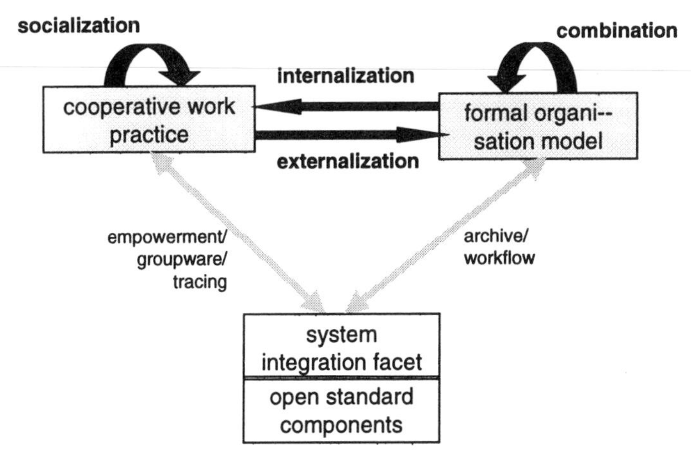

Figure 10: Organizational KM adapted from [NT95]

This approach is, in its initial version, IT-free. The FOQUS project, conducted in cooperation with engineering institutes and companies, has studied information technology support for such an inter-organizational collaborative knowledge creation process, using the example of industrial failure management [KJ98]. The service quality of global failure and reclamation management for complex products has become an important competitive factor in which companies invest significantly through setting up call centers, helpdesks, and similar distributed organizations.

As the ovals in the middle of figure 11 show, distributed failure management can be organized according to what we call the *escalation principle.* As a callcenter agent goes through the classical cycle of problem capture, analysis, and correction, she may be forced to escalate steps to back-office agents, possibly all the way to the initial product and process designers. An escalation step is managed by the workflow system as a speech act [MW+92], consisting of the phases of request, commit, execute, and accept. The growing content information about the failure in question is attached to the workflow in the form of electronic circulation folders.

The knowledge management goal of helpdesk systems is to shorten the response time by maintaining an *organizational memory (OM)* of similar cases and situations, thus reducing over time the involvement of back-office agents. Besides the workflow models which support combination and internalization of formal models captured as a side-effect of case handling, the FOQUS prototype also supports socialization and internalization based on informal media-based advice linked according to product or process characteristics, following the AnswerGarden approach [AM90], cf. the example of industrial gear production in the lower right of figure 11. In the background, all of these aspects are linked through a repository based on metamodels of the products and processes supported.

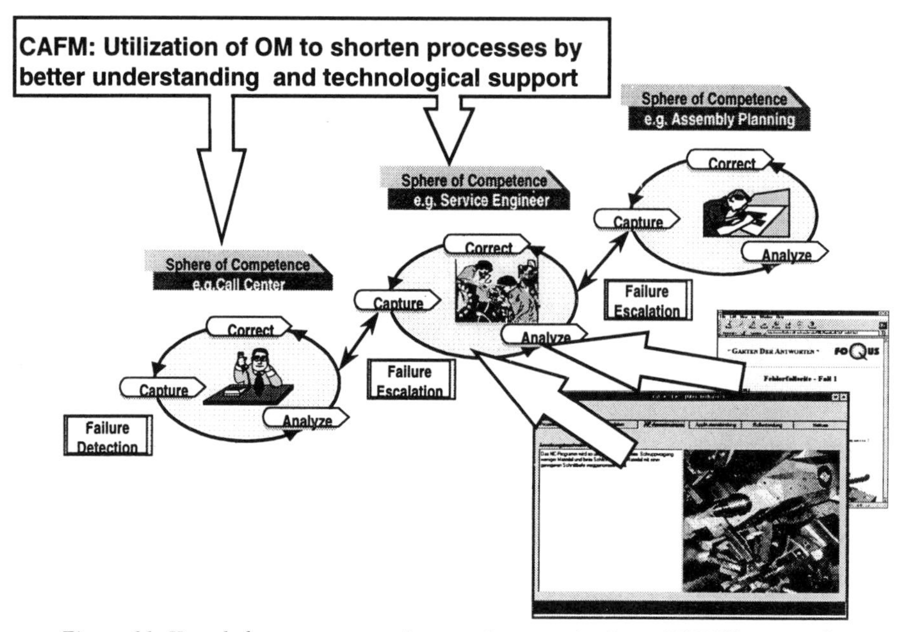

Figure 11: Knowledge management in a service organization – FOQUS approach

5 Summary and Conclusions

Organizational learning and organizational memory are emerging as key competitive strategies in the age of electronic business but a unified view of how the many individual information technology approaches to this topic might fit together has been lacking. As a first step towards a more comprehensive understanding, we presented three different approaches to experience-based knowledge management under a three-faceted framework of cooperative information systems.

The first approach involves the pragmatic usage of novel information technologies which mitigate ad-hoc problems faced by experience sharing and experience abstraction in distributed work settings. The second approach emphasizes the role of goals, scenarios, and full traceability in order to bring an experience-oriented component to a model-centric approach to organizational information management. The third approach draws these aspects together under our adaptation of one of the best-known organizational knowledge creation frameworks, thus suggesting integrated IT support for the full improvement cycle in networked organizations.

The cooperative information systems framework suggests that there might be a much deeper integration of the three approaches than the ones we have studied so far. We have illustrated the approaches with examples from our own work, partially done in joint projects with Janis Bubenko's group whose ideas have been a frequent source of inspiration for my work since the time of my Ph.D. thesis in the late 1970's. Experience-based knowledge management moves the question of how to integrate conceptual modeling formalisms with industrial practice – a constant theme of Janis' research –into yet another exciting direction : getting away from normal-case modeling to capitalize on failures and missed opportunities.

Acknowledgments. This work was supported in part by the Deutsche Forschungsgemeinschaft under projects PRIME and TROPOS, and by the European Union under the projects CREWS, CAPE-OPEN, and a European-Canadian cooperation project on cooperative information systems. Thanks are due to many partners in these projects, as well as to my colleagues at RWTH Aachen and GMD-FIT, especially to Ralf Klamma for discussions about the KM framework presented here and to Wolfgang Prinz for information about the NESSIE project.

References

[AJ98] Appelt, W., Jarke, M. Interoperable tools for cooperation support using the world wide web. Final Report, EU Telematics Engineering Project CoopWWW, Aachener Informatik-Berichte 98-12.

[AM90] Ackerman, M., Malone, T.W. AnswerGarden – a tool for growing organizational memory. Proc. ACM Conf. Office Information Systems, Cambridge, Mass, 31-39.

[Bub93] Bubenko, J. Extending the scope of information modeling. Proc. 4th Intl. Workshop Deductive Approach to Information Systems, Lloret de Mar 1993; 73-98

[DD+98] DeMichelis, G., Dubois, E., Jarke, M., Matthes, F., Mylopoulos, J., Papazoglou, M., Schmidt, J.W., Woo, C., Yu, E. A three-faceted view of information systems: the challenge of change. Comm. ACM 41(12):64-70

[Dh98] Dhar,V. Data mining in finance: using counterfactuals to generate knowledge from organizational information systems. Information Systems 1998; 23, 7: 423-437

[HPW98] Haumer, P., Pohl, K., Weidenhaupt, K.: Requirements elicitation and validation with real-world scenes. IEEE Trans. Software Eng. 1998; 24(12):1036-1054

[HD98] Heymans, P., Dubois, E. Scenario-based techniques for supporting the validation and elaboration of formal requirements. Requirements Eng. 1998; 3(3/4): 202-218

[JBC98] Jarke, M., Bui, X.T., Carroll, J.M. Scenario management: an interdisciplinary approach. Requirements Eng. 1998; 3(3/4): 155-173

[JB+93] Jarke, M., Bubenko, J., Rolland, C., Sutcliffe, A., Vassiliou, Y. Theories underlying requirements engineering: an overview of NATURE at genesis. Proc. 1st IEEE Symp. Requirements Eng., San Diego 1993

[JB+99] Jarke, M., Becks, A., Tresp, C., Köller, J., Braunschweig, B. Designing standards for open simulation environments in the chemical industries: a computer-supported use case approach. Proc. 9th Intl. Symp. Systems Engineering – Sharing the Future (INCOSE 99), Brighton 1999, 69-76.

[KJ98] Klamma, R., Jarke, M. Driving the organizational learning cycle: the case of computer-aided failure management. Proc. 6th European Conf. Information Systems (Aix-en-Provence), 1998, 378-392

[MFS95] Mark, G., Fuchs, L., Sohlenkamp, M. Supporting groupware conventions through contextual awareness. Proc. 4th Europ. Conf. CSCW 1995; 253-268

[MW+92] Medina-Mora, R., Winograd, T., Flores, F., Flores, C. The action-workflow perspective to workflow management technology. Proc. 4th CSCW Conf., Toronto, Ca, 1992; 281-288.

[NT95] Nonaka, I., Takeuchi, H. The Knowledge-Creating Company, Oxford Univ. Press, 1995

[O'Le98] O'Leary, D. Knowledge management systems: converting and connecting. IEEE Intelligent Systems 1998(5):30-33.

[RT99] Ramesh, B., Tiwana, A. Supporting collaborative knowledge management in new product development teams. Decision Support Systems 27 (1999); 213-235.

[Sch94] Scheer, A.-W. Business Process Engineering. Springer 1994.

[WP+98] Weidenhaupt, K., Pohl, K., Jarke, M., Haumer, P. Scenario management in software projects: current practice. IEEE Software, March 1998; 34-45.

A 2025 Scenario and Vision on Stream Data Management

Alfonso F. Cardenas
University of California,
Los Angeles, U.S.A.

Abstract

Technological advances happen as a result of technological challenges posed by visions of what we are determined to achieve. The vision of two major multimedia data management scenarios that touch most of us is described: firefighting in a suburban environment and the digital medical patient. The scenarios are described as if one were already living them in the future 2025 year, recalling problems we now face as we enter the 21st century. There are sufficient indications today of advances that will be furthered, and of the strong possibility of new advances, in computing, communications, power sources, sensors and electronic data interchange needed to achieve the vision of the two scenarios described. Following the scenarios we present the multimedia database management features towards which we are now embarking to enable the data management and access needed to support such scenarios.

1. A Firefighting Scenario -- Vision

It is the year 2025. A large fire has erupted in a brush and hill area in the vicinity of major housing developments in the outskirts of the city. Firefighter crews in 4-wheel vehicles and larger firefighting trucks rush to the scene to control the fire from spreading and to extinguish it. Fires in such rough terrain of deep canyons and hills, particularly in dry and windy conditions, spread quickly and unexpectedly, thus endangering property and lives. Modern computer and communications technology assists in this challenge. The firefighters have an advanced communications system and hand-held devices to permit them to get up to date information about the terrain, inhabited building layouts, and situation and character of the fire that they are to extinguish. Each firefighter group carries such a portable device that is a combination of the telephone, computer processor and data gathering or sensor device. It has a small screen of around 4x4 inches arranged in a double/triple panel that can fold out and snap to be 8x8 or 12x12. The device provides displays of: the terrain with various characteristics such as elevation and type of vegetation shown in various colors, layout of roads or major trails, a representation of where the fire is on a real time basis, the layout of where dwellings are, icons showing where firefighter and various firefighting equipment are deployed, sources of water, etc.

The device has voice communication, like a telephone with voice activated telephone connections, to other similar devices, or to command and control centers. The device includes GPS type of instrumentation identifying its location that is transmitted on a real time basis to a command and control center for access by others. Voice output can communicate various types of information in the vicinity of the device, such as dangerous conditions, distance to closest fellow firefighters, instructions from command and control centers, etc. One can of course use it as a telephone to communicate as a normal wireless phone and to establish communication with any of the other designated similar devices.

These devices are also available to non-firefighters to communicate. It has become a consumer product available like the wireless telephone, although more expensive of course.

Technological advances early this 21st century have permitted the deployment of such devices. There is multimedia data base support for all the pictorial/image data needed, such as the terrain and its characteristics and the dwelling layout. The data base and integrated data access and display facilities support various operations such as the ability to zoom in and out of various displays so as to see them as if one were far above the ground or practically standing on the ground. A pertinent subset of the terrain and dwelling data base stored in command and control centers is transmitted to the portable device's local data base. This subset is refreshed almost on a real time basis depending on where the device is located. There is information about dwellings such that one can practically take a tour of any dwelling layout and see in almost three dimensions and fish tank virtual reality (VR) perception of what the outside and even rooms of the house inside look like. This is supported by advanced versions of older products such as Apple's Quicktime VR and Microsoft Direct Show, which have been integrated with data base facilities. The data base stores video or 3D representations of views of dwellings and any desired area which can be seen on the hand held devices carried by the firefighters. One may practically walk through and see up and down and to the side from outside or inside a dwelling with 360 degree angular motion from a fish eye. A property owner determines to what extent data about his house is to be made available in such data base, thus protecting the privacy of the owner.

Data communications advances support the bandwidth and reliability needed to communicate the large volumes of data when needed. Large volumes are involved in video displays or 3D or VR oriented visualization. Streaming video works such that one can see the video without downloading the whole video as was needed in the past. Communications are highly reliable such that there are no dead space areas as there was with wireless phones in past years.

There is voice support to indicate driving or walking instructions to whatever destination one indicates via voice. This facility is an advanced version of what started to appear in automobiles in the early 2000's based on GPS to direct drivers to their destination. In the early days one had to type in the destination and read the directions on the screen – a cumbersome process.

Satellites monitor major areas of concern on a continuous basis. Fire detection devices transmit to the pertinent satellites the triggers to start gathering the image data of the endangered areas. Such real time information gathered from satellites and other data gathering devices is consolidated in a data base and processed in command and control centers to produce the images showing the location of the fire. Satellites also monitor the deployment of firefighters and equipment should these be hurt or destroyed in the firefight.

A major recent functionality provided is the ability to see a display of the predicted evolution of the fire should it not be put in check. A mathematical model has been developed and continues to evolve to predict this fire movement as a function type of vegetation, dryness and humidity conditions, fire brakes, the slope of the terrain (calculated from topographical elevation), wind data, etc. This model is run every so many minutes in a command and control center producing the information to then compose the video images with the fire movement superimposed on the images of the terrain and dwelling areas in 2D or 2.5D. This information is transmitted to the pertinent devices on the field along with instructions from firefighters directing the firefight. Firefights can thus take appropriate decisions locally with such information.

2. The Digital Patient Scenario -- Vision

As a patient, parent of a patient, or doctor in the year 2025, one has today much of the digital patient record not available in the early 21st century. Here are highlights the digital medical record.

A wrist watch, and alternatively a more powerful and larger palm device, is available that provides pertinent medical information about a person at all times. It provides personal information that one frequently forgets, such as what blood type, medicine and other medical information one has – important information needed in case of emergency. Furthermore, now there are micro sensors in the watch and other sensors that may be placed in our bodies to gather important medical information, such as diabetic conditions, blood pressure, temperature, etc. that is sent to the wrist watch for consolidation and display to the person and for transmission to a doctor's office and hospital of choice. Some of the probes are non-invasive and others are invasive. This information (e.g., stream of blood characteristics through time, electrocardiograms, etc) is stored as streams locally and then transmitted on a wireless basis periodically upon demand for longer-term storage from the wrist watch to one's home computer and to a doctor or hospital data base. As one grows older or is afflicted by health challenges, additional kinds of probes may be placed in the body to measure and monitor particular conditions.

The medical wrist watch has an option that allows one to receive via wireless communications various such information from another person with a similar wrist watch. Thus, one can monitor a child's or spouse's condition at all times wherever the person may be. A stream of data for each condition measured in a person's body can be transmitted to a home computer and/or a doctor's office.

There is a professional doctor's wrist watch and a more advanced hand held patient palm size device that may show images of the data streams of selected patients, for example, the electrocardiogram of the patient on an almost real time basis; thus doctors may readily receive critical information from several patients.

Management by exception principles are utilized. The situations under which a medical condition needs to be flagged are programmed and thus one may receive an audio warning of this event from the medical watch or palm device. Advice on what to do may be programmed and available in text in the display for some of the more straight forward conditions (e.g. Abnormal heart beat level, call doctor immediately).

These advances have prevented many premature deaths and disabilities due to early detection of problems and faster medical attention than what was possible in the early 21st century. In the past, the check up of one's body was primarily done when one visited personally the health care provider after waiting for sometimes critical hours of travel time to see the doctor or days after calling for an appointment to see the doctor. Now, critical information can be gathered on a real time basis and communicated to the health care provider. Of course, one still has to go physically to a medical facility for personal consultations and various data gathering procedures, such as the scans to produce various types of images (X-rays, MRIs, etc.)

Each type of medical image taken such as MRIs and CTs is stored as a stream of images for viewing. A stream may be visualized in 2.5 or even 3D on a large monitor by special software, as one may want to see for example a 3D image of what one's heart arteries and wall blockage looks like from inside. For a number of such compute intensive tasks, one uses workstations rather the medical watch or palm computer that is limited in compute and storage capacity.

We have now much of what started to be called the "integrated digital patient" at the beginning of this 21st century. Each patient's alphanumeric, long text, image and audio data is digitized and stored in the data base: (a) in the wrist watch the most recently monitored data and selected samples of longer term data are stored, and (b) in the data base in the health care provider's offices and possibly in the data base at the patient's home all data gathered is stored. Most such multimedia data is integrated and may be accessed by and transmitted to those authorized. Thus, although a patient may undergo various treatments at different places by different doctors, the data is integrated and readily available (all this facilitated due to common medical information data transfer protocols agreed on by the medical industry after years of efforts since late last century). We avoid now the data redundancies and associated costs of retaking of images so common in the old days: frequently image scans such as X-rays were taken over and over again by different health care specialists as the data was on film and not readily available and shareable electronically.

With such digital patient data bases available, we can better manage the various payment and insurance coverage for each patient. The insurance data base systems can now communicate electronically with the insured patient data bases and expedite the costly and awkward payment process of the old days. Most medical systems software in doctor's offices can now communicate with other systems such as those of insurance companies via medical industry wide protocols for electronic data interchange. Thus, a doctor's invoice goes electronically to insurance company adjusters for payment approval or negotiation, as well as to the patient electronically if requested – paper invoicing is still sent to the patient unless the patient indicates otherwise.

One can now access electronically a wealth of information via a workstation: the patient data for a family including streams of images, medical test results, medical measurements, etc; the doctors' and pharmacy charges related to each particular stream and stream element, e.g., particular laboratory test; and the insurance company record of payment to the doctor or pharmacy, or reimbursement to the patient, tied to the particular entry in the streams of medical information.

Advances in medical informatics provide now a more visual interface to the digital patient record. An example of the highly visual access to the digital record in a timeline paradigm is shown in the next section.

3. Multimedia Stream Concept and Data Management

Now back to the present earlier time. Various major advances as we start the 21st century particularly in computers, communications, sensors, power supplies, and data interchange protocols and collaboration among various organizations, enable the prior firefighting and digital patient scenarios. We address herein enabling advances needed in multimedia data management.

A *stream* is a chronologically ordered set of objects or entities, such as a series of medical images or a series of clinical laboratory reports of blood analysis taken at different points in time, shown below Figure 1(a) and 1(b) below. Streams are usually the inputs to many visualization methods or modules, for example a 3D rendering module for objects usually takes as input a stream of 2D images of the objects.

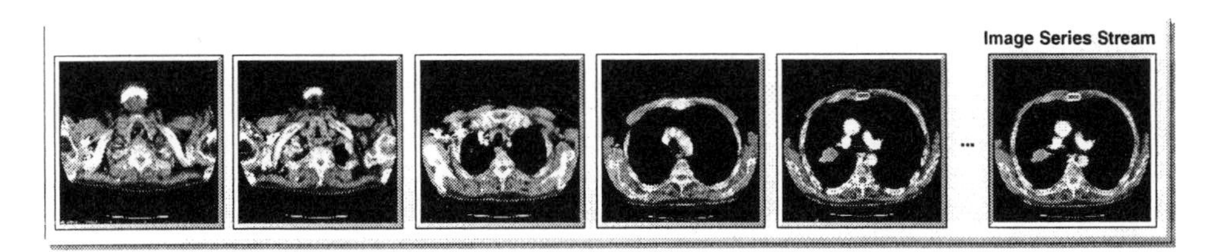

Figure 1(a) A stream of medical images

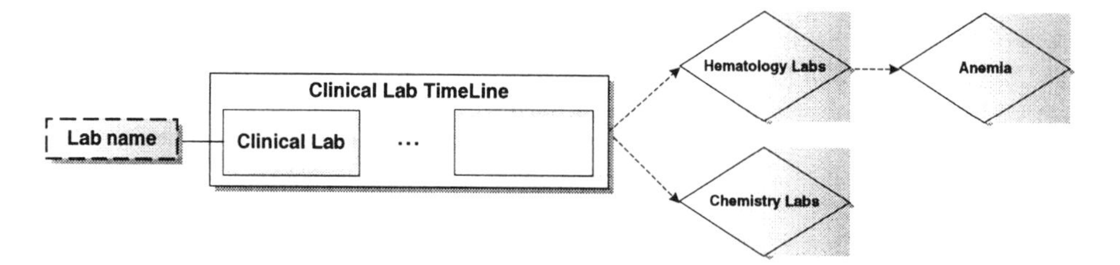

Figure 1(b) A stream of text clinical laboratory reports, from which substreams may be obtained, e.g., a substream composed of the lab reports dealing with hematology and including the subset indicating anemia conditions.

This major building block in multimedia data base support complements the other more traditional data base constructs entities or classes, attributes and relationships. Multiple streams of data are involved in the firefighter and digital patient scenarios above. Each sensor gathers measurements that are generated as streams of alpha-numerics, pictures/images, or audio. The data base at both the hand held devices and the larger more permanent computer devices have many streams of information. The ability to electronically and cost-effectively gather, store, share, relate and communicate streams of information is a major advance.

Multiple streams may be related. In fact, we spend much time in many domains dealing with these relationships. For example, as illustrated in Figure 2, a stream of insurance payment for two patients A and B may have a payment that covers all the costs be related to a set of streams for Patient A, while the other payments may be related to single stream or even specific elements in a stream (e.g., a laboratory blood test) for Patient B. We would like to see easily such streams and the relationships involved. The data base relationships of DBMS such as CODASYL and relational object DBMS and the document relationships proposed by XML late last century were the precursors for this advanced capability to relate information.

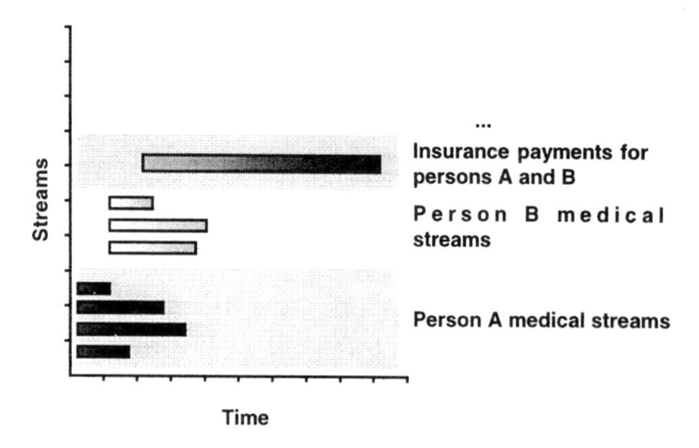

Figure 2 Multiple streams - Example

Advances in electronic data transfer protocols and support for streams enable much of the data management needs in the two scenarios. This flexibility enables the users to see on demand the various streams of information. The timeline access paradigm in the digital patient scenario and the firefighters device are good illustrations calling for different data gathering sources and different organizations to use common data formatting and exchange protocols to allow presentations of data as if it were all in one place, while in reality it is distributed but logically integrated or related by various means. Advances in the World Wide Web are expediting this connectivity.

Figure 3 shows a timeline display of a prototype system being developed with the vision of the digital patient and the underlying multimedia stream data base support briefly outlined above. For a patient, one may see the text report by the doctor for the particular examination or visit, along with a graphical representation of each stream of information from left to right with the date of the information shown next to the corresponding stream. The example shows several streams of different areas of medical concern: images of the evolution of a chest tumor (area indicated by the cross bars), a graph showing quantitatively the white blood count (WBC) through time, and streams of icons showing the existence of information at the indicated point in time for each of several streams such as Pulmonary, Hemo/Oncology and Renal.

Clicking on any of the dated icons in the stream of interest will show the detail of that medical concern. One may get, for example, the report of blood tests at that point in time; or the shown series of lung tumor images at a point in time, or, by invoking an option, a video showing the 3D visualization of this lung tumor image stack. A similar timeline interface may be used to show the timeline of payments and reimbursement by the insurance company, that is, what insurance payments have been made for what specific medical activity shown in the timeline (a payment icon appears whenever a payment was made, and a click on the icon displays the detail of the particular payment).

The timeline display in Figure 3 assumes the use of a large display screen. A small subset of the information, particularly that being gathered via sensors in the patient body, will be stored in the medical watch or smart card and displayed on a small watch screen or palm device -- and subsequently uploaded to larger analysis and archival storage.

These data management advances are essential to achieve the firefighting and digital patient visions of 2025.

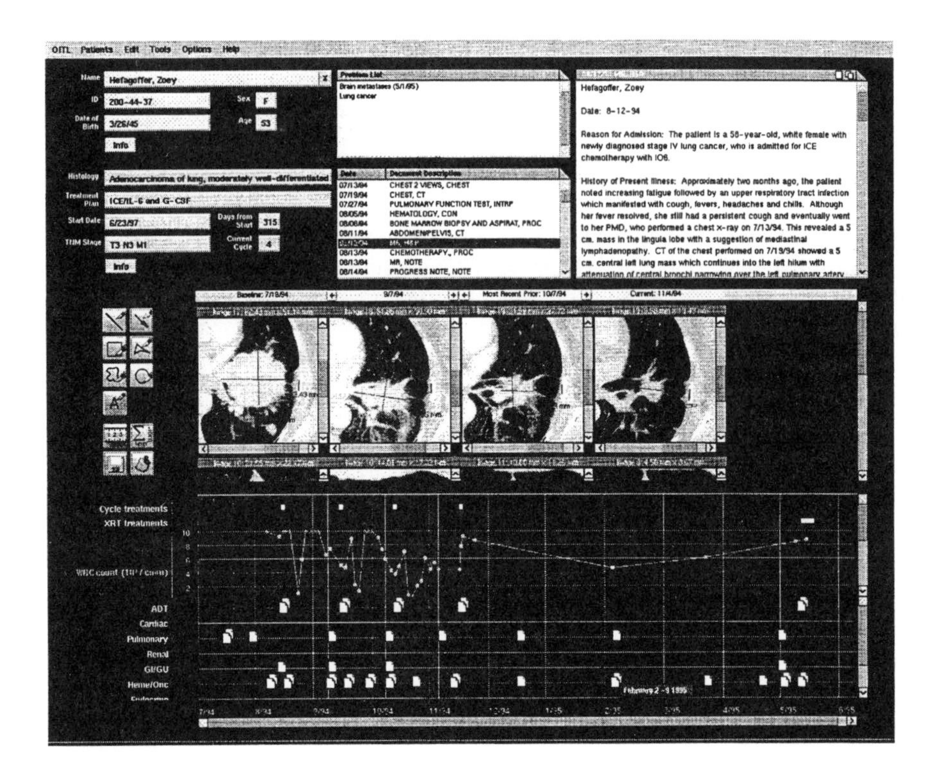

Figure 3 Timeline display of a digital patient record, with several substreams of multimedia information (images, graphs, icons denoting documents and other types of information)

Access as Value Addition

Tomas Ohlin
Adjunct Professor
Department of Computer and Information Sciences
Linköping University, Sweden

Abstract

Value added information services often stress content that is given, or that generates added value. But also form plays a role. This text will comment on the value and consequences of simplified access possibilities for information systems users. Effects of disappearing borders within information systems will be discussed. These effects are likely to have broad consequences in many parts of society.

Background

It may be allowed in this context, as a point of departure for this discussion, to remind ourselves the time span we are referring to when we analyze expanded access to information resources. In an early contribution to the analysis of computer operating systems principles (Bubenko, Ohlin, 1971, chapter 10), it is noted:

> "Today´s .... computing systems .... can hardly be regarded as complex, considering what we may expect in the future, with networks of different size computers, databases and terminals ....
> ..... a development- and user-friendly operating system may be very hardware resource consuming. We are left to evaluate possibly decreased personnel costs, increased "user access value" and possible gains concerning system starting up time, against possibly increased hardware costs".

The discussion about balance between user access systems demand, and system performance, apparently has roots at least from the early 70s. This paper expands further on the effects of change in system access.

Over the years after the time when the above statement was presented, it has come to be clear for many that increased user convenience and decreased total system cost go together. This is partially because of decreased hardware costs, something which the competitive market has had as a consequence. But it is especially the fruit of a value change in the minds of many. The position of the user of information resources today is simply considered to be more important than was the case a few decades ago. This is partly because user communities today are larger and more active than they used to be, but it is also the fruit of a qualitative change of opinion.

Borderless systems

At present, around the change of the century the statement is often heard that "borders" connected to and situated around information systems are being observed to be less sharp than they used to be. The word "seamless" system is also increasingly used. This situation to some observers seems "evident", and is simply taken for granted. "Information systems make it possible to cross over borders", it is said. And that is that. The effects of this are being noted with increasing interest by many, and consequences are starting to be discussed. This situation has not only technical, but also organizational, social, economical, legal origin, and they are, to this authors opinion, not at all concentrated around information systems only. It is worth elucidating some of the reasons behind. Possibly there are lessons to be learnt about the expanded organizational, economical etc applications of this development. Will its importance increase, or will it change, perhaps with a result in an opposite direction, so that there is less concern?

We argue here that a decrease in application importance over time is not likely to appear. Let us discuss the reasons for this.

First, it is easy to note that the economic and technological development to increasingly low-cost communications technology, at the change of the century makes it ever cheaper to use border-overlapping "methods" in information systems. Telecommunications facilities of today make this possible, cheap and accessible. It is simply easier to communicate because networks are easily accessable, and physical borders are easy to overcome through efficient programming. These new bridges are usually defined to be stable and firm.

There are also other reasons behind the changed view on the decreased importance of borders in information systems.

Conceptually, there exist unlimited different views on the concept of a systems "border". What is regarded as a border in one view is an inner systems characteristic in another, or an outer in a third. Thus, the discussion on borders here does not refer to a fixed, well defined and standardized view on the concept. Rather, it is applied generously, on the concept "border" as it is seen in many actual applications.

It is possible that some roots for "border dissollution" have their origin in the information science field. Mathematically and logically, a border is infinitesimally narrow. However close you examine it, it is just as narrow. In "real life", however, socially, legally or organizationally, a border often is quite substantial, more like a thick hedge, a hurdle. You can even measure its thickness.

Borders in information systems are narrow, sharp. They are logical. The closer we look, the more details around the border we can find, but the border itself is untouched. The border stands there like a statue, invariant, unreachable by rain and storm. We can build a bridge over it, but we cannot touch it. When we refer to and discuss this border, we refer to its origin, the presumed border "core", that nobody really has seen or touched. We can build a bridge over that core, a small but dramatic bridge, by defining an adequate core bridge concept and include that in the actual system. It may be a simple construction, but it overlaps the border core. We know it is there, although we don't see it very clearly. As we depart from this close view, this characteristic is still there, we know there is a bridge in there, but we don't quite know where. It suffices to know that there is a bridge there somewhere, available to be used for somehow crossing the border. The border is not as relevant any longer. Crossing over it has been made possible.

A purpose for this discussion is to make plausible that even small border crossings may have large effects.

Let us look at a more market oriented technological background for the decreased border importance.

The legal and organizational system around information services in many countries has traditionally been different for "database" services than for "telecommunications" services. The roots are different. Database services were born on a market, a market where influence from western countries, especially USA, was strong, sometimes dominant. In the 1970s, the supremacy of IBM was so intense that this company was almost considered untouchable. There existed a market de facto monopoly. The functioning of market principles even around this exceptional monopoly was proved when it turned out that even this position could be shaken, moved. Sometimes only a limited market mistake is enough, like the understanding for the importance of the personal computer. The change to today's market dominance by another market player (in this case Microsoft) was considered to be almost impossible two decades ago. However, the facts of today prove that hardware and software development was born on a real market where competition is a reality.

The situation for telecommunications services was different. In many countries, there were natural or legal monopolies around basic telecommunication services, and especially telephony. In Sweden, this monopoly at the start of the former century was born as a private and natural monopoly, but it grew to be placed closer and closer to public ownership. This was partly for strategic and, later, military reasons. As computing systems were expanded in the 1960s and the 70s, those systems that wanted to use telecommunications facilities had to cooperate with a public structural monopoly for telecommunications. This monopoly in the beginning definitely was not applying competitive principles for its presence on what was to become an information systems market. But this reluctance was changed when the need for increased resources became evident, and when the principles of open markets from the computing services came to be putting increasingly stronger pressure on the telecommunications systems.

As computer technology was developed to be more and more flexible and efficient, it came to be obvious for telecommunications systems engineers and managers that more general purpose computing equipment was natural, and soon necessary, as building blocks inside many types of telecommunications systems. This was an effect of availability of increased hardware and software efficiency, both born on the computer market.

As these computing systems elements technologically became integrated into the expanded telecommunications systems, it became evident that also the principles for their market existence had to be integrated. It turned out to be impossible to implement public or private monopoly control over integrated systems where many subsystems were organized after market principles. Development and delivery was simply principally too competitive. One side had to give up.

The fact that most information systems then as well as today integrate both database and telecommunications technologies lead to the situation that the legal and organizational forms for one of the two principal sides mentioned had to give up. This turned out to be the monopolistic one - a "market decision" that was not evident in the beginning, and which caused much questioning at the time.

However, as a result, telecommunications systems organizations were moved from a monopolistic environment to a market. The divestiture of AT & T is an early but notable effect of this change. This demonopolization is a shift that has generated many followers. Still, in the beginning of the new century, in many systems this change turns out to be more of a complicated cultural shift than it is a technological and organizational one.

With a common market principle behind, it is today natural with business motivated border crossings inside many information systems that use both data base oriented and communications subsystems. A system convergence is taking place. There are many effects of this. Some important ones are cultural.

Also legally, in many countries, this convergence is demandingly difficult, because telecommunications systems often carry much of the mass media - especially TV and radio - who nationally often are both legally and organizationally regulated in quite different forms than other, industrial and educational etc, information systems.

A third effect behind this cultural change concerns the position and the desires of the information systems users. Users traditionally have felt and expressed an unsatisfactorily limited influence over information system structures. This has included the system structure decision making. The system providers and equipment deliverers from the beginning have been placed at the steering wheel. Users have demanded better access to the decision making. This has been supported by organizational efforts, and increasingly strong user opinions have been presented to the information system constructors. Concerning communications systems, the users have found monopolistic telecommunications organization less flexible, and demanded a greater influence through plurality.

As a result, the border between market and monopolistic structures has been challenged.

Summing up the reasoning that has been referred to, we find ourselves with a clear result about border importance: borders are not as apparent any longer inside information systems. There may also be prejudicates. Those borders that do appear, are not considered and being observed as seriously as was the case a decade ago. It is increasingly easy to build bridges, whether real or virtual.

It is interesting to note that this, partly theoretical and to its applicability primarily limited, conclusion seems to be having exceptionally strong effects in environments outside of the information systems themselves. The effects of this have shown to be of many different types. They are social, economical, legal, and cultural, and more. They have also shown to be close to organizational and political changes. National borders are no longer as important as they used to be. It is simple to transfer data, but - more important - a change of communications culture has taken place.

User distance

Dissolved communications borders are relevant for information flows. But they also imply a change of player roles. Players who traditionally were active inside a system, now find themselves related to external activities that they often earlier simply did not know of, or were familial with.

It is of interest to consider how the decreased importance of borders affect the distance between information systems and their users.

Traditionally, when the information system referred to well defined separate pieces, hardware/software and communications equipment, there were evident borders between the system and the user. Early, in the beginning, this was especially evident, as there were physical and logical glass doors and windows around the "computer room". You had to book time for access to these amazing machines.

As time has passed, focus, or the heart, of the information systems, has been transmitted to positions closer to the user. The rational parts of a system are being standardized to form, and more concern is placed on the non-rational, emotional, parts, "the soul in the system".

How is the user affected by this?

As the communications functions in many systems grow to be more important, the users find themselves at a physical distance from each other. But logically this naturally is not so. Users become part of a larger system, a system that often grows to be more and more distributed in structure. Then, as systems borders and boundaries are lowered and overcome, as they even disappear, with an exaggeration perhaps we may say that also the users' distinct appearances almost disappear from sight. They find themselves somewhere, out there, in the outer parts of the system. The system takes the shape of a continuum, where there are no barriers or discontinuities, and where the users are active in the outer but integrated systems parts.

New systems tend to become increasingly continuous in shape. Their parts float. In this, the user is no longer as well defined as she was before. She takes the role of a direct partner, with the original meaning of the word.

Naturally the user wants access to the fundamental parts and functions of the system. She demands not only access but also possibilities to take part in the decision making about the system, the strategic positions to be taken on what next should be given priority. This is, at least potentially, made easier through a borderfree structure.

Broadened empowerment

With this in mind, we may consider the division of information access, or "power", inside a system. Lowered borders invite to division of access, to keep the system going, to keep it "happy" by inviting all systems elements and functions to take part in the systems' work, aiming towards the overall systems goal.

With lowered borders, the user takes a step towards being active not only as a consumer, but also in the production of information, or, to follow Toffler, she becomes a "prosumer".

The closer information systems mirror real life, the more their administration and maintenance may also mirror the type of social structure that we have built in our real world. Users are systems citizens, and the systems basic software is the public administration.

Within a system that is characterized by a lack of borders between processing facilities and input and output resources, the system usage is a continuation of the system itself.

The digital divide

The concept of digital divide often refers to unbalanced access to the Internet and subsequent applications resources. Socially, it is generally considered desirable to spread information access to all users, all citizens. This opinion is shared by many. We may formulate fundamental democratic reasons for this.

In any system, because of lack of manufacturing precision and for other reasons, there are discrepancies, functions that more or less temporarily do not work "up to specifications", parts that do not behave. As a consequence, users who are in touch with these functions at times find themselves at a disadvantage, being left without adequate access to important system resources. There appears a "digital divide". This also may be the fruit of unbalanced user knowledge, or incomplete systems planning.

Let us consider the Internet.

A recent package of supportive actions in the US has, at the turn of the century, been suggested for spread of access to the Internet. President Clinton has revealed the details of a multibillion-dollar proposal to ensure that all Americans have equal access to the Internet. Clinton's plan to bridge the digital divide offers $2 billion in tax breaks to technology companies in exchange for their participation in the effort, $150 million in technology-training funding for teachers, $100 million for the creation of 1,000 technology centers in low-income areas, $50 million to help low-income families purchase computers, and $45 million to fund the creation of technology projects in low-income areas. In addition, the plan contains $25 million to help the industry provide broadband services to rural and other areas, and $10 million to help train Native Americans for careers in technology. The hope is said to be that the plan will make Internet access as common as telephone access in America.

This last statement does not get the same meaning in the US as it gets in Sweden and Finland, with their already close to universal telephony access. Comparing to these two countries, it is an even stronger statement. Similar plans, if not of the same quantitative size, are presented in other countries.

These measures against a digital divide are supported by the evolution mentioned earlier, with decreased borders in fundamental information systems. The efforts are aiming at common access to fundamental information and communications resources. Reasons may be looked upon as democratic rather than market oriented. For instance, providing Internet access to all users of fundamental public information systems, citizens as well as companies, refers to a type of value addition that increases the common resources for everybody in a country. A smooth society.

We argue here that, because of their borderless nature, this development is more natural for information systems than for other types of systems. With an example from the telecommunications organizational concepts, value addition in the form of expanded public systems resource access may be looked on as a type of expanded "universal service" within an environment. Lowered systems borders no doubt will support distribution of such services.

Value addition forms

The conditions for value addition for the information systems discussed are changing as a consequence of the structural change that follows lowered borders. Value addition is looked on as a central concept for market acceptance of new ideas, products and services. The type of value addition that now appears or may appear, is built on a continuous platform.

Traditionally, the existence of friction is a basis for successful market activities. The successful business woman markets products that successfully treat relevant types of systems friction. What happens on an arena where there are less and less borders and friction? Border bridges, the increased continuums, overbuild friction. If friction may help create the basis for successful market transactions, this may result in decreased competition possibilities.

The smooth and continuous market then perhaps is not as inviting for innovative development, and not potentially as profitable as the market where there are bumps, mountains, valleys, and friction. Smooth future information systems may then, at least in principle, turn out to be less inviting for profit seekers than today, because of lack of discontinuities and borders.

But then, there naturally are markets where one-dimensional friction is not the only parameter of importance. There are markets where success, where the function to optimize, is multi-facetted. There, smooth and soft information systems may find "profitable" natural environments. These may be looked upon as democratic fields. There, the soft and smooth systems will have good prerequisites and possibilities for success. Satisfactory systems access naturally is valuable for efficient system usage. Valuable access generates a high degree of value addition.

As access to information systems becomes increasingly integrated, borderfree and smooth, such value addition that is related to this access, also becomes less visible, apparent. There is an almost quantitative connection between access and this type of value addition. The smoother we make the system access, the smoother we also find the degree of value addition that is related to the access.

Systems resources, in this case like access to data bases and communications facilities, in short systems power, will likely be distributed more equally and democratically in systems with low boundaries. If, on the other hand, we fear that user inequality, for example for social reasons, will appear, there is a need - at least temporarily - to build borders around these inequalities, in order to define specific counteractions there. However, over time also such artificial borders will be overbuilt by natural smoothening efforts.

The position for innovation

Difficult situations invite innovation in order to overcome the difficulties. Borders create difficulties. If the systems development implies decreased amounts and forms of borders, and increased systems continuums, we may find less room for innovation. This may lead to more static systems, systems that are not dynamically being developed as fast as before, and also to less creative systems. In this respect, smoothness can be balanced against the bumpy road, ground for innovation.

Smoothness thus is a positive factor for user friendly systems access. A smooth system also makes moving around inside the system easier. It invites flexibility. On the other hand, smoothness may counteract innovation. Will future information systems be less innovative and more stable because of their user friendliness?

Summary

Activities that aim towards increased access to basic communicative resources, Internet for everybody, are desirable as a democratic platform for users, citizens and companies. This type of smooth value addition forms basic prerequisites for efficient information services of many types. Without doubt, this is for many democratically desirable.

However, there is a price to pay. The smoother we make the platforms, the harder it may be to find room for creativity and development. Market oriented motives often are one-dimensional, but democratically oriented motives often are more multi-facetted. As more capable basic resources are being considered important for users, increased interest is being placed at the multi-facetted motives for future information systems. It is surely more challenging to optimize towards these, but resulting systems may likely turn out to be more sustainable in the long term.

As users are integrated in the systems of tomorrow, systems structures will likely be stronger, and user-less systems will certainly not appear as useless systems. Smoother access to many concerned will likely be considered to have positive value for large groups of people concerned.

Increased smoothness in information systems is likely to generate major organizational effects in large parts of society. The effects from these systematically and technologically motivated merges of telecommunications and data base systems have so far been amazing. They have reached far longer than expected, and in short time. Furthermore, it is not likely that these merging and smoothening effects yet have found their limits, and lost their power. On the contrary, they likely will continue to influence not only technological but also social, cultural and political information systems on all levels.

References

1. Becker, T & Slaton, C, The future of Teledemocracy, 2000
2. Castells, M, The Information Age, vol I, 1996
3. de Sola Pool, I, Technologies without Boundaries, Harvard Univ Press, 1990
4. Farrar, C, "You left out the Italians!": Difference and Democratic Citizenship, Stockholm, Sept 1999
5. IA Newsletter, Internet Alliance, USA, Feb 2000
6. Masuda, Y, Information society, Tokyo, 1980
7. Ohlin, T, Sprid medborgarmakten,IT i demokratins tjänst, SOU 1999:117
8. Toffler, A, The third wave, 1980
9. Tsagarousiano, R, Cyberdemocracy. Cities and civil networks, 1998
10. Wresh, W, Disconnected, 1996

From Information Modelling to Enterprise Modelling

Pericles Loucopoulos
Department of Computation, UMIST
Manchester, UK
pl@co.umist.ac.uk

Abstract

An organisation's knowledge has always been critical to its competitive success; efficient operations come from shared knowledge of how things work and how they could work if the organisation were to change; market share grows with better knowledge of customers and how to serve them. Such knowledge is normally implicitly managed. However, rapid organisational change, knowledge- intensity of goods and services, the growth of virtual enterprises, and the drive for 'customer-centric' business processes, have intensified the need for better use of enterprise knowledge. In the context of this paper the term enterprise knowledge refers to a set of structured models that represent an integrated view of both user-defined application-related concepts and discovered patterns of behaviour in data.

1 Introduction

In the closing session of the IFIP WG 8.1 CRIS conference in 1986, Janis Bubenko presented a paper on a research agenda for information systems [Bubenko 1986]. He first took a retrospective look on information systems and then proceeded in identifying what were for him a set of key research issues that needed to be addressed by the information systems research community. The issues at the time were centred around the question of "how best to develop information systems?" and included topics such as appropriate description languages, support tools, ways of working, architectures and such like. To a large extent this question has nowadays been sidelined by what many perceive as a much more relevant question in today's fast changing business environment, and that is "how can information systems facilitate business engineering?". Indeed Bubenko identified this area as a potentially important area for research stating that "... methods and principles for the early stages of the IS development process such as organisational goal-, problem-, change- and activity analysis, strategy planning etc. is a

research area dominated by informal approaches, less well defined ideas, concepts, and methods and sometimes, by very strong beliefs". Bubenko argued in the same paper for "**early stage systemeering**". This paper looks at some current approaches to 'early stage issues' considered under the banner of *enterprise modelling*, an area to which Bubenko himself has made significant contributions.

Naturally, information systems continue to serve 'back-office' business needs such as co-ordination of production and enhancements of services offered. But, a new and important role has emerged namely the potential for such systems to adopt a supervisory and strategic support role for supporting changes to organisational structures and the improvement of business processes. The motivation for this new role of information systems comes from the need of enterprises to deal with rapid and turbulent changes such as electronic commerce, deregulation, mergers, globalisation and increased competition, and customer-centric business developments. These causal forces manifest themselves in the need for on one hand, *integration* of both business processes and support systems and on the other hand, *externalisation* of business practices.

These needs demand the adoption of appropriate techniques for dealing with enterprise knowledge. Rapid organisational change, knowledge- intensity of goods and services, the growth in organisational scope, and information technology have intensified organisational needs for knowledge. In addition virtual organisations that are made up of complementary allied entities place greater demands on knowledge sharing [Ruggles 1997].

Unstructured business knowledge is important for a company's performance, but cannot be systematically used and is not an asset a company can own. Clearly there is a need for support in terms of conceptual frameworks for structuring and managing enterprise knowledge so that it is clearly defined, controlled, and provided in a way that makes sure that it is available and used when needed. The activity that deals with such aspects is called *enterprise modelling*.

The term 'enterprise modelling' refers to a collection of techniques for describing different facets of the *organisational domain* e.g., goals, structures and work roles, flow of information etc, as well as the rationale behind these operations, the *operational domain* e.g., the business processes, business rules etc and the *informational domain* e.g., the database structures, the mappings between different (possibly heterogeneous databases) and the rules for and from discovered patterns of behaviour in data [Bubenko 1994; Loucopoulos and Kavakli 1995; Yu and Mylopoulos 1994]. It is a natural extension to information modelling whereby the models target not just the information system requirements but also the enterprise objectives, the work processes, the customer needs and processes that create value for customers. The models take a holistic view and cut across traditional functional boundaries.

The focus of enterprise modelling therefore, is to provide the means for dealing with the management of change [Kardasis and Loucopoulos 1998; Kavakli and Loucopoulos 1998; Loucopoulos, Kavakli, et al 1998]. Indeed the key to successful business change is knowledge shared by multiple enterprise stakeholders about: (a) where the enterprise is currently; (b) where the enterprise wishes to be in the future; and (c) alternative plans to effectively bringing about desired transformations.

This knowledge, is often introduced as the organised, high-level knowledge pertaining to goals, objectives, organisational structures, personnel and competencies, tasks and routines. However, this is not sufficient for understanding all aspects that may influence radically the future functioning of an enterprise. A more appropriate view, espoused by the author, is one that considers enterprise knowledge as the confluence of knowledge from both the business application (and indeed the business domain) and the data supporting these applications. The interplay and integration between these two knowledge sources, as shown schematically in figure 1, could result in a rich picture of the enterprise under change.

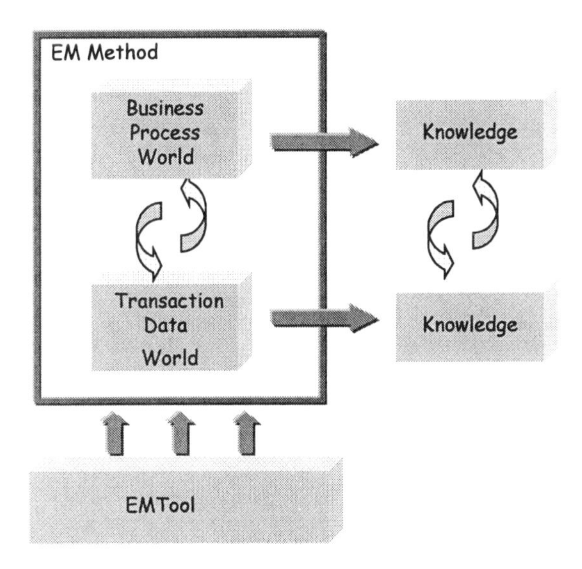

Figure 1: An integrated view of enterprise knowledge

This paper discusses both methodological and tool-oriented issues of enterprise modelling. The background to this discussion is the development of the *Delos* enterprise modelling approach and support tool. Section 2 introduces a methodological framework. It discusses enterprise modelling in the context of supporting the management of change. Section 3 covers the modelling views of *Delos*. Section 4 concludes with a few observations on the use of enterprise modelling in the change process .

2 Enterprise Modelling in the Change Process

In order to be able to systematically model and ultimately improve the change process, a framework is necessary that will enable the understanding, analysis, and tracing of the decisions involved. Research in the area of organisational reform [Morton 1991; Scheer 1991; Vidgen and Loucopoulos 1997] suggests that modelling of organisational change encompasses the following intentions:

1. The intention of *understanding the current enterprise situation*: Any type of change whether it involves the development of a computerised system or the re-engineering of business processes involves many assumptions about the embedding enterprise domain. As discovered by empirical studies [Yu 1994], poor understanding of the domain is a primary cause of project failure. In order to obtain a deep understanding of an enterprise, one needs to understand the current goals of the enterprise and how these are achieved through the involvement of organisational actors in enterprise processes.
2. The intention of *exploring change* from the different perspectives of the interested parties: The need for change is typically stated in a simple manner, sometimes called the change vision. However, even if the primary goal for change is given, it does not reflect the way the need for change is understood by the enterprise stakeholders or the way change is contextualised in the particular enterprise situation. Such understanding requires the articulation of the change concept in the context of the enterprise and its social environment and the deployment of these change goals in terms of appropriate changes of current enterprise structures and processes.
3. The intention of *designing the future enterprise situation*: Change goals form the requirements upon which the re-engineered enterprise structure will be based. This task concerns the mapping of change requirements onto a future enterprise model, which in turn involves the modelling of the future enterprise goals and how these will be realised in terms of operational enterprise components.
4. The intention of *evaluating* enterprise models against the criteria of the involved parties: The previous intentions concern the formulation of distinct enterprise models with respect to the current, change and future enterprise states. The aim of evaluation is to deliver an enterprise model, that is consistent with the stakeholders' experience and/or expectations. Often, alternative enterprise models may be possible (e.g., there may be multiple change models, leading to alternative future solutions). The appropriateness of a model depends on a number of criteria (termed *evaluation goals*) both qualitative and quantitative in nature. Such criteria are not pre-existing but need to be defined within the context of the particular change application.

Summarising, it is possible to make the distinction between four different enterprise models with respect to organisational change, namely:

1. knowledge about the current enterprise goals and how they are achieved through the current enterprise behaviour (**As-Is** model);

2. knowledge about the stakeholders' change goals and how they can be satisfied in terms of alternative change scenarios (**Change** model);
3. knowledge about the desired enterprise situation, i.e., future enterprise goals and how they are achieved by the re-engineered enterprise behaviour (To-Be model); and
4. knowledge about the stakeholders' evaluation goals concerning the appropriateness of an enterprise model (Evaluation model).

Therefore, in order to model organisational change one needs to reach the following 'knowledge states', namely: (1) the **As-Is model defined** state; (2) the **Change model defined** state; (3) the **To-Be model defined** state; and (4) the **Evaluation model defined** state. For simplicity we refer to these states as As-Is, Change, To-Be and Evaluation states respectively.

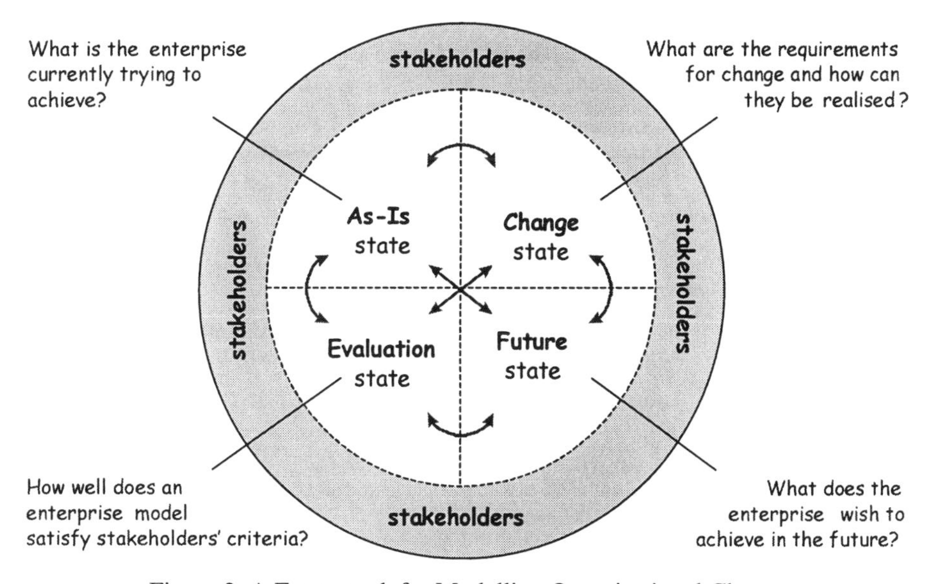

Figure 2. A Framework for Modelling Organisational Change

For example, in a business process re-engineering project one may start by understanding the current situation (reach the As-Is state) and proceed with exploring alternative change scenarios (reach the Change state), continuing with the evaluation of alternative scenarios (reach Evaluation state) and finally, the design of the re-engineered business processes according to the selected change plan (reach the To-Be state). Alternatively, one might start with the analysis of current problems and the setting of corresponding change requirements (reach the Change state) then proceed by designing the new enterprise models that satisfies these requirements (reach the To-Be state) and finally, evaluate alternative designs (reach the Evaluation state). This view of change modelling is illustrated in figure 2.

This framework defines the set of applicable knowledge states that need to be reached in an organisational change project. However, it should not dictate any particular ordering between these states, i.e., there is no unique *route* for navigating the framework. Instead, each state to be reached is dynamically selected in the course of the change management process. The sequencing of states as well as the particular manner of reaching each state is not prescribed but depends on the enactment context of the particular change project. In other words alternative processes can be followed by change engineers in order to manage organisational change. Thus, it is a *non-deterministic* model of the change management process.

The way of using enterprise modelling can be visualised as a directed, bipartite graph, whereby places in the net correspond to knowledge states whilst transitions correspond to strategies (see figure 3 as an example). The directed nature of the graph indicates the way one might progress from one knowledge state to the next. This represents a *navigational structure* in the sense that it allows users of enterprise modelling to determine their route between the different organisational knowledge states. None of the routes included in the roadmap is recommended 'a priori'. Instead the approach suggests a dynamic construction of the actual path by navigating in the roadmap. In this sense the approach is sensitive to the specific situations as they arise in the process.

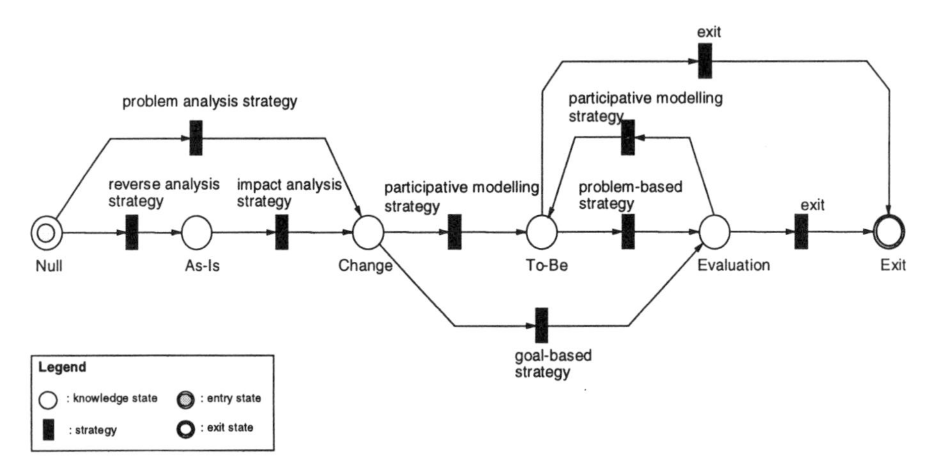

Figure 3: An Enterprise Modelling Roadmap

3 Enterprise Ontology – The *Delos*[1] Approach

Enterprise modelling in *Delos* can be used in any business or information system engineering settings including both forward (from business requirements to data and process development) and reverse engineering (from legacy systems to business policy). The *Delos* modelling framework serves as a way of having different partitions on possible enterprise knowledge models that examine an enterprise and its engineering or re-engineering requirements from a number of interrelated perspectives.

In summary, the *Delos* models are logically grouped in terms of three view:

- The **organisational** view. This provides a definition of business goals, business actors and their roles and business rules that constrain the functioning of the business.
- The **operational** view. This provides a definition of how the business realises its objectives in terms of its daily routine operations.
- The **informational** view. This provides a definition of the information, its structure and allowable operations that ultimately support the users in carrying out their daily operations.

Each one of these three views is facilitated by specific modelling orientations that address specific concerns. Specifically these involve: enterprise goals (strategy view), enterprise actors their roles and goals (structure view), legislation and high-level policy (policy view), triggers, interrelated roles and their activities (processes view), business happenings (events view), business objects (objects view), operational rules (business rules view), data organisation (database view), collections of heterogeneous data (warehouse view) and finally business behaviour patterns (discovered rules). These are shown in figure 4.

All these different views conform to a standard set of conceptual aspects (the horizontal level in figure 4). They have semantic structures (shown as model concepts), i.e. the meta-models, they have one or more external notations (model presentations), they may represent pre-defined specifications of 'best business practice' knowledge (model solutions) and finally in any given situation the views may correspond to existing AS-IS situation (source models) or to future TO-BE situation (target models). This framework has proven to be very useful in a variety of project contexts including change management [Grosz, Loucopoulos, et al 1998; Kavakli and Loucopoulos 1998; Kavakli and Loucopoulos 1999; Prekas, Loucopoulos, et al 1998; Rolland, Grosz, et al 1998] and customer profiling [Kardasis, Loucopoulos, et al 1998; Wangler, Holmin, et al 1999].

---

[1] The name *Delos* is used to signify the orientation of the approach towards integrated, co-operative enterprise modelling. It is the name of a Greek island which according to ancient history, at 478B.C., gave its name to a treaty for a common, co-operative defence policy between different Greek cities.

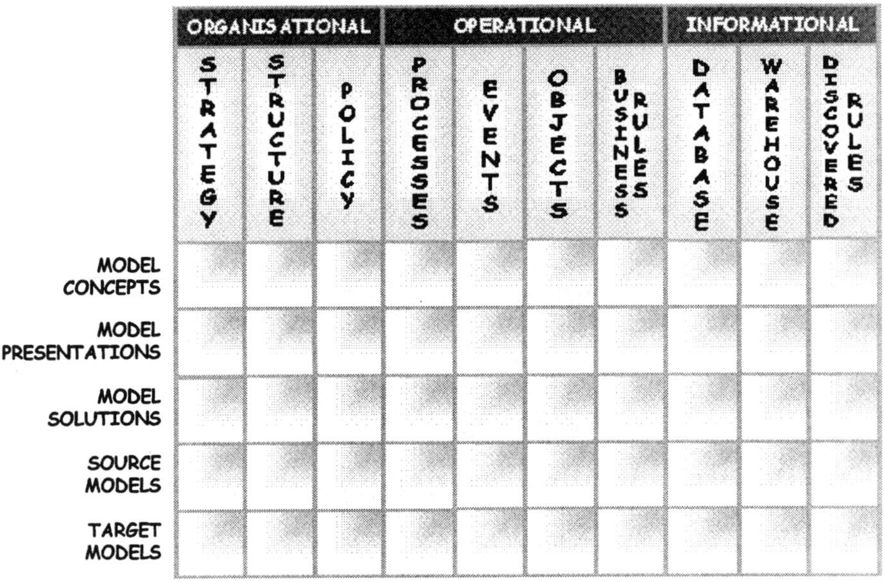

	ORGANISATIONAL			OPERATIONAL				INFORMATIONAL		
	STRATEGY	STRUCTURE	POLICY	PROCESSES	EVENTS	OBJECTS	BUSINESS RULES	DATABASE	WAREHOUSE	DISCOVERED RULES
MODEL CONCEPTS										
MODEL PRESENTATIONS										
MODEL SOLUTIONS										
SOURCE MODELS										
TARGET MODELS										

Figure 4: An Enterprise Modelling Conceptual Framework

In an ideal situation therefore, enterprise knowledge will be represented in a way that encompasses all these aspects in a fully integrated manner. Indeed there are causal relationships between all these views. For example, an enterprise goal model represents the teleological perspective, the purpose behind business processes and systems; a process model represents the way that enterprise goals are realised together with the business rules that define constraints on these operations; a data warehouse may define the collection of enterprise data needed for analytical processing and these together with discovered rules will serve as a continuous updating of the business rules and if necessary the re-engineering of the business processes.

This framework integrates enterprise knowledge from both the business and information systems domain. It has the advantage of bringing what have hitherto being considered as technical tasks to the attention of domain experts and through this to enhance the potential for effective system development and utilisation.

The table below summarises the way that the different views may be used in an integrated fashion.

Description	Models	Results
Customisation and extension of **model solutions** (patterns).	• Patterns for the business domain. • Enterprise application models.	A repository containing the complete set of application-specific business models.
Development of the enterprise data **warehouse**.	• Models of the source and target data objects. • Mapping rules.	The enterprise data warehouse.
Identification of **measures** to be estimated and understanding of the data to be mined.	• Goal models. • Business object models. • Business rules.	Specifications for the data mining experiments (runs on the data mining tool) and generation of results
Integration with the business knowledge.	• Discovered rules (from DM results).	Revised enterprise models.

The integration of business domain knowledge, business application knowledge and system application knowledge is an important factor in supporting the process of change. The management of the different models requires an appropriate tool. The *Delos* tool uses repository technology (the Microsoft Repository environment) for editing, browsing, querying, importing and exporting enterprise models and well as linking these to other environments. The architecture of the Delos tool is shown in figure 5.

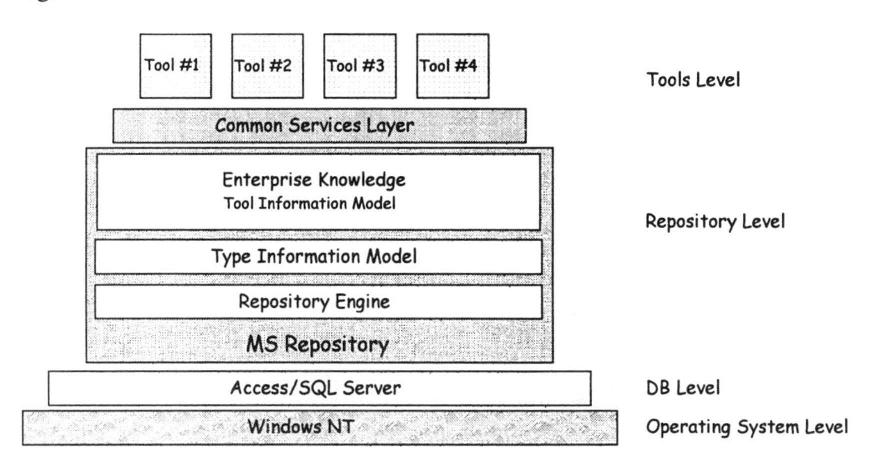

Figure 5: The Architecture of the *Delos* Tool

For every modelling aspect of the *Delos* method, the environment provides a specialised tool with graphical or form-based editing and verification facilities, together with storing, retrieval and visualisation facilities. Special attention has been given to the management of reusable patterns of business knowledge, which can be easily customised to enterprise-specific models. An important feature of the *Delos* tool is that it treats models as collections of objects. It supports uniqueness, i.e. no matter how many models contain a certain object, the object is stored in the repository only once. It also supports multiple and localised views, i.e. an object may appear to have different appearance and properties in different modelling views, however the repository contains only one reconciled version of it with all its properties.

The repository contains both meta-level descriptions and their instances, the former being enterprise independent whereas the latter being confined to a specific organisation. The metamodel layer considers both Delos method concepts (the different views) and pre-defined concepts such as the UML and DBM metamodels that come bundled with the repository.

4 Concluding Remarks

An organisation's knowledge has always been critical to its competitive success. This paper has put forward the premise that there is a need for support in terms of conceptual frameworks for structuring and managing enterprise knowledge so that it is clearly defined, controlled, and provided in a way that ensures that it is available and used when needed. The paper argues that the structured business knowledge should constitute the confluence between domain knowledge, developed through the use of requirements engineering techniques, as well as discovered knowledge developed through the use of knowledge discovery techniques applied on operational data.

The challenge in today's fast changing organisational environments is the rapid development of descriptions that may be used for evaluating alternative scenaria to change. Some approaches generally view change management as a *top-down* process. For example, BPR approaches assume that the change process starts with a high level description of the business goals for change. These descriptions constitute a very abstract representation of the future reality at the intentional level. The initial goals are then put into more concrete forms during the development process, progressively arriving at the specification of the future system requirements that satisfy these goals. Other approaches (e.g., TQM) advocate a *bottom-up* orientation whereby the need for change is discovered through analysis of the current organisational situation and reasoning about whether existing business structures satisfy the strategic interests and concerns of the involved stakeholders.

In the first case the goals for change are *prescribed* and may not explicitly link the need for change to the existing organisational context, rather they reflect how change is perceived from the strategic management point of view or is codified in the organisation's policies and visions; there is the likelihood therefore, that such goals may not always reflect reality [Anton 1996]. On the other hand, in bottom-up approaches goals for change are *described* i.e., they are discovered from an analysis of actual processes; however, descriptive goals tend to be too constrained by current practice, which can be a serious drawback especially when business innovation is sought. A more 'liberal' approach that recognises that the approach to change should be dependent to the situation in hand, is arguably more appropriate. This approach would describe change both in terms of understanding the current situation and designing future situations, whilst change management would be viewed as the process of discovering business goals for change and analysing the impact that these goals have to existing business structures and practices. To achieve we need appropriate reasoning tools and enterprise modelling provides an important foundation for developing such tools.

References

Anton, A. (1996) *Goal-Based Requirements Analysis*, ICRE '96, IEEE, Colorado Springs, Colorado USA, 1996, pp. 136-144.

Bubenko, J. (1994) *Enterprise Modelling*, Ingenierie des Systems d' Information, Vol. 2, No. 6, 1994.

Bubenko, J.A. (1986) *Information System Methodologies - A Research View*, IFIP WG 8.1 Working Conference on Comparative Review of Information Systems Design Methodologies: Improving the Practice, T. W. Olle, H. G. Sol and A. A. Verrijn-Stuart (ed.), North-Holland, Noordwijkerhout, The Netherlands, 1986, pp. 289-318.

Grosz, G., Loucopoulos, P., Rolland, C. and Nurcan, S. (1998) *A Framework for Generic Patterns Dedicated to the Management of Change in the Electricity Supply Industry*, 9th International DEXA Conference (DEXA98), Vienna, Austria, 1998.

Kardasis, P. and Loucopoulos, P. (1998) *Aligning Legacy Information Systems to Business Processes*, 10th International Conference on Advanced Information Systems Engineering (CAiSE'98), B. Pernici (ed.), Springer-Verlag, Pisa, Italy, 1998, pp. 25-39.

Kardasis, P., Loucopoulos, P., Scott, B., Filippidou, D., Clarke, R., Wangler, B. and Xini, G. (1998) *The use of Business Knowledge Modelling for Knowledge Discovery in the Banking Sector*, 2nd IMACS International Conference on Circuits, Systems and Computers (IMACS-CSC '98), Hellenic Naval Academy, Athens, Greece, 1998.

Kavakli, V. and Loucopoulos, P. (1998) *Goal-Driven Business Process Analysis: Application in Electricity Deregulation*, 10th International Conference on Advanced Information Systems Engineering (CAiSE'98), B. Pernici (ed.), Springer-Verlag, Pisa, Italy, 1998, pp. 305-324.

Kavakli, V. and Loucopoulos, P. (1999) *Goal Driven Business Process Analysis: Application in Electricity Deregulation*, Information Systems, Vol. 24, No. 3, 1999, pp. 187-207.

Loucopoulos, P. and Kavakli, E. (1995) *Enterprise Modelling and the Teleological Approach to Requirements Engineering*, International Journal of Intelligent and Cooperative Information Systems, Vol. 4, No. 1, 1995, pp. 45-79.

Loucopoulos, P., Kavakli, V., Prekas, N., Dimitromanolaki, I., Yilmazturk, N., Rolland, C., Grosz, G., Nurcan, S., Beis, D. and Vgontzas, G. (1998) *The*

ELEKTRA Project: Enterprise Knowledge Modelling for Change in the Distribution Unit of the Public Power Corporation, 2nd IMACS International Conference on Circuits, Systems and Computers (IMACS-CSC '98), Hellenic Naval Academy, Athens, Greece, 1998, pp. 330-336.

Morton, M.S. (1991) The Corporation of the 1990s: Information Technology and Organisational Transformation, Oxford University Press, Oxford, 1991.

Prekas, N., Loucopoulos, P. and Dimitromanolaki, I. (1998) *Developing Models of Reform for Electricity Supply Industry Companies*, 2nd IMACS International Conference on Circuits, Systems and Computers (IMACS-CSC '98), Hellenic Naval Academy, Athens, Greece, 1998, pp. 337-342.

Rolland, C., Grosz, G., Loucopoulos, P. and Nurcan, S. (1998) *A Framework for Encapsulating Best Business Practices for Electricity Supply Industry intoGeneric Patterns*, 2nd IMACS International Conference on Circuits, Systems and Computers (IMACS-CSC '98), Hellenic Naval Academy, Athens, Greece, 1998.

Ruggles, R. (1997) *Why Knowledge? Why Now?*, Perspectives on Business Innovation, Centre for Business Innovation, Ernst and Young I.I.P., No. 1, 1997.

Scheer, A.-W. (1991) Architecture of Integrated Information Systems: Foundation for Business Modelling, Springer Verlag, 1991.

Vidgen, R. and Loucopoulos, P. (1997) *Toward an Enterprise Process Architecture: The Contribution of the Viable System Model to Business Process Modelling*, BCS Confefence on Information Systems Methodologies, Springer-Verlag, Preston, UK, 1997.

Wangler, B., Holmin, S., Loucopoulos, P., Kardasis, P., Xini, G. and Filippidou, D. (1999) *A Customer Profiling Framework for the Banking Sector*, 9th European Japanense Conference on Information Modelling and Knowledge Bases, E. Kawaguchi, H. Kangassalo, H. Jaakkola and I. A. Hamid (ed.), IOS Press, Iwate, Japan, 1999, pp. 57-73.

Yu, E. (1994) Modelling Strategic Relationships for Process Reengineering, Ph.D., University of Toronto, 1994.

Yu, E. and Mylopoulos, J. (1994) *Understanding 'Why" in Software Process Modeling, Analysis and Design*, 16th International Conference on Software Engineering, Sorrento, Italy, 1994, pp. 159-168.

Horizontal and Vertical Integration of Organizational IT Systems

B. Wangler
Department of Computer Science, University of Skövde
Sweden

S.J Paheerathan[1]
Department of Computer and Systems Science, Stockholm University/KTH
Sweden

Abstract

Legacy information systems are usually tailored to support particular business functions, such as payroll or purchasing, and are as a consequence usually difficult to integrate. However, companies of today strive to streamline their business processes (intra- as well as inter-organizational). This gives rise to a strong need for much better integration of systems (existing as well as new in-house developed systems or packaged ERP solutions) across functional boundaries and in line with the business processes. This paper surveys and discusses methods and technique to achieve such horizontal integration as well as vertical integration between different operational and management levels in an organization.

1. Introduction

Organizations and their IT-support have traditionally been structured around business functions where each function is supported by its own information systems. These applications may be recent or old. It is not rare to have application systems that are 10 – 20 years old, if not more. As a consequence, they have been built using different technologies, such as different languages, different protocols, even running on different platforms, and all aimed to support well the business within the functional area they were originally designed to support.

Competitive businesses of today need to have computing support to business processes that are not restricted to a single business function, but usually cut across many such functional areas and are liable to change over time. Furthermore, such business processes are often required to support running the business electronically

---

[1] On leave from Institute of Computer Technology, University of Colombo, Sri Lanka

over the Internet not only for presenting the data but also for enabling electronic commerce transactions. These trends demand for the integration of current functionally oriented systems in the lines of business processes. For many companies this integration needs to be delivered within short time in order to ensure the competitiveness of the businesses.

The need to reduce the cycle time of business transactions premeditated to gaining competitive advantages also encourages inter-organizational co-operation. Globalization of businesses and mergers between enterprises of related interests are some other factors that drive the interests toward business processes beyond the interest of a single institution.

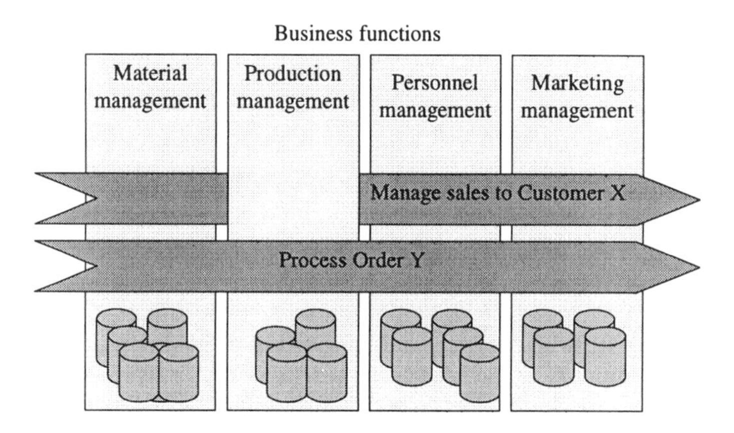

Figure 1. Process Orientation

These trends give rise to new requirements on software applications and their integration, including the incorporation of existing legacy applications and Commercial Off-The-Shelf (COTS) products in the formation of new business processes. In addition, systems on different management and control levels within an organization need to interact. For instance, process control systems need to be fed with control information and to forward data to higher level systems such as enterprise resource planning systems and data warehouses.

2. Approaches to application integration

Though there are different approaches adopted by organizations for the application integration, a more complete categorization based on the market segmentation of software tools and technologies for application integration can be defined with the following six levels [1]:

- Platform integration using technologies such as messaging, Remote Procedure Calls (RPC), Object Request Brokers (ORB) and mechanisms such as publish and subscribe etc.
- Data integration using database gateways or data warehousing technologies.
- Component integration with transaction management, application services, business logic integration facilities achieved through application servers
- Application integration through adapters, rules and content based routines, and event based data translation
- Process integration with process and workflow modeling,
- Business to business integration, with EDI/XML and supply chain integration and on-line trading brokers

The approach applicable for a particular integration project depends on the need for integration, positioning of the business in its competitive market place and the maturity of the technology used.

The lowest in the category is the platform integration, which provides connectivity among heterogeneous hardware, operating systems and application platforms. Technologies that provide platform integration range from messaging, which can only facilitate asynchronous connectivity between applications, through Remote Procedure Calls (RPCs) providing synchronous connectivity to Object Request Brokers (ORBs) that, combined with messages, can provide both types of connectivity.

Data integration is generally of two types. The first includes database gateways enabling data access to heterogeneous data stores using Structured Query Language (SQL). This approach demands for the knowledge of underlying database schemas during application development. The second category in data integration is with tools for extracting, transforming, moving, and loading (ETML tools) data originally designed to support data warehouses after bypassing logic of the original application designed to access the database. These solutions are even found enhanced through messaging support and to provide meta-data management and data cleansing capabilities and may be associated with graphical user interface tools to draw the relationship between applications.

Component integration enables new functionality to be combined with ERP packages, and legacy systems with provisions for load balancing and fault protection, connection pooling, state and session management, security and data access to relational and non-relational sources through application servers.

Application integration is achieved with a reusable framework, which combines a collection of technologies. This collection usually includes underlying platform integration technology, event integration through message brokers facilitating data translation, transformation and rules based routing, application interface integration provided to leading ERP packages through application adapters etc. The frameworks are added value by providing facilities to abstract the complexity of creating, managing and changing the integration solution.

Process integration approaches while requiring all the integration services mentioned so far, provide the highest level of abstraction and adaptability by enabling business managers to define, monitor, and change business processes. Business processes model the information flows across systems and organizational boundaries. Middleware products supporting this level of integration enable business managers achieving this through a graphical modeling interface and easy to handle declarative languages. Changes to generated integration solutions can be made, simply by changing the respective business model via the same graphical interface and declarative language and regenerating the solution.

Business to business (B2B) integration facilitates the extended enterprise concept by enabling the integration of systems beyond the corporate boundaries of an enterprise. For example, systems at suppliers' sites and customers' sites can be combined with systems at enterprise's site to create an extended business process for practicing e-business. Though the EDI was the traditional approach to make conversations between enterprise systems, now trends point towards using XML as a solution for providing the lingua franca of e-business. B2B integration solutions are commonly provided using two approaches [1]. One is to take the process integration approach to define business processes across the organizational boundaries and the other is through the creation of technologies for brokering exchanges in on-line trading communities.

There are many commercial middleware products available in the market to help in application integration. The more advanced and business process oriented of these tools are usually referred to as Enterprise Application Integration (EAI) software [2].

EAI products completely or partially automate various aspects of the process that enables custom built and/or packaged business applications to exchange business level information. A complete EAI product offers the connectivity services provided in middleware products, the data transformation services provided in ETML products, and the process management services provided in workflow products [3].

3. Realities in enterprise application integration

In addition to the categorization of integration approaches based on a market segmentation of tools and technologies for application integration as presented above, we see three different forms of enterprise application integration:

- between different systems supporting different functional areas of a business. We refer to this as horizontal intra-organizational integration.

- between systems on different control and managerial levels of an organization . We refer to this as vertical intra-organizational integration.
- between systems of different organizations. We refer to this as inter-organizational integration.

Whatever the tools and technologies adopted during integration, it is problems having to do with these dimensions we face in integrating different systems already running in an enterprise or at collaborative sites.

Though the common problems to be addressed in every application integration project include handling differences in data formats used by different candidate systems, resolving the protocol differences during communication between them, there are specific problems and issues associated with each specific integration dimension.

3.1. Horizontal intra-organizational integration

A typical example of horizontal integration is Supply Chain Management, in which an organization tries to optimize the complete set of activities of order entry, purchasing, production, shipment etc. in order to minimize the lead-time and costs for production, and at the same time maximize value for the customer.

Figure 2 shows four business processes each spanning a number of applications or processes. The gray area depicts the organization in question. Rectangles depict applications or sub-processes. Process no 1 takes place completely inside the organization. No 2 starts inside and finishes outside. No 3 starts outside and finishes inside. No 4 leaves the organization at one point but comes back before finishing such as may be the case if a separate company takes care of delivery before the original company checks that the invoice has finally been paid. In the course of the process, queries and one-way messages are sent, answers and one-way messages are received, timers are set and run-out.

The current trend towards componentization of software causes functional systems to be split up into smaller pieces, each responsible for its own task or service. By combining these task-oriented components in new and different ways, organizations are able to design and introduce new services or service combinations. Moreover customers should be able to specify themselves the service combination they want. For example, it should be possible in the ordering process to specify which combination of services as regards e.g. delivery and payment methods one wishes. This calls for the ability of enterprise application integration software to dynamically introduce or restructure business processes.

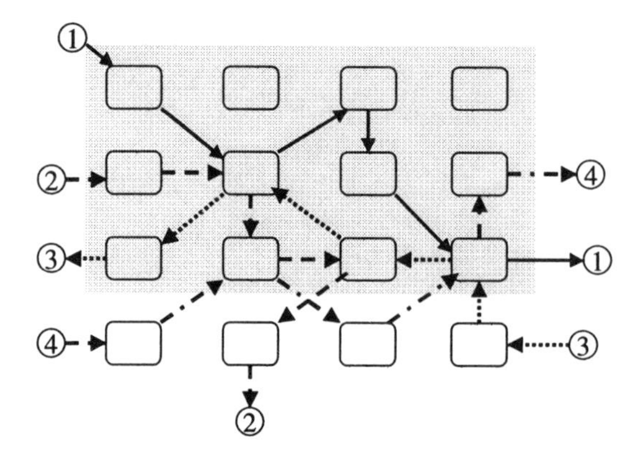

Figure 2. Business processes affecting various applications and processes within (dark area) and outside an organisation.

Many enterprises have taken an expensive ERP path for integration by acquiring packages from vendors like PeopleSoft, SAP, Baan, etc., with increased hopes for inter-process chain integration. Some enterprises have seen Y2K as a catalyst to replace an ageing IT infrastructure with modern ERP systems [4]. Further, it is common to find enterprises with ERP applications supporting a part of some process chains together with legacy systems serving other requirements in the same domain e.g. by incorporating functionality that refers to multiple organizational levels.

Replacing the entire information systems portfolio of an enterprise with ERPs is neither possible nor economical due to the following reasons. Firstly, there is no ERP solution that will provide all the functionality an organization requires and therefore some custom applications will be present among the candidate applications for integration. Even if it were possible through a combination of different ERP packages to complement the inadequacy of coverage, not every enterprise would be willing to depend on such solutions. Secondly, there is always a tendency to maximize the returns of the past investments on information systems. Throwing away some expensive items for some fancy reasons is not an attractive action in business, which is usually aimed to reduce its spending unless the spending promises profitable outcome. Furthermore, businesses are always alert to changes and the amount of change that may be introduced by a complete replacement will simply be intolerable in many cases. These facts prove that the presence of legacy is unavoidable during an integration projects.

Though many ERP packages provide standard interfaces or connectors to connect systems, they are usually built as monolithic solutions [5]. It follows, that the support provided by them is often inadequate and that a main challenge during horizontal integration will be to connect different systems that include the categories of legacy systems, newly in-house developed applications, and COTS, into more or less complete processes and to do so within the timeframe permitted by the business concerned. The solution should favor quick changes to implemented business processes, as business may demand for such changes due to the stiff competition it faces.

A feasible way of achieving horizontal integration at the level of sophistication we are seeking will be with EAI products supporting the application integration level and levels above it as referenced in the categorization of application integration approaches presented in Section 2. Though many middleware products enable integrating applications built using recognized ERP packages and legacy applications with recognizable interfaces, and there are many legacy wrapping tools for preparing legacy applications for integration, a complete plug-and-play situation is still impossible. In these circumstances, the only option available is to handle the task through extensive coding work.

As indicated before, a recent trend in integration is that the business processes that should be compiled do not require the incorporation of the complete applications. Instead in many cases only selected functionality of existing applications are the ideal candidates for incorporation. Though achieving this level of integration will greatly enhance the effectiveness of integrated solutions, the technological support for this task and support from current legacy systems is found to be limited.

3.2. Vertical intra-organizational integration

Vertical intra-organizational integration is aimed to integrate systems implemented at different administrative levels of an organization. Though it is not uncommon to incorporate the functionality at different levels of an operational system into a single application , the normal is to see different systems implemented to address business functions at different management levels.

As an example, a typical function in any company is “production”. In manufacturing industry this function may, at the lowest level, be controlled by process control systems and computerized NC machinery using proprietary formats of data and messages (See Fig. 3). Frequently they use different operating systems and utilize different networking technologies. These systems need to be fed with control data stemming from higher level planning and scheduling systems while the lower level applications need to collect data and pass them upwards. For example, a company implementing some modules of some ERP package still requires using its current MES (Manufacturing Execution System) and EMTL systems.

Hence, vertical intra-organizational integration may demand the flow of data between candidate applications in both directions. While the feed control data are to flow downwards from systems at higher levels to operational control systems, systems at the lower level need to collect information about results and send them upwards in order to support decision making at higher level management systems such as for purchasing. In data warehousing oriented environments this is to be done periodically in order to feed data to a data warehouse. In this process data may have to be cleaned such as to equalize definitions and to improve the quality of data. Furthermore, data may have to be pre-aggregated along several dimensions. This calls for a separate data staging activity as part of upwards transfer of information. Whereas the data staging files may be fed with data as it is created, the final update of the data warehouse would take place periodically, e.g. once a day. In this process several steps involving user interaction may be needed in order to analyze, improve, and complement the data.

A wide range of approaches and technologies as well as products based on them is available for vertical integration. These range from relatively comprehensive process brokers to message brokers and such things as single function data transformation tools or even data warehouse oriented solutions. Whatever is used, a product, a combination of products and programming, the integration solution is also influenced by the economy and the level of expected sophistication of the solution.

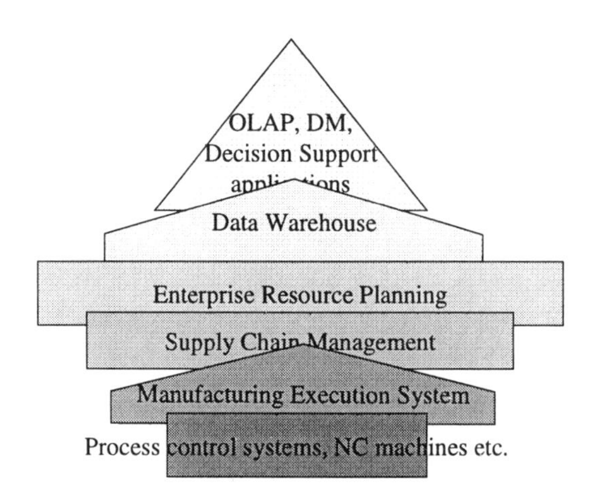

Figure 3. Different layers of systems needing integration.

3.3. Inter-organizational integration

The trend among organizations towards making commerce electronically tends to change the original interaction patterns between enterprises, in particular in their roles as customers and suppliers. Enterprises need to exchange information electronically in order to collaborate and negotiate. The influence of Internet, and the availability of middleware technologies now enable the businesses to exploit the opportunities to connect applications spreading across multiple enterprises. Supply-chain logistics and fulfillment companies integrating with their customers' fulfillment and shipping systems, or financial services firms integrating with underwriters and retailers of financial products, are a few examples of this multi-enterprise integration.

The key issue in inter-organizational integration is to get the data operations of one application of an organization matching with another application of another organization. Though the Electronic Data Interchange (EDI) has now been used for this purpose, the complexity and restrictions of e.g. the EDIFACT standards limits the use of EDI for large organizations. XML stands a better chance of becoming standard for transfer of data and meta-data between applications with its loosely coupled document model and with the new XML based transaction servers facilitating document transfers back and forth between businesses. XML is also proving its ability in letting the enterprises to structure and exchange information without rewriting their existing systems or adding large amounts of heavyweight middleware.

But the actual state of inter-organizational integration today falls short of what is really needed to achieve an efficient multi-enterprise business process. Inter-organizational business operations are still mostly handled through call-center operators keying e.g. in the relevant back-office order-entry system. Large volumes of business transactions over the Internet are received today via relatively unreliable FTP, requiring handcrafted error-handling programming. Furthermore, many of the application-integration solutions today rely upon proprietary technologies and are therefore not well adapted for use in an e-business context.

The shifting focus of application integration from inside the enterprise to multi-enterprise integration introduces a different variety of requirements and priorities. Some requirements are refinements of approaches that have evolved for EAI, but there are still many differences. One such is the stronger emphasis that is put on security issues. Another is the increased need for an integrated support for Internet application architectures, such as application servers that support Enterprise JavaBeans (EJBs), CORBA etc. The distinction between EAI solutions and application servers becomes fuzzier in the context of inter-organizational integration, to such an extent that a new culture referred to as Internet Application Integration (IAI) is emerging.

4. Process Oriented Solutions in Application Integration

Traditional strategies to achieve application integration are through hard-coding these services within applications to be integrated. These approaches while resulting at heavy programming work to application developers, is liable to extensive maintenance when the integration solution require changes.

Alternatively, an approach using a middleware software layer providing various broker services between the different information resources has gained appreciation. In consequence, a whole lot of commercially available middleware tools to help in this integration have appeared in the market.

At the same time, customer requirements for application integration are also maturing and increasingly becoming (a) decentralized, as distributed computing architectures and technologies complement the increasing IT influence of lines of business managers, and (b) coordinated, as batch connections are increasingly replaced with real-time systems to support the industry move from the management of transactions to the management and re-engineering of business processes [3].

EAI products completely or partially automate various aspects of the process that enables custom built and/or packaged business applications to exchange business level information. EAI is not middleware. EAI is not workflow. EAI is not data transformation as found in data warehousing (ETML products). Each of these types of products offers a solution to a specific piece of the general business problem. Rather, EAI is a combination of the technologies employed in these kinds of products. A complete EAI solution employs the connectivity services also provided in middleware products, the data transformation services also provided in ETML products, and the process management services also provided in workflow products [3].

In the long run, EAI will ease the burden and lower the costs of application integration by providing a level of automation not traditionally available for the integration of applications. This may be achieved by further eliminating the requirement for programming during integration and help achieving the integration at the business process level in contrast to the data or functional levels.

Current trends in the EAI space are driven by three major forces according to industry specialists [6]. The first is the move towards doing things at a higher level i.e. at a business process level, hence facilitating process integration [2]. The second is an increasing proliferation of packages which makes the integration opportunities more predictable. The third is extending the process orientation beyond the corporate walls and delivers the full promise to e-business by integrating customers, suppliers and partners while transferring the controls to parties outside the organization enabling the business to enjoy the benefits of an extended business.

Industry experience demonstrates, that process orientation allows organizations to solve more complex business problems and be more competitive. Business process integration provides mechanisms for transferring content between two business processes that act as sender and receiver. The content can be of any type and may consist of several different combined contents, e.g. a physical product together with paper or electronic information. Process integration interfaces of an EAI product can be managed by a Business Process Broker

A Business Process Broker will provide functions for synchronizing events into one business transaction, handling parallel business transactions, mixing email (person to person) and application (machine to machine) messages in one business transaction. It will, hence, be possible to build Process Management Systems that align the IT-support to the business processes. Advantages of this alignment are that people and systems can be linked together in the processes, and that decisions based on business rules may be facilitated and even automated. The impact of process design can be increased, and business processes can be changed and managed in a simpler way through graphical interfaces while accommodating run-time modifications to business processes. A Business Process Broker facility of an EAI product provides a higher level of abstraction than a Message Broker provides under similar conditions, by giving users the ability to define integration requirements through workflow and business process models [7].

The management of business processes is a key issue in process orientation and will greatly improve the effectiveness of such integration efforts when it is combined with facilities to simulate the use of process models before the implementation. Simulation of business processes will enable business managers to identify potential bottlenecks before the system is implemented. Collection of real-time data on the performance of processes for use in analysis, and facilities to re-implement a previous solution quickly in case it is found to be more favorable, will further improve the support to today's businesses.

One of the key players in the EAI market place, Viewlocity, aims to improve their EAI products through extensive research in these directions [7]. The ProcessBroker project [8] run by Viewlocity and SYSLAB of the DSV particularly takes further the process-oriented aspects of EAI in order to address the requirements engineering, modeling and architectural needs of the concept.

5. Concluding remarks

Ending up, we may conclude that differences between the requirements posed by horizontal, vertical and inter-organizational integration are smaller than one perhaps would expect. The differences there are concern such things as the fact that vertical processes may involve a more disparate collection of applications, i.e. systems for detailed control of low-level processes and machinery mixed with batch-oriented system for loading data into databases aimed for strategic and tactical management. Inter-organizational processes are special in that they involve autonomous organizations thereby posing specific demands concerning e.g. security and the fact that process integration should still allow for some independence in the way that individual companies prefer to make business

While the possible disparities between applications to be integrated in a horizontal integration effort can be considerable, the opportunities for this kind of integration are brightening due to vendors such as CrossWorlds, Forte, Extricity, HP,

Viewlocity, and Vitria. However, it is not only novel software solutions that are needed. These have to include graphical interfaces allowing for comprehensive and easy-to-understand ways of describing, designing and managing processes including the various actors, systems and people that are involved. They also need to be complemented with methods and guidelines helping the user in the analysis and design of processes and data models. Business analysis methods such as EKD (Enterprise Knowledge Development) [9] by Bubenko et al. offer a good starting point for this work. Work by Johannesson and Perjons [10] represents a first attempt to enhance such methods with design principles specifically targeting application integration.

Acknowledgement

This work was performed as part of the project ProcessBroker [reference to web], a joint project between SYSLAB and Viewlocity. SYSLAB's participation in the project is sponsored by Swedish National Board for Industrial and Technical Development, NUTEK.

References

1. Beth Gold-Bernstein, "EAI Market Segmentation", EAI Journal, July/August 1999.
2. Torun Lidfeldt, White Paper: Middleware, Datateknik 3.0, nr 7, 25 nov 1999 (in Swedish).
3. Katy Ring, Enterprise Application Integration: Making the Right Connections, White paper, 1999, http://www.ovum.com/ovum/news/apiwp.htm.
4. Jeanne W. Ross , The ERP Path to Integration: Surviving vs. Thriving, EAI Journal, February 2000, http://www.eaijournal.com/ERPIntegration/ERPPath.htm.
5. David S. Linthicum, Integrating SAP R/3, EAI Journal, February 2000, http://www.eaijournal.com/ERPIntegration/IntegratingSAP.htm
6. Colleen Frye, Middleware Moves Up a level, Application Development Trends, Oct 1999, http://www.adtmag.com/pub/oct99/fe991002.htm
7. Processbroker consortium, A Process Broker Architecture for Systems Integration, white paper, Nov. 1999, http://www.dsv.su.se/~pajo/arrange/Publications/publications.html.
8. Processbroker Consortium, Project Proposal, Nov. 1998, http://www.dsv.su.se/~pajo/arrange/index.html.
9. Bubenko, J. A., jr, D. Brash and J. Stirna (1998). EKD User Guide. Dept. of Computer and Systems Science, KTH and Stockholm University, Electrum 212, S-16440, Kista, Sweden.
10. Paul Johannesson, Erik Perjons, Design Principles for Application Integration, Proceedings of 12th Conference on Advanced Information Systems Engineering, Stockholm, June 2000.

Information Systems Research Need for New Methods of Work

Eva Lindencrona
Swedish Institute for Systems development
Kista, Sweden

Abstract

Information systems have been an area of R&D for at least four decades. Research in Information Systems started in response to evolving industry needs. The results from Information Systems R&D are of many different kinds. A wealth of knowledge, methods and tools has been created. It is interesting how these important results have come to use in practice.
R&D in information systems has been carried out in Computer Science departments, in Informatics departments, in Business Administration department as well as in several other new university departments. R&D in information systems does involve not only different disciplines but also different kinds of activities such as research, engineering, and development.
Four decades ago only very few specialists were active with analysis, development and use of information systems, and although there was a gap between R&D and practice, the need for R&D in the area seemed to have been well recognised by practitioners. Today, almost everyone frequently finds herself in contact with computers and information systems, directly or indirectly. One would think that this might lead to an increase in recognition and demand for research in the area of information systems. However, not only has the gap between the state of the art and the state of practice increased, but the recognition and demand for R&D concerning information systems seems to be lacking behind.
This paper discusses issues related to R&D in information systems. It questions whether information systems research, as conducted today, will be successful in responding to the Information Society needs for new technologies and methods. Research is changing in many areas in terms of role and style. This paper advocates that such a change is necessary also for research in information systems.

1. Introduction

Things are not what they used to be! Our society is changing, taking steps into the Information Society. There are different views on what is meant by "The Information society". The Information Society is sometimes described as knowledge based society where Information and Communication technologies are the engines. The changes imposed on today's society when entering into the Information Society are not yet well understood or can be clearly foreseen. Different trends and possible development directions for the new Information Society are described in futuristic literature, articles and foresights. Substantial changes in all aspects of society are a common denominator, although different authors foresee different positive or negative lines of development.

Globalisation and internationalisation are characteristics of future business. New models for competition, as well as for co-operation are evolving. Hierarchical structures are decomposed and transformed into network structures. New business opportunities and business models are emerging as the technological development offers completely new possibilities. The product cycles are increasingly shortened and the time between innovation and new products or services is decreasing dramatically. These changes are not limited to industry or business but apply equally to public service. The changes to business and public services are fundamental in the sense that they impact not only how things are done but also what things that can be done. It is hard to anticipate that R&D in the area of information and communication would not be affected by such global changes.

For a long time there has been a trend in favour of applied and short term R&D. Although there are signs of a trend in the opposite direction and new programmes for basic and long term research are launched, as for example in Japan and in the US, the general trend pointing towards R&D in support of explicit applied industrial and societal goals is clearly dominating. Research funding is being adjusted. National R&D funding is complemented by or changed into international R&D funding. The European framework programmes are examples of this.

There is a trend from so called "curiosity driven" research where researchers freely choose their research issues and are driven primarily by scientific goals towards research driven by industrial or close political goals and where research issues are chosen based on their immediate relevance for industry and society.

The developments in computing-, information- and communication systems are fundamentally changing the daily life of most people in Europe and elsewhere. The number of people in contact with information technology and the daily activities supported by information systems is increasing exponentially. The people actively taking part and using the technologies of the Information Society in it self create new opportunities and a base for still new ways of using the technologies. The use of Internet is an example of this and a "critical mass" effect can be recognised. Although the Internet and its forerunners have been around for at least 30 years it is only in the last decade that its use has exploded due not only to the introduction of new technologies, such as graphical user interfaces, hyperlinks protocols, security protocols etc. But also due to this critical mass effect which makes the Internet an increasingly attractive platform for new products and services.

Four decades ago there were few academians among the practitioners in industry or among the "users". They were at that time primarily working in the IT departments inside big companies. The academians were often physicists or mathematicians. Computer- or information sciences did not exist as university curricula at that time. Today this situation is totally different. Although many more are needed, there are today an increasing number of academicians among users, and among the "consumers" of information systems R&D.

The contexts in which information systems are being developed and in which they will serve will be much more complex than earlier. They will be more complex primarily because they will exist closer to human behaviour and to human needs than earlier information systems applications. Also, the new information systems and services will to a larger extent be built on other systems and be embedded in real or virtual information environments.

The users of information systems are no longer limited to male IT-professionals but are people of all ages, sexes and professions. Information systems usage will not be restricted to getting information, writing letters or doing administrative work, not even to e-commerce or use of public services on the net. The interaction between people for all kinds of purposes as well as the interaction between people and machines will be the base of the next generation of information systems. This may require change of focus of today's research in information systems.

For decades, the area "new methods of work" has been an important area for information systems research. Information and computer scientists have contributed substantially in areas as CMC (computer mediated communication), CSCW (computer supported co-operative work), CML (computer mediated learning) etc. In the subject of Informatics and in Business Administration, over time, different "schools" for business development have been dominating. One example of such a school was BPR - Business Process Reengineering. Researchers in information systems have been fast to move into new areas and to adopt the ideas of different management "schools", including BPR, in order to develop information systems support for activities of relevance to the new schools. However, the idea that such new methods of work would apply also to research work itself has been less promoted.

New methods of work are not restricted to the use of ICT in the research work itself but should be interpreted in a wider sense. It is likely that new methods of work, new modes and models of R&D procedures will change the way research is carried out. It will not only change the research organisation, but more important, it will have an impact on the kind of results that are created in R&D activities. And, it will probably change the image of research. All these changes will challenge existing structures. R&D in information systems is an important area for the development of the Information Society and it will have to respond to all these changes. The question is how.

2. Discussion

2.1 Research context

Innovation is going to be the key to success in the years ahead. This is specifically important for information systems and services. Innovation in systems development techniques - including improvements in systems management - will distinguish success from failure. Success will also depend on understanding the application contexts well enough and to apply the technologies appropriately. Again, this points in the direction of application driven research. From technical points one gets quicker to key issues which might not have been addressed otherwise. In addition, in this way of working the potential for collective learning is high.

For many companies, advanced and qualified customers are considered as the main driving force behind product development and innovation. The same reasoning can be applied also to research.

In the new competitive regime, commercial successes require the ability to generate knowledge using external resources and working together with them. It is the people, not papers that transfer technology and generate new knowledge leading to new products and services. Industry researchers and managers will benefit from better understanding of what to expect from their external research partners.

Turning a prototype into a product is a complex process. New technologies can fail in the marketplace because of false assumptions about factors totally unrelated to technical merit - pricing models, distribution channels, support processes, and of course customer needs and perceptions. By better understanding of the product or service development cycle, researchers will gain a more realistic view of their roles in this process.

Europe shows excellence in research (as measured by number of publications per unit invested) but its technological and commercial performance (as measured by patents issued per unit investment) is low. Still, in many research institutions, advancement is based primarily on publication and the system does not envisage the possibility of implementing reward schemes aimed at risk, innovation and possible exploitation. Innovative thinking often meets resistance from those whose ideas are dominant. An application driven approach may not change the difficulties that innovative thinking can meet but embedded in an application context the process of acceptance may be shortened.

New models of incitement for exploitation of innovations are required to raise researchers´s efforts to obtain patents and property rights for inventions and discoveries. Recognition for the researcher or inventor responsible for an innovation is a key aspect. This is difficult already in traditional disciplinary research at universities but might be even more complicated in application driven transdiciplinary research [7].

Standards development is another area where researchers could contribute with their knowledge and methods. Standards promote a common technical "language" for industry, confidence in products and services to consumers and therefore requires careful investigation of its foreseeable technological, as well as socio-economic implications. R&D involvement in standardisation could aid in producing better standards. New models for prestandardisation could support redefinition of relationships between research and standardisation [9].

2.2 Research is changing

The research community is growing. The number of scientific papers published annually has doubled every 10-15 years. In parallel, the market for research is growing, and there is an increasing demand for research. For research in computing, communication and information, the application areas are wider than ever. The market for research increases when researchers are willing to work together in temporary multidisciplinary teams and on problems representing well-defined customer needs.

The use of research in support of economical growth and economic policies is today a dominating goal. This is very obvious for R&D in areas such as computing, communication and information. For many big companies related to those areas, in-house research is not cost effective due to shortened product lifecycles and global competition. These companies are establishing new kinds of links with universities and research institutes. For many years now, company sponsored university research or cost shared research programmes have been a norm rather than the exception. As a response to such changes there is also an emerging market for specialised, private research companies operating internationally and specialising in niche competencies.

In later years there is a change towards greater focusing of public R&D on more socially oriented objectives. Areas of concern are those addressing citizens needs such as environment, health, transport and public services.

The reasons for private companies as well as for public services to relate to research and to share research funding are often described as a guarantee to avoid investing in old fashion technologies and to anticipate new trends and directions in knowledge and technology.

The cultural and social milieu in which research is carried out is changing. Scientific, technological and industrial research is becoming more closely connected. This is specifically true for R&D related to ICT.

R&D become part of a larger process of knowledge production and play a vital role in the general changes of the society and of the new economy.

The number of scientists in the ICT area is still limited and as there are comparatively few scientists working in companies and public administrations there is a risk of misunderstanding of the potential of research work.

Even today, in the ICT area, it is not unusual with unrealistic expectations of what researchers can do. Sometimes researchers are expected to be gurus able of solving almost any problem. Sometimes researchers are regarded primarily as additional development resources. Sometimes there are expectations that researchers could be switched to any topic as company requires change [2].

As time between research and commercial products is shrinking dramatically as for example in the communications area, research is less far ahead of product development and its commercial value is faster assessed. The more researchers working outside research departments the more qualified users and consumers of research and the better use of existing research resources and the public perception of research can change in a constructive direction.

The limited number of researchers in the ICT sector also has an impact on how researchers perceive themselves and their role. One can still find researches who consider themselves above research criteria related to society or business needs, who at the same time believe that they themselves are best able to set research priorities and who do not want to be bothered with potential exploitation of their results. Also it may not be too difficult to find examples of research projects that are not ended - even after they have paid off - because the researcher do not want to stop. As long as this is the case the gap between researchers and the general public continues to increase and the topics chosen for research may appear more and more esoteric. At the same time, much of the stress experienced today by scientists is caused by this pressure to explain and justify how their work fits into a much larger context. With the increased competition in research this is sometimes felt to be contradictory to the need for greater specialisation and leading edge competence.

The organisation of research is changing. A few decades ago, independent individual researchers or at most small groups, was the norm. Today this is changing towards big institutions and collaborative research efforts.

Research institutions are becoming players on a market leading to new challenges in terms not only of collaboration but also in terms of competition.

R&D are short for research and development. The results from research is knowledge expressed in terms of concepts and theories and documented in scientific papers, reports, books, lectures e.t.a. The results from development are artefacts such as technology, prototypes, demonstrators, and methods, tools e.t.a.

In the ICT community R&D is often treated as generating the same kind of results and it is not unusual that a project which does not result in a product or at least a prototype of a product is regarded as a failure. At the same time there are many examples of the opposite i.e. projects which develop a prototype without documenting what was learnt in the development process as for example unresolved design issues or solutions that did not work.

The time elapsed from R&D to results in terms of new methods, products or services is often used to classify different types of R&D. In average, research refers to up to 5 years, advanced development to 1-3 years and development to 2 years from market. Long term research is often used synonymously with basic research and refers most often to more than 5 years to market.

These kinds of classification are dangerous, as they are misleading. As the demand for fast results from R&D increases it may lead to shortsighted priorities. Long term R&D is an investment in the future independent of whether it is basic or applied research or even development. Research can have an impact within timespans usually reserved for development. The degree of uncertainty and risk might be a better measure than time in order to characterise different types of R&D.

In many research organisations there is an ongoing debate about the balance between researcher-driven research and consumer driven research. Different wording is used in different contexts in this ongoing debate. For example, researcher-driven research is also called curiosity-driven or unfettered research. Consumer-driven is used synonymously with user-driven or commercially funded. Researcher-driven research has been practised primarily at universities and, in some countries, in governmentally funded research institutes.

In many universities research seems to have a higher priority than teaching and research has become an area for - national as well as international - competition among universities. In most countries, public funding is decreasing at the same time as the number of researchers is increasing. In order to cope with this situation and often in response to political demands universities are taking on more and more externally funded and consumer driven research. The "pure" researcher-driven research where the researcher freely defines the research problems without any need to justify their relevance for the financier and where the results are judged by peer researchers based on purely scientific criteria, and without other relevance criteria, is extremely rare today.

Although academic research may seem to be comparatively researcher-driven, it may in essence be driven by research policy or by consumer's demands. Many senior researchers are depending on grants or project funding from external partners and have to adapt to the areas, priorities and requirements of the financing institution or company. Some of the European research programmes, for example, are very detailed on the areas and problems they want researched and where they want technology to be developed and projects are selected on criteria constituting a mix of scientific and other relevance criteria.

Research problems are much more complex today. As information systems come closer to human behaviour their complexity increases rapidly. Also the cultural and social contexts in which the new information systems operate are much more integrated in social life or in strategic business models than earlier generation information systems, adding to the increased complexity.

The increased complexity leads to requirements for transdiciplinary research. In many research institutions this demand for transdiciplinary research is recognised although the systems - the way in which research is organised in Universities - does not easily adapt to the new requirements. Many researchers today recognise that the most dynamic areas of research are in the borderlines of the existing, traditional disciplines. There is an urgent need for reinterpretation of the norms and values traditionally associated with disciplinary research. Such a change would support dynamic new areas of science where disciplinary knowledge and competence are combined in novel combinations and configurations.

The need for change is recognised by most research organisations as well as by most research funding organisations, public or private. R&D and its evolution are in itself becoming an area for research. For example, in Gibbons et al [1] different modes of research are identified and discussed. Mode one is characterised by problems being set and solved in a scientific context, and in the interest of academia, by homogeneity in discipline, by hierarchical structures and by preventives. Mode two is characterised by problems being set in the context of an application, by being transdiciplinary and characterised by heterogeneity and by being transient.

Gibbons et al [1] propose that "It is cognitive and social norms that determine what shall count as significant problems, who shall be allowed to practice science and what constitutes good science".

2.3 New applications and new technologies

The information systems and services of the information society will be different from the systems of today. There will be a shift from business orientation to people orientation. The way business and people co-operate and behave is changing and will require new kinds of information systems and services. Information systems of tomorrow will be addressing new areas; it can be foreseen that they will focus on communication between people as well as on communication between people and machines. The information systems and services of tomorrow will be much more integrated in a systems context as well as in an applications context. In many cases the roles of information providers and information retrievers will be reversed and information systems development must be flexible and adaptable to the new needs and systems implications that may arise in such situations. The need of simple and intuitive interfaces to the systems, needs to be stressed over and over as with the new categories of users, systems and services with poor interfaces simply will not be used. Deep understanding of the application domains and the contexts in which the new systems and services are to be used will be a base for identification and formulation of research issues in the area of information systems. A continuing trend seems to be information systems that serve different user groups or communities and new services specifically addressing the needs of communities will be developed. The communities may be real or virtual, they could correspond to interest groups, families, pressuregroups, groups of partners, professionals, shareholders, neighbourghs, carpool members, political groups, support groups etc, etc who have a need for collecting and distributing ideas, who wish to share views, who want to take decisions etc. The basic characteristic of these systems is the need for communication between people where the users have equal status as communicators, and where each one can take the role of information provider as well as information user.

The information systems developed today are no longer primarily serving the industry or addressing administrative efficiency. They are serving the citizens and they are doing things that could not have been done before. The next generation information systems will serve the individuals in their homes, cars and offices, in their private as well as in their professional lives. Research in the area called "Things that think" where, principally, all electrical apparatuses are connected to the net will form the base the next generation information systems with a potential for tremendous impact on daily life for everybody.

Many of the new applications will be built on existing information and systems resources. Value addition will require integration of new information systems with existing information resources and such integration will be typical characteristics of tomorrow's information systems. Earlier and ongoing R&D in information systems as for example in areas as databases, digital libraries, metadata, information architectures do address issues that will be fundamental components also in next generation information systems. These technologies will allow users to store, retrieve, share, and manipulate all kinds of data in order to create added value and to build exciting new information services. Development of such applications in real contexts will require substantial changes in our approach to the design of information architectures, databases and storage systems. Also user interfaces based on all human senses as well as on cognitive and affective theories may lead to the need for very different methodologies and tools. Again, the information systems R&D community has a very important mission to fulfil and their area of R&D could play a key role in the development of the Information Society. In addition to this, research related to better understanding of the effects of the technologies that are being created will be urgently needed.

Internet based information systems are more important than earlier information systems as they will have a much higher impact on peoples lives, simply because the Internet has the potential of reach a much wider audience than any platform has had before. The Internet allows for completely new kind of applications and for reaching out to new categories of users and to new kinds of usage. The Internet and the Web are not only new platforms for information systems and services; they are primarily new platforms for communication between people and between people and things. Internet based information systems are a different kind of information systems thus requiring people to think about them much differently than of traditional systems. Inexpensive, seamless and reaching out to a wealth of information and leading to new kinds of organisations and new ways of interacting with existing organisations. Internet based information systems are sufficiently different form traditional information systems to require new theories as well as new approaches to information and systems engineering.

Companies are faced with time-to-market pressure combined with the requirement of higher quality and low cost systems development and management. For information systems research, advanced applications that include qualified and challenging R&D problems may be the most useful driver. It can stimulate more ideas for the researcher to work on. Application driven research may be faster than traditional researcher-driven research in contributing to innovation in information products and services. When research takes place in the context of an application, it involves a variety of actors and skills and it will contribute not only to science, to the application but also to knowledge creation in a more general sense.

2.4 Methods and tools

Research carried out in the seventies in the area of databases and data modelling has had a strong impact on the products and systems available today and with all probability this will be true also for systems to be developed in the near future.

Research in the area of information systems development was dominating in the eighties. Many new methods were developed and meta methods and method architectures were presented. What happened to the results from that time? Some of the findings did get into commercialised methods and were used primarily by big companies. Case tools were developed that did support systems development according to some method. Still however, these methods and tools do not seem to have had an equally visible impact on products or on the day to day practice of developing information systems and even less so in the development of today's new information services Why is that so? Are the methods and tools not good enough, are the tools not supporting the methods in a proper way? Or are they too limited in their scope? Some of the problems seem to be lack of simplicity, very little support for the creative parts of the systems development, not adapted to new kinds of applications and too narrow in scope.

Many researchers have challenged the waterfall paradigm. Component-based methods have followed. Even if there are new methods they are still often "closed" methods in terms of the theories on which they are based, in terms of goals and contexts. They still refer to traditional technology and most often do not take into account the context in which they are to be used and the context of the applications whose development they are to support

Open methods are required which take into account theories from different disciplines, that aim at supporting the creative aspects of systems development, that considers not only text, graphics and sound but other senses as well and that take into account the complex nature of tomorrows information systems and services.

New methods and tool must also take into account that their users are. The traditional roles of professionals building systems for "end users" may well be turned upside down, with "end users" being the Information providers and "professionals " being ad hoc users of the services being developed. This will not be true for all kinds of information systems but for many enough to challenge the traditional perspectives and to illustrate the need for change of focus for the future methods and tools.

The volume of research in information systems has increased over the last decades. There are many competing methods, tools and languages for development of information systems available today. The "same" problem can be solved by different approaches; the "same" application can be built using different technologies. Better understanding of the qualities of different technologies as well as better understanding of the contexts and applications where the technologies are to be used will lead to better use of existing, alternate technologies.

2.5 Technology transfer

Traditionally, research, development and technology transfer are seen as three sequential activities. Research creates new ideas, concepts and theories, development is used for " proof of concept" and for building prototypes in order to verify and validate ideas generated during research. After that, technology transfer can start. According to this scheme, technology transfer is an activity where R&D results are presented and demonstrated to practitioners who then accept the new ideas by implementing them in their organisations. Technology transfer is thought of as a "from - to" activity. Typically technology transfer is carried out as seminars, courses, demonstrations i.e., in essence as different forms of one-way communication. The expression "technology transfer" in itself transmits the idea of moving knowledge from someone to someone else, in this case to move knowledge about technology from the R&D community to private or public business. There are several reasons why this sequential model needs to be questioned.

Has the sequential model for technology transfer been successful? There seem to be little evidence of this. The gap between R&D and best practice in the information systems area is widening. The technology transfer from universities and institutes to application development in private and public business has been rather poor. Only very large companies seem to benefit from R&D in information systems. Two decades ago, results from R&D in information systems were quickly picked up by intermediaries as consultancy or software companies thereby being indirectly beneficial to "user" companies. Today this seems to be much more rare. Most of the Internet consulting companies have not adopted information systems development methods and tools originating from R&D. One reason for this has been the inability of R&D to apply information systems engineering and management principles in the right context.

The sequential technology transfer model has difficulties in coping with the demand for fast results. New models for technology transfer will be requested in the Information Society. Cyclic models based on closer collaboration, on networks and again, on research being carried out in the context of advanced applications are needed.

Information systems research is primarily carried out in universities and institutes. Some very large companies do research in the area and do develop their own methods and tools for building information systems. The majority of companies building information systems for others or themselves are not involved in research in information Systems. However, new technologies are introduced and used in companies. This may be new architectures, languages, and data definition tools e.t.a. If these new technologies are not developed in the Universities where are they developed? The answer is in other contexts such as industry consortia (f. ex OMG, W3C,) joint R&D ventures, industrial research e.t.a. The big difference between the two contexts are that in traditional way of working the new knowledge is produced by researchers in the University and thereafter transferred to business. In the other ways of working, knowledge is produced in collaboration with industry and technology transfer has become much more integrated in the knowledge production. EU R&D programmes attempts to build consortia for collaborative production of knowledge and where typically there are consortia members representing research, technology providers, problem owners, and users. The risk however with these kind of projects is that they are more concerned with application of existing knowledge to industrial uses than with the creation of new knowledge.

As has been discussed above, the information systems developed today are characterised by higher complexity which gives rise to new research issues. This also makes it much more difficult to apply the traditional sequential way of thinking of technology transfer. As complexity increases the collaborative way of creating new knowledge is requested both from a competence point of view as well as from a technology transfer point of view. The technology transfer is not finished before the new technology is exploited in one way or the other. Again, exploitation is more complex that envisioned in the sequential model where science leads to technology and technology satisfies market needs. Information systems technologies normally are not a commodity available off the shelf; it is something, which is developed to solve the problems of a specific business, service or product.

3. Conclusions

Information systems R&D is an essential component in the creation of knowledge for the coming Information Society. Looking at what has been achieved in information systems research during the last decades there seem to be a huge potential for applying already existing results in new processes, products and services.

Not only the application of existing results but the continuation and expansion of R&D in information systems will be a key factor for the developments in the near and the future years.

Impact of information systems R&D and enhancement of the understanding of information systems R&D could be speeded up by the adaptation of new methods of work. Different aspects of new methods of work have been discussed in this paper. Main points made can be summarised as:

Research in the context of advanced applications

The debate on the balance between basic research and applied research as well as the balance between research and development will, and rightly so, continue. The position taken here is that it does not exist a "correct" balance between the different categories. What constitutes a "correct" balance will vary over time, will vary with the area of R&D and in relation to the current needs of the society.

For the near future, it is suggested that R&D in information systems, to a larger extent, should be carried out in the context of advanced applications. For information systems, R&D agendas should not be defined unilaterally by either research nor development or unilaterally by researchers or consumers.

Transdisciplinary research

The complexity of the next generation information systems will require closer co-operation between different disciplines and areas of R&D. Information systems are complex systems involving artefacts as well as human beings. With new application areas for information systems there will be, not only a need for new methods and tools for systems development, but also a need for better understanding of the use and integration of artefacts in everyday life. The traditional borderlines between disciplines can be questioned and new transdiciplinary areas supported. Close collaboration between researchers from different areas of R&D should be supported for the mutual development of areas involved. Again, the context of advanced applications can highlight the need for transdiciplinary research and stimulate close collaboration between researchers from different disciplines.

Cyclic rather than linear knowledge production and technology transfer

The traditional sequential model for technology transfer is based on the assumption that there is a natural sequence of research, development, and thereafter the transfer of the technologies developed. New methods of work are cyclic rather than sequential. In such a cyclic model the research and development are much closer integrated and the technology transfer is, to some extent, integrated in the process. In order to keep pace with the needs of the Information Society a cyclic way of working will be required.

Incorporate more risk

In the near future, it will be a challenge for R&D in information systems to conduct research that will contribute to the Information Society. There is a need to implement measures in support of this effort. In later years R&D in information systems has resulted in

- incremental scientific advancements
- new technologies where market success seem to be related more to de facto standards and interoperability than to technological innovation

Information systems research is a comparatively new discipline and there has been a need to develop new concepts and theories. It seems that it is now time for information systems as a research area, to open up for expansion and to question its recent borders. One way to do this is to take on more uncertain problems where the solutions are not primarily a matter of time or resources but of innovation.

Continuous networking

New models for co-operation between industry and research institutions are being tried and developed. There are models based on cost-shared research programmes or projects, on "technology watch" models, on empirical studies etc. Also between research organisations, different models for closer co-operation are established. Such models could aim for example, at co-ordination of research efforts in order to jointly develop substantial contributions to an area. Universities are building networks for closer co-operation. Student or researcher exchange as well as co-operation in development and implementation of curricula is becoming frequent. The use of ICT for distributed teaching is another interesting model for such co-operation.

Networking based on win-win models will be sustainable and profitable for the partners involved. Information systems research should continue to get involved in networking and take initiatives to create new models for networking.

References

1. Gibbson et al : " The new production of knowledge" Sage Publications Ltd, ISBN 0-8039-7794-8, London 1994
2. Foley J.:" Technology transfer from University to Industry" CACM, September 1996, Volume 19, Number 9

3. Isaacs E. A.., Tang J.C:"Technology transfer: So much research, so few go products", CACM, September 1996 - Volume 19, Number 94
4. Stein A.:"Openness in Scientific Advisory Committees" IPTS report 39, Nov 1999
5. Fielding R.T. et al:" Web-based Development of Complex Information Products",CACM August, 1998, Volume 41, Number 8
6. Castells M.:" The rise of the network Society", Blackwell 1997
7. Moncada P. et al.:"The management of intellectual property in Public Funded research", IPTS report, July 1998, No
8. Vonortas N.S.:"Research Joint Ventures: The use of Databases", IPTS No 24, May, 1998
9. Vinard D.:"Linking R&D to standardization: Getting more from the results of industrial research", IPTS Report, No 35, June 1999
10. Roussel P,A. Et al.:" Third generation R&D:Managing the Link to Corporate Strategy" Harvard , Business School press, 1991
11 Odlyzko A.:" The decline of Unfettered research", AT&T, 1996
12. Isakowitz T. Et al,:"Web Information Systems" CACM July 1998, Volume 41, Number 7
13. Bohanec M-,et-al.:" Knowledgebased portfolio analysis for project evaluation", Information Management.,vol 28,no.5,1995
14. DePiante Henriksen A.,Jensen Traynor A.:"A practical R&D project-Selection Scoring tool", IEEE, Transactions on Engeneering Management, Vol.46, No.2 May 1999
15. Barré M.:"Public research programmes:Socio-economic impact Assessment and User Needs", IPTS Report, No 40,dec 1999

II.

Approaches to Information Systems Engineering

A Framework for Component Reuse in a MetaCASE Based Software Development

Kalle Lyytinen, Zheying Zhang
Department of Computer Science and Information Systems
University of Jyväskylä, Finland
kalle@jytko.jyu.fi, zhezhan@cc.jyu.fi

Abstract

Reuse forms one mechanism in improving system development productivity and quality. In this paper, a systematic framework for reuse processes and situations is suggested. The framework enables us to think reuse in a comprehensive manner in a metaCASE environment. It covers both design and method components and consists of a system development process model, type level hierarchy, reuse situation types, component granularity classification, and reuse operations. Based on an analysis of MetaEdit+ tool support, we demonstrate how reuse process can be incorporated into a metaCASE environment.

1 Introduction

System development is becoming increasingly market driven. This demands domain knowledge and better interface design. It also requires system development methodologies to accommodate new features of application domains. Meanwhile, the shortening of lifecycles and demand for higher productivity require that tool supported methodologies should be developed at an increased speed. In this paper we argue that a metaCASE environment is necessary to bring productivity and quality increases. To overcome incompatibilities and redundant implementations in the current methodology specification and system development process, reuse forms a baseline of a metaCASE based solution for this problem.

The concept of reuse has been widely discussed for about forty years in software engineering. It is estimated that over 50% of the predicted cost of software development in the next decade, assuming current practices, can be avoided if software reuse is widely practised [1]. Much of the software reuse activity to date however has been limited to the reuse of parts, especially code components. Yet, reusable artefacts include other design artefacts. Therefore we consider expanding reusable assets from code to all design artefacts generated during the system development process. We thereby need more advanced techniques and better management support. One aspect of them is technological integration which demands a use of advanced system development support environments, i.e. a metaCASE environment [2].

While introducing reuse into a metaCASE environment, we need a systematic framework that helps us organise reuse activities. This is challenging because reuse techniques have not matured to the point where a single conceptual framework is established for organising reuse research and practice [3]. There have been several attempts to define such a framework (e.g. [4], [5], [6], [3], and [7]), but most of them concentrate on a specific aspect of reuse. For example, Biggerstaff et al. [4] characterise software reuse technologies; Prieto-Díaz et al. and Liao et al. [7, 8] present a review of library management; and Moore, Krueger, and SRI [3, 5, 6] focus on reuse activities. A general framework synthesising different types of reuse in the system development process is lacking.

In this paper, we discuss how to improve component reusability in a metaCASE environment. The paper seeks to demonstrate how reuse "ideas" can be implemented in an industry strength environment called MetaEdit+ [9]. We will develop a general framework for metaCASE component reuse. This framework will consider reuse from the perspectives of system development lifecycle, modelling levels, types of reuse, component granularity, and reuse activities[1]. The structure of this paper is as follows. The next section clarifies basic concepts such as a metaCASE environment, a reusable component, and reuse concepts. Section 3 develops a general architecture for reuse in a metaCASE environment. Section 4 presents the reuse process on a more detailed level. Section 5 concludes the paper and discusses future research.

2 Reuse and MetaCASE Environments

2.1 Reuse

Reuse applies existing software and design artefacts to deliver new applications, or to maintain the old ones [5]. In the beginning, the interest in reuse stemmed from the realisation that one way to increase software productivity was to produce less new software for the same functionality [10]. In recent years, software reuse has been "enabled" by object-oriented concepts and its component concept, and distributed network technologies which allow run-time reuse by late binding and utilisation of object service, like e.g. CORBA. These make it possible that in future components can be identified and retrieved for reuse across different systems.

A *reusable component* is any component that is a specifically developed to be used, and is actually used, in more than one context [11]. Today, reusable component research concentrates mostly on the source code. Although code reuse forms the most common type of software reuse, it is the least productive form [12]. Moreover, reusing design artefacts at other stages of system development has greater potential leverage because of their greater expressive power. This can further trigger source code reuse at the implementation stage. We thereby assume that component reuse occurs in all stages of the system development process and thus components

---

[1] This study forms a part of the RAMSES project whose goal is to develop and evaluate component reusability techniques on the application level as well as the method construction level. RAMSES stands for Reuse in Advanced Method Support Environments. This project is funded by TEKES, MCC, and Nokia NMP.

cover all kinds of software artefacts: descriptions, formal specifications, graphic design documents, source code, test cases, user manuals, and so on.

Generally, a *reuse process* consists of four basic steps: abstraction, selection, specialisation and integration [5, 13]. Abstraction forms a process that standardises components by assigning them into uniform profiles. The profile describes interfaces and specific features of each component. The abstracted components are stored in a repository so that users can search and retrieve them. Normally, selected candidate components can not satisfy all requirements. They need to be adapted to cater for the specific functions that were recognised during the specialisation process. Finally, the integration step integrates revised components into a new application or service platform.

Reuse leads to increases in productivity and quality and decreases in maintenance costs, and helps organise company's experience [14, 15]. However, despite its merits, reuse has not succeeded. Some reasons for the failures are technical (e.g., technical immaturity of environments [16]), while others relate to management practices, financial disincentives, and legal constraints [3]. Another impediment has been a persistent mindset of favouring greenfield development [17]. Therefore, common software processes do not take reuse adequately into account. In a similar fashion, current CASE or metaCASE tools provide weak methodical support for reuse, although such tools embed a large number of method components and design artefacts.

2.2 MetaCASE and MetaEdit+

Instead of providing just one fixed CASE (computer aided software engineering) environment, a metaCASE environment provides facilities for method engineering which involves design, construction and adaptation of methods, techniques and tools for various system development contingencies [18]. Therefore, a metaCASE tool can generate a variety of CASE tools [19].

MetaEdit+ is a fully configurable metaCASE environment that provides functionality for both CASE and CAME (computer aided method engineering) [9, 20-23]. As a CASE tool, MetaEdit+ offers a versatile and powerful multi-tool environment that enables flexible creation, maintenance, manipulation, retrieval and representation of design information among multiple developers. As a CAME environment, it offers an easy-to-use, yet powerful, environment for method specification, integration, management and reuse [9]. MetaEdit+ has been widely used in research institutions. Over the years it has been used to implement system development methodologies and tools for several different design domains [23-27] and it is also currently used in large industry strength development tasks.

The use of MetaEdit+ as a metaCASE environment can be clarified by an example. Here we take a simple phone interface design as an example.

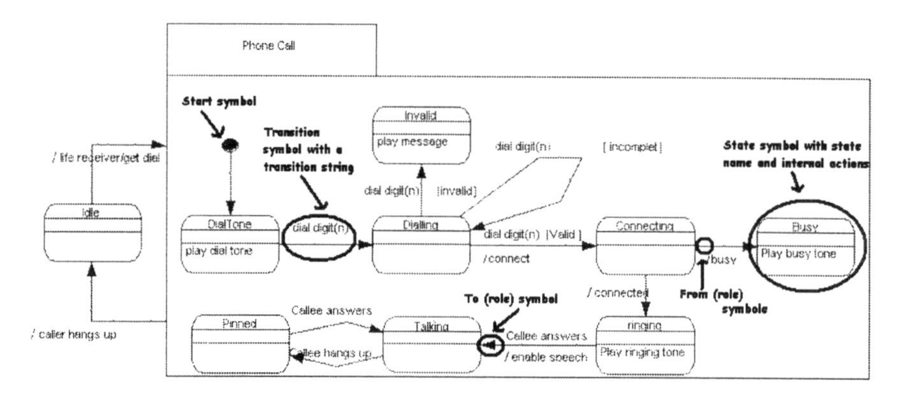

Figure 1 Phone Call State Diagram specified in MetaEdit+

Figure 1 illustrates a state transition model of a phone call process designed in MetaEdit+. Besides the state transition diagram, we could use other methods to model the phone interface from other perspectives, such as a class diagram to specify the static structure of the system. In MetaEdit+, these heterogeneous methods are specified and customised using GOPRR [20] metamodelling language. GOPRR stands for the acronym of Graph, Object, Property, Relationship, and Role, which are the primary meta data types. The conceptual GOPRR modelling constructs and their instances are shown in Table 1.

Meta Data Types	Description	Type Example	Instance Example
Graph (G)	A graph is a specification of a method (technique). It is an aggregation concept that contains all other GOPRR meta types.	State Diagram Definition	Phone Call State Diagram
Object (O)	An object is a conceptual thing in the universe around us.	Start	DialTone
		State	Dialling, Connecting, Ringing, Talking, …
		Stop	Idle
Property (P)	Properties are describing/qualifying characteristics associated with other meta types.	State name	DialTone, Dialling, …
		Internal actions	Play busy tone, Play ringing tone, …
		Guard condition	Incomplete, Valid, Invalid, …
Relationship (R)	A relationship is an association between two or more objects. It connects objects through roles.	Transition	Dial digit(n) [valid], Connected, Callee answers, …
Role (R)	A role is a link between an object and a relationship to specify how an object participates in a relationship.	From, To	From, To

Table 1 GOPRR metamodelling language

Table 1 illustrates the concept of five types of meta data used in metamodelling methods and their instances in designing an application. We can see that while using

GOPRR metamodelling language there are around 10 meta data type definitions in a state diagram and around 20 instantiated components in a phone call state diagram. This is but one small and simple diagram in the whole phone interface design task. We can imagine that it may include hundreds of meta data type definitions, and thousands of instantiated components to design the phone interface from different perspectives. Furthermore, many components have common features. Introducing reuse into such a design environment is thereby possible and necessary.

3 An Architecture for Reuse in a MetaCASE Environment

When integrating reuse into the metaCASE environment, we need to specify a general architecture for reuse processes to derive the functionality of such an environment and specify the role of reuse in use. The functionality should cover most situations where reuse takes place. As shown in Figure 2, a metaCASE environment can involve reuse situations in two dimensions. The horizontal dimension illustrates the basic stages of system development, i.e. analysis, design and implementation. The vertical dimension presents the information type levels. The terms of information type level are derived from the data levels of Information Resource Dictionary System (IRDS) framework [28] where they are called the IRD schema level, the IRD definition level, the IRD level, and the application level respectively. Accordingly reuse can take versatile forms at different development stages and information type levels. In general, we will distinguish between three types of reuse: vertical reuse, horizontal reuse, and instantiation reuse.

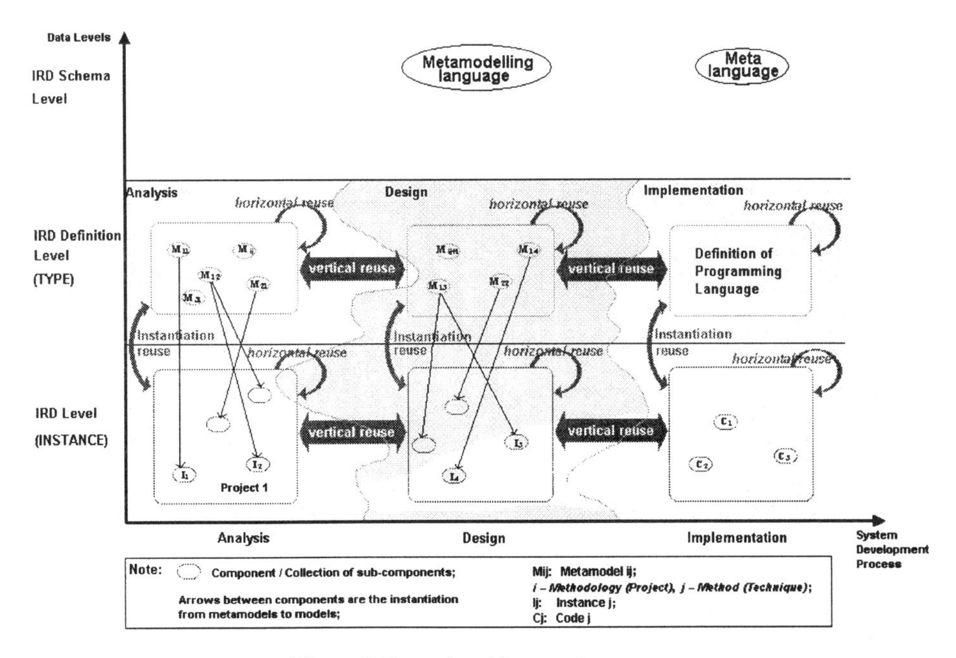

Figure 2 General architecture for reuse

3.1 Information Type Levels

Information type includes the IRD schema level, the IRD definition level, the IRD level, and the application level. The application level deals with the specific instances, for example, the data pertinent to a phone, such as phone number "12345", is recorded on the application level. Normally, the application level information is not represented in a metaCASE environment. We integrate it into the IRD level. Therefore, in a metaCASE environment we have three levels of information type: the IRD schema level, the IRD definition level, and the IRD level.

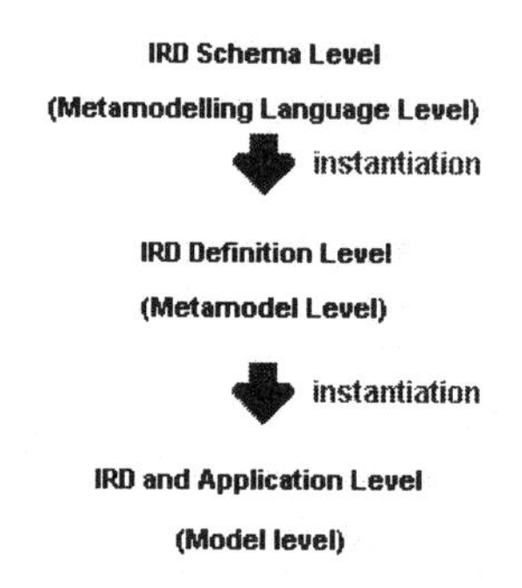

Figure 3 Information type levels and the relationship

Within MetaEdit+, these three information type levels have more technical names: meta-metamodel level, metamodel level, and model level, as shown in Figure 3. Instantiation relationships exist between models at two continuous levels. For example, the model of the IRD level must be an instance of one metamodel on the IRD definition level, and metamodels on the IRD definition level must be instances of the metamodelling language on the IRD schema level.

The IRD schema level (meta-metamodel level) has the most abstract information type. It prescribes types of objects about which data may be specified on the IRD definition level [28]. It is composed of a metamodelling language that contains a set of primitive types and rules to specify a methodology and a set of frames that justify the use and the logic of constructing a methodology. In MetaEdit+, GOPRR forms this kind of metamodelling language. Since the GOPRR metamodelling language is appropriate for building metamodels for system analysis and design, we have added a normal meta-language as a complementary abstract concept to specify the syntax of programming languages at the IRD schema level, e.g. BNF (Backus Naur Form).

The IRD definition level contains IRD definitions [28]. On this level (metamodel level) a methodology engineer specifies the methodology in terms of a set of metamodels using the metamodelling language. Activities of this level belong to the method engineering process. The method engineer constructs methodologies and populates them into a metaCASE environment. For example, methodologies such as OOAD, OMT, UML, as well as "local" methodologies can all be specified, customised and implemented at this level.

The purpose of the method engineering process is to develop methodologies that support system design through models on the IRD level (model level). It enables designers to develop systems, e.g. an interface design, a business application, etc. by instantiating metamodels. For example, the phone call state diagram is on the IRD level. Different metamodels can support model construction at different system development stages.

3.2 System Development Process

Another dimension of the architecture involves basic stages of system development. There are several process models that specify the system process, such as Waterfall, Rapid Prototyping, or Spiral [29] models. No matter which model is applied in the development process, some basic activities, such as analysis, design, and implementation, are indispensable. Typically, representations in these stages are transformed from informal to formal [30]. MetaCASE tools provide facilities to support these kinds of activities. For example, metamodels of methodologies help describe the target system during the system analysis and design tasks. Based on the specified models, code generation takes place at the implementation stage.

Reusable components may have different representations at each stage. At the analysis and design stages, most components can be semi-structural documents such as the software requirement specification, and the requirement dictionary [29], represented in notations like an object diagram and a state transition diagram. At the implementation stage, the components are represented as formulas or code segments, in a particular programming language. In Figure 2, for example, collections of components at different data levels are represented by small ovals. On the IRD definition level, four components M_{11}, M_{12}, M_{13}, and M_{14} comprise the metamodels of Methodology I. On the IRD level, their corresponding instances are I_1, I_2, I_3, and I_4. Models I_1, I_2, I_3, I_4, and code segments C_1 and C_2 form Project 1.

The boundary between stages is not clear because of a blurred distinction between analysis and design, as reflected in current methodologies. Reusable components at each stage thus need not be distinguished strictly.

3.3 Types of Reuse

Since a metaCASE environment includes components at different information levels and system development stages, the purposes of reuse and the type of reusable information can be different. We have accordingly three types of reuse: horizontal reuse, vertical reuse, and instantiation reuse. In Figure 2, we represent these three types of reuse by different kinds of arrows.

Horizontal Reuse refers to the reuse which exploits similarities across two or more application domains [31]. In a metaCASE environment, it takes place on the same information type level and at the same development stage, as the annular arrows show in Figure 2. There are two forms of horizontal reuse. The first refers to the exploitation of functional similarities across different domains. An example might be the "Person" class, which is the superclass of the "Lecturer" and "Student". It can be reused horizontally in many domains. The second refers to the exploitation of similarities in technical domains, which are independent of application. An example might be the algorithm reuse, such as a sort function. The horizontal reuse in technical domain can often be more easily achieved, since components are standardised and independent of the application domain.

Vertical reuse exploits functional similarities within a single application domain [31]. In a metaCASE environment, it usually takes place between development stages, but at the same information type level. Vertical reuse captures the domain knowledge that is specific to an application and presents a value added to the development process. It is presented by bi-directional arrows between stages in Figure 2. In the method support environment, since none of the methodologies offer adequate support for developers to all stages and for all aspects of the system development process [32], it is possible to expand the methodology functionality by vertical reuse.

Instantiation Reuse is a possible reuse approach in a metaCASE environment. It exploits semantic similarities across information type levels to reuse mainly design knowledge. When a new method is implemented by reusing the existing metamodel, it implies that its instances can be constructed by means of reuse. Since design rationale is recorded between the original metamodel and the target metamodel, such as adding, deleting, revising specifications and representations, version information, instantiating knowledge, and experience, we can construct instances of the target method by an instantiation reuse. The instance components of the original method can be reused and modified according to the design rationale to generate new instances by adding new features supported by the target method. This will ease the new instance creation process. Meanwhile, the same process can take place from the model level to the metamodel level.

Although the design rationale across information levels enhances reuse, we can observe difficulties in the conversion process. It is not a physical addition or deletion process, since the syntax, semantics, and rules are distinct between the information levels. The original modification record or design rationale can not be applied directly. Finding out potential relationships and a conversion schema between the original knowledge and the target information and providing facilities to support such a reuse process is not available in the current metaCASE environments. However, it indicates one direction to improve reusability.

So far we have discussed different types of reuse on the IRD definition level and IRD and application level. We have ignored the most abstract level: the IRD schema level. Even on this most abstract part of the framework, reuse takes place. On this level, the reusable component is not a metamodel, but a metamodelling language. Reuse can be a means to enhance the functionality or description ability of the metamodelling language. A good example is the extension of GOPRR to GOPRR-p metamodelling language that supports process modelling in MetaEdit+ [33].

4 Reuse Framework

The reuse process and its components need to be also specified. In Figure 4, we present a reuse framework concerning reuse process-related facets.

This framework forms a three-dimensional cube: the abstraction level, the component granularity, and the reuse activity. Abstraction levels represent two levels of reusable information. Component granularity and reuse activity aspects represent the basic building blocks that should be considered in developing any type of reuse process.

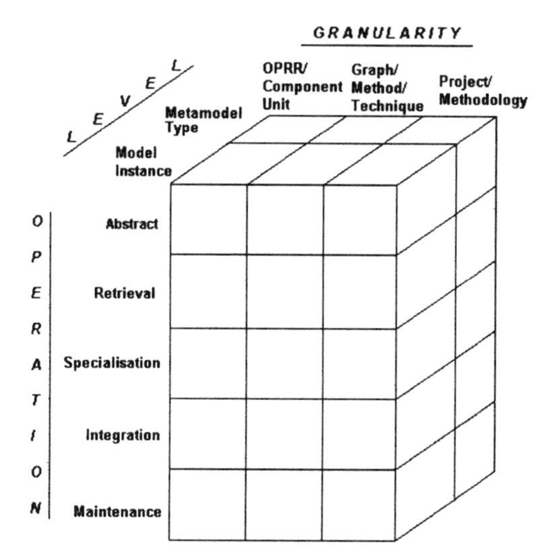

Figure 4 Reuse framework

4.1 Information Type Levels

Information type levels are the same as in Figure 2, but only the two lower levels, i.e. model level and metamodel level, are represented in the framework. In MetaEdit+, each level has its own users: method engineers are interested in the reuse on the metamodel level, while software engineers are more concerned with the reusable components on the model level. We aim at improvement of the reusability throughout the system development lifecycle, and thereby ignore the highest abstraction level here, the GOPRR metamodelling language level.

4.2 GOPRR based Component Granularity

On the component granularity dimension, we have different rules to classify components. In MetaEdit+, using the specification of GOPRR metamodelling language, we can distinguish between a component unit, a graph, and a project.

- Component units are non-property primary data types of MetaEdit+. They are object, relationship, or role types and their instances. Component units are the smallest reusable components in MetaEdit+. If we further decompose these component units, the benefits we would obtain would not be worth the effort of retrieval, specialisation, and integration. A new implementation would be easier and cheaper. Therefore, component units are taken as the smallest components used in analysis and design stages. Examples of metamodel level component units and their instances can be found in Table 1 in section 2.2.

- A graph forms a collection of objects, relationships, roles and properties. It provides a representation of a technique on the metamodel level. It results in diagrams describing specific tasks within the problem domain or a solution domain. For example, a metamodel of a state transition diagram and the model of a phone call state diagram (or its subsets), are two graph components on the metamodel level and the model level, respectively.

- A project is a design product, or a plan to produce it. Like a graph, it has a dual meaning. On the metamodel level, a project is a methodology, or an approach including a set of methods and rules to guide the system development, such as a UML notation that includes a class diagram, a state diagram, and other techniques. On the model level, it is a system development project, and its instantiated models and code, e.g. all aspects related to interface design of a specific phone.

Generally, the study of component unit reusability will be conducted on a fine-grained granularity level, and the project level reusability on a coarse-grained granularity level. The reusable information of every component includes both a conceptual specialisation, and a representational definition. Normally, graphic notations are relatively straightforward and easily reused, but the conceptual specialisation needs more consideration due to difficulties in defining the semantics, and resolving conceptual conflicts.

4.3 Reuse Activities

Every component can be involved in horizontal reuse, vertical reuse, or instantiation reuse. No matter which type of reuse is applied, we always need a set of operations to implement the reuse process: abstract, select, specialise and integrate. Furthermore, one more operation, maintenance, has to be added into the reuse lifecycle. Through reuse processes, more and more components will be added into the component repository. We should consider also the repository maintenance issues while enlarging the size of the repository. Although most reusable components have a long life span, some components may not be used any more and should be deleted from the repository to decrease its complexity. Therefore, from the abstraction to maintenance operations, we have the process to introduce components to the repository, to query and locate candidate components, to modify and specialise selected components, to integrate components into the target artefacts, and finally to maintain the repository. Figure 5 shows the reuse process.

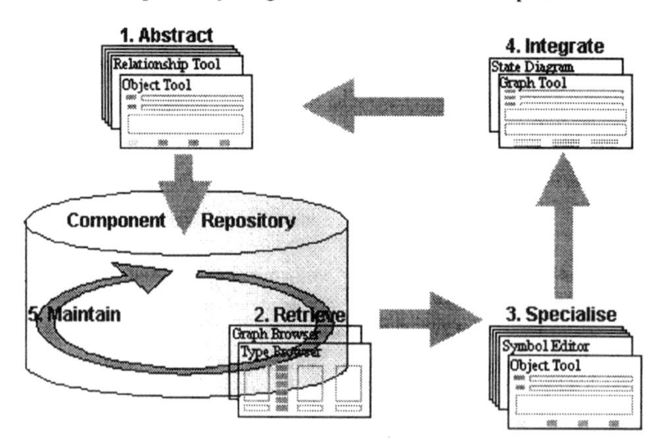

Figure 5 Reuse lifecycle

In MetaEdit+, some tools have been implemented to support reuse. For example, features of a component are specified in tools like the Object Tool, the Relationship Tool, the Role Tool, or specific editors. For example, the "State" object in a state transition diagram can be specified in the Object Tool in the form of its name, its owner, and its properties. If we model the phone interface design using the state transition diagram, it is not easy to present the complicated state that includes several internal actions. We thereby consider using state diagram in UML methodology. To specify such a method in MetaEdit+, we retrieve and reuse the existing definition of a "State" in the metamodel repository, and then specialise it to "State(UML)" in the UML methodology. There are two distinct ways to reuse "State": copy and reference. Since the properties of "State(UML)" are consistent with the properties of "State" except for the additional property of internal activities, there is no serious conflict between these two concepts. We thereby reuse "State" by reference, add new properties by using the object tool, and create a symbol for "State(UML)" by reusing the existing symbol library. Finally, "State(UML)" is integrated into a new state diagram by using a Graph tool. In the process of integration, conflicts such as an inconsistent specification and representation, homonyms and synonyms must be removed.

Table 2 presents a checklist that evaluates the reuse capability in MetaEdit+. In Table 2, symbol ✘ represents that there is no facility to support the reuse operation, while symbol ✔ shows that there are facilities in MetaEdit+ to carry out the required operations. Although we have for example tools to support component abstraction, or browsers to support component retrieval, many existing facilities need further development. For example, browsers lack efficient query tools to search the components based on a predefined facet description. MetaEdit+ lacks also metrics for component reusability measurement. Detailed descriptions of existing and expected functionality to support reuse operations for every component granularity level in MetaEdit+ are shown in appendix 1.

	Component Unit		**Graph/Diagram**		**Methodology/Project**	
	Metamodel level	Model level	Metamodel level	Model level	Metamodel level	Model level
Abstract	✔	✔	✔	✔	✔	✘
Retrieval	✔	✔	✔	✔	✔	✔
Specialisation	✔	✔	✔	✔	✘	✘
Integration	✔	✔	✔	✘	✘	✘
Maintenance	✘	✘	✘	✘	✘	✘

Table 2 Reuse support in MetaEdit+

While evaluating the reuse capability of MetaEdit+ we can observe that the basic functionality exists for the methodology specialisation. However, auxiliary tools to support efficient reuse process at the model level are largely missing including a comprehensive component retrieval tool, a version management tool, and a traceability support tool. Moreover, maintenance of reusable components should be improved.

5 Conclusion

In this paper, we have examined component reuse in a metaCASE environment holistically and presented a framework that systematically guides the evolution of reuse activities. The framework differs from other software reuse frameworks [3-5] in that it introduces reuse in a metaCASE environment that consists of both methodology construction process, and a system development process. Reusable components are thereby expanded from the source code to any type of design artefacts generated at different system development stages, and to metamodels generated during the method implementation process. The framework expands reuse processes by shifting the focus from code design reuse, to conceptual abstraction in a metaCASE environment. At the same time, it classifies the reusable components into different granularity levels according to the data types and semantics defined by the metaCASE environment. Each component can be further decomposed into the representational part, and the conceptual part. Such a classification represents components in a comprehensive manner and supports systematically component management.

This reuse framework not only focuses attention on new technical challenges posed by software reuse processes, but also acknowledges the organisational challenges implicated by such environments. It demonstrates that a combination of systematic reuse techniques within the metaCASE technology helps enhance the system development process. The reuse techniques avoid greenfield development in both the software engineering process, and the method engineering process. It accumulates the development experiences and saves the efforts and costs in the methodology/system development. The reuse in a metaCASE environment effectively combines methodology construction process with the system development process, which can benefit the "family" based product development. Furthermore, with the increasing storage of the reusable components, system maintenance work will become simpler due to the use of standardised components.

Improving component reusability in a metaCASE environment is not fully accomplished. We have not implemented all solutions that can support effective reuse in the metaCASE environment. Specifically, facilities to manage and retrieve candidate components are now being researched, and a novel framework to specify and use design components is being established. Moreover, reuse support tools, such as facilities for modification tracking or semantic conflict checking, have not been developed. It seems that some facilities will be difficult to implement. For example, the semantic checking during the component integration and maintenance is difficult to accomplish automatically, as we know from semantic view integration.

References

1. DDR&E Software Technology Strategy. Department of Defense: Washington, 1991
2. Lyytinen, K. and V.-P. Tahvanainen Introduction: Towards the Next Generation of Computer Aided Software Engineering (CASE). In: K. Lyytinen and V.-P. Tahvanainen (ed) Next Generation CASE Tools, IOS Press, 1992, pp 1 - 7
3. SRI DOD Software Reuse Initiative Technology Roadmap (V2.2). Software Reuse Initiative, Department of Defense, 1995

4. Biggerstaff, T.J. and C. Richter Reusability Framework, Assessment, and Directions. In: T.J. Biggerstaff and A.J. Perlis (ed) Software Reusability Volume 1: Concepts and Models, 1987
5. Krueger, C.W. Software Reuse. ACM Computing Surveys 1992; 24(2): 131 - 183
6. Moore, J.M. Domain Analysis: Framework for Reuse, In: R. Prieto-Díaz and G. Arango (ed) Domain Analysis and Software Systems Modeling, IEEE Computer Society Press, 1991, pp 179 - 203
7. Liao, H., M. Chen, and F.-l. Wang A Domain-Independent Software Reuse Framework Based on a Hierarchical Thesaurus. Software- Practice and Experience 1998; 28(8): 799 - 818
8. Prieto-Díaz, R. and P. Freeman Classifying Software for Reusability. IEEE Software 1987; (1): 6 - 16
9. Kelly, S., K. Lyytinen, and M. Rossi MetaEdit+: a Fully Configurable Multi-User and Multi-Tool CASE and CAME Environment. In: Advanced Information Systems Engineering, Proceedings of the 8th International Conference CAISE'96, Springer-Verlag, 1996, pp 1 - 21
10. Neighbors, J.N. Draco: A Method for Engineering Reusable Software Systems. In T.J. Biggerstaff and C. Richter (ed) Software Reusability, 1989, pp 295--319
11. Karlsson, E.-A., ed. Software Reuse: A Holistic Approach. In: C. Tully and I. Pyle (ed), John Wiley & Sons, 1996
12. Ambler, S. A Realistic Look at Object-Oriented Reuse. Microsoft Corporation, URL: http://msdn.microsoft.com/library/periodic/period98/html/Object-Oriented_Reuse.htm, 1998
13. Sutcliffe, A. and N. Maiden Supporting Component matching for Software Reuse. In: Lecture Notes in Computer Science, 1992, pp 290 - 303
14. Grande, C.D. Software Reuse Overview and Rediscovery. IBM Corporation, International Support Organisation, 1994
15. McClure, C. Software Reuse Techniques: Adding Reuse to the System Developent Process, Prentice Hall, 1997
16. Bell, J.L. Reuse and Browsing: Survey of Program Developers. Centre Universitaire d'Informatique, University of Geneva, 1992
17. Hooper, J.W. and R.O. Chester Software Reuse: Guidelines and Methods. In: R.A. DeMillo (ed) Software Science and Engineering, New York and London: Plenum Press, 1991
18. Brinkkemper, S. Method engineering: engineering of information systems development methods and tools. Information & Software Technology 1996; 38(6): 275--280
19. Kelly, S. MetaCASE Tools. URL: http://www.metacase.com/, 1997
20. Smolander, K. GOPRR: a proposal for a meta level model. University of Jyväskylä: Finland, 1993
21. Lyytinen, K., *et al.* MetaPHOR: Metamodelling, Principles, Hypertext, Objects and Repositories. Technical Report TR-7. University of Jyväskylä: Finland, 1994
22. Rossi, M. Advanced Computer Support for Method Engineering: Implementation of CAME Environment in MetaEdit+. PhD Thesis, University of Jyväskylä: Finland, 1998
23. Tolvanen, J.-P. Incremental Method Engineering with Modeling Tools: Theoretical principles and Empirical Evidence. PhD Thesis, University of Jyväskylä: Finland, 1998
24. Cronholm, S. and G. Goldkuhl Meanings and motives of method customisation in CASE environments - observations and categorizations from an empirical study. In: Proceeding of the fifth workshop on the next generation of CASE tools, University of Twente, 1994, pp 67--79
25. Hillegersberg, J.v. Metamodeling-based integration of object-oriented systems development. PhD Thesis, Amsterdam, 1997
26. Hillegersberg, J.V., K. Kumar, and e. al Using metamodeling to analyze the fit of object-oriented methods to languages. In: Proceedings of the 31st Hawaii International Conference on System Sciences, IEEE Computer Society, 1998

27. Oinas-Kukkonen, H. Improving the Functionality of Software Design Environments by Using Hypertext. PhD Thesis, University of Oulu: Finland, 1997
28. ISO Information processing systems: Information Resource Dictionary System (IRDS) Framework. International Standard ISO/IEC DIS 10027. 1989
29. Pressman, R.S. Software Engineering: A Practioner's Approach. McGraw-Hill, Inc. 1992
30. Pohl, K. Process-Centered Requirements Engineering. Research Studies Press, John Wiley & Sons, 1996
31. Ezran, M., M. Morisio, and C. Tully Practical Software Reuse: the essential guide. 1998.
32. Solvberg, A. and D.C. Kung Information Systems Engineering: An Introduction. Springer-Verlag, 1993
33. Koskinen, M. A Metamodelling Approach to Process Concept Customisation and Enactability in MetaCASE. University of Jyväskylä: Finalnd, 1999

Appendix. Tables of reuse capability evaluation in MetaEdit+

Model level reuse	Component Unit	Diagram	Project
Abstract	✓ Component Unit Properties	✓ Graph Property	× Abstraction Tools
Retrieval	✓ Graph Browser to locate appropriate component units; In Graph Browser → "Info" to navigate through the related components × Facet-based retrieval tools	✓ Graph Browser to locate appropriate diagram; In Graph Browser → "Info" to navigate through the related diagram and component units × Facet-based retrieval tools	✓ Graph Browser to locate appropriate project × Facet-based retrieval tools
Specialisation	✓ Reuse by reference (limited, without specialisation) × Support to copy/paste reuse; Modification tracker	✓ In Diagram Editor, reuse by Exporting/ Importing diagram (copy/paste) and specialising in Diagram Editor, Matrix Editor, or Table Editor × Reuse by reference; Modification tracker	× Tools to specialise the project
Integration	✓ In Diagram Editor (simple) × Integration related tools (such as conflict/overlap inspection, modification tracker,...)	× Integration related tools (such as conflict/overlap inspection, modification tracker,...)	× Tools to support project integration
Maintenance	× Maintenance support	× Maintenance support	× Maintenance support

Note: ✓ **Existing facilities** × **Expected facilities**

Metamodel level reuse	Component Unit	Graph	Methodology
Abstract (Need abstraction tool for reusable component creation)	✓ Definition tool for OPRR meta data types such as Object, Class, Transition, … ✗ Abstraction tools	✓ Definition tool for Graph such as Class Diagram, State Diagram, … ✗ Abstraction tools	✓ Definition tool for Methodology such as UML. ✗ Abstraction tools
Retrieval (Need a comprehensive retrieval tool)	✓ In Metamodel Browser to locate appropriate OPRR components; In Type Browser → "Type Info" to navigate through the related OPRR components ✗ Facet-based component retrieval tools	✓ In Metamodel Browser to locate appropriate graph; In Type Browser → "Type Info" to navigate through the related OPRR components and Graphs ✗ Facet-based retrieval tools	✓ In Metamodel Browser to locate appropriate methodology ✗ Facet-based retrieval tools
Specialisation (Need facilities to better support reuse and traceability)	✓ In Type Edit Tools, e.g. Object Tool to specialise the component in both concept and representation (symbol) ✗ Instructions to choose between reference reuse or copy/paste reuse; Modification tracker	✓ In Graph Tool to specialise the graph in conceptual aspect ✗ Instructions to choose between reference reuse and copy/paste reuse; Support to detect and remove conceptual/semantic overlap or conflict; Modification tracker	✗ Tools to specialise the methodology
Integration (Need facilities to retrieve and solve conflict, support traceability as well)	✓ In Graph Tool to integrate the components into a graph level component ✗ Homonyms and synonyms inspection; Support to detect and remove conceptual/ representational overlap or conflict between each component unit; Modification tracker	✓ In Graph Tool to integrate a graph into one project ✗ Tools to integrate graphs to a new graph; Homonyms and synonyms inspection; Support to detect and rid of conceptual/ representational overlap or conflict between each graph; Modification tracker	✗ Build relationship between methodologies
Maintenance (Need reuse metrics to measure reusability and examples)	✗ Support to maintenance; Facilities to track changes	✗ Support to maintenance; Facilities to track changes between metamodels and reports	✗ Support to maintenance

Note: ✓ **Existing facilities** ✗ **Expected facilities**

Method Engineering with Web-enabled Methods

Sjaak Brinkkemper
Software Engineering Process Group, Baan Company R&D
Barneveld, the Netherlands

Abstract

Method engineering is the engineering discipline to design, construct, and adapt methods, techniques and tools for the development of information systems. The advent of web and intranet technology gives rise to dramatic innovations in systems development methods due the fast company-wide deployment capabilities and the efficient methodical support on each developer's workstation overcoming the drawbacks of methods in paper format. The R&D department of Baan Company implemented a web-enabled software development method as one of the first companies in the world. About 1500 software engineers distributed in development offices over the world have now on-line access to procedural and technical support on their workstation through the Baan Development Method (BDM) intranet site. Web-enabled methods bring about new method engineering research themes that are discussed with illustrations taken from the BDM approach.

1 Introduction

In his keynote address at the third IFIP WG8.1 conference on the Comparative Review of Information System Design Methodologies (CRIS-86 [13]), Janis Bubenko stated:

> *"The state-of-art of information system methodologies is characterised by hundreds, if not thousands, of more or less similar academic as well as practical methodologies. There is a conceptual as well as terminological gap between principles and methods developed in academic environments and those developed and applied in practice. ..."*

That was 1986. Has anything changed since then? We observe that nowadays there exist tens of thousands, if not hundreds of thousands more or less similar methodologies. We have witnessed the coming of series of object oriented methods, methods for workflow systems design, and at the moments all kinds of methodologies for web applications and e-commerce systems are popularised. On the other hand we can argue that the conceptual and terminological gap might have been diminished, as the academic world and industrial worlds have grown towards each other. Terminology from practitioners was purified, obscure academic terms never made it into the real world. Whoever is still using "infological", "systemeering", or "NOLOTs"?

The abundance of methods, techniques, and tools is still there. Few attempts for generalisation and integration of methods were reported, and therefore have limited reducing effect on the creativity of method inventors. This unsatisfactory situation has lead to the research area of Method Engineering [2], [4], which aims at the identification of generic development principles in methods, the unification and homogenisation of concepts, and the optimal suitability of methods in development circumstances. There are still many research questions open, but the emergence of web technology has boosted the activity as we will see in this paper.

In 1992 Kuldeep Kumar and Dick Welke [11] proposed the notion of situation specific engineering of methodologies, later baptised with the name *Situational Method Engineering*. Their proposal consisted of four intertwined strategies:

1. *Modular construction of methods*. As methods consist of phases, containing some hierarchy of steps, each producing intermediate or final deliverables, it is possible to identify core components in methods that can be reused in various development situations. These method components, or method fragments, are to be collected, described, categorised and stored in a Method Base for future use. Method fragment descriptions are based on meta-modelling specification [1] [7].
2. *Situational method assembly*. First, suitable method fragments are selected based on a careful characterisation of the development situation using factor categories application domain, system contents, external factors such as law and norms, and technical platform, and development expertise [16],[9]. These method fragments are then assembled into one overall development approach using an extensive system of assembly rules and well-formedness heuristics for methods [5].
3. *Automated method engineering*. Method fragments are stored in a Method Base being part of a Computer Aided Method Engineering (CAME) tool. CAME and MetaCASE technology allows to generate tools based on their meta-models [6],[10].
4. *Organisational embedding*. Proper implementation of Method Engineering (ME) in a development organisation is only then realistic if an adequate support team is established. This team is a, usually central, department in charge of maintaining the Method Base, evaluating method usage in actual projects, providing method and ME training, and accumulating development experiences and new method technology into the Method Base [9],[12].

Various research teams are active in the area of Method Engineering. The team of Rolland et al [15] works on method description and assembly in the process dimension. Tools for procedural support of the requirements engineering process are generated based on hierarchical trees consisting of <Situation, Decision, Action> triplets. A metaCASE tool, called MetaEdit, is being developed in the research unit of Kalle Lyytinen [10]. The kernel of MetaEdit is the Graph-Object-Property-Relationship-Role visual specification language for the meta-modelling and tool generation. ConceptBase [8] supports the stratification of knowledge bases in meta-levels based on the Telos language. Each meta-level defines the language constructs for the level below, thus allowing for flexible domain specific concepts. The laboratory of Motoshi Saeki is working on the Method Base providing various method assembly tools [17]. At the Open University in the Netherlands, Karel Lemmen has investigated the educational perspective of ME, based on a comprehensive ME course [12]. Method fragments on the one hand, and project situations on the other hand are mapped upon a 32 cells Aspect/Level framework, thereby enabling the proper method fragment selection. An extensive empirical study with more than 140 experienced software developers justified the validity of ME in the development practice.

All these approaches do hardly take the web into account, where we claim that this technology is rapidly changing the role of methods in the development practice. In this paper we want to elaborate on the consequences of a web enabled method. The next section reviews the basic concepts of ME, and presents the Method Management System as the central web-site for method usage *and* method engineering. The 3rd chapter presents the realisation of a web-enabled method at the Baan Research and Development department as an illustration of the solutions provided to method engineering in practice. We end with a discussion of major experiences and several new research themes for web-enabled method engineering.

2. Method Engineering and web technology

2.1 Basic Concepts

Method Engineering evolved from a new vision on development methods to a complete discipline in the overall area of information systems development. In order to have a good understanding of this area we have to identify the basic concepts. We reformulate Method Engineering as it was defined in [1]:

> *Method engineering* is the engineering discipline to design, construct, and adapt methods, techniques and tools for the development of information systems.

To understand the notion of Situational Method we have to define (1) method for information systems (IS) engineering, or just, shortly, method, and (2) method fragment:

> A *method* is an integrated collection of procedures, techniques, product descriptions, and tools, for effective, efficient, and consistent support of the IS engineering process.
>
> A *method fragment* is a description of an IS engineering method, or any coherent part thereof.

According to these definitions UML [3] is an example of a method, as well as a method fragment. Class diagrams, and Data Flow diagrams, being specification techniques part of UML are also method fragments. Decomposing UML to its concepts leads to atomic method fragments, such as Class, Association, Process, and Data flow.

A *situation* is the combination of circumstances at a given moment, possibly in a given organisation, or in a given role.

A *situational method* is a IS engineering method tailored and tuned to a particular situation.

This implies that the situation drives the method and not the other way round. No method description can ever be applied step by step to the very letter. Adaptation of the method to the circumstances is always needed. The principles and formalisation of this practice is the core research objective of Method Engineering. Observe, however, that in contrast to what has been formulated in [1] or [6], we extend the concept of situation to fit both the situation of the organisation or project, as well as the situation of the individual participant in the IS engineering process, i.e. the various roles of a project, e.g. project manager, software engineer, tester, etc. The latter is needed to accommodate the needs for individualised methodical support for the various roles active in the development project.

2.2 Web technology

Web technology, such as hyperlinks, browsers, HTML, web-sites, fire-walls, intranets, navigation aids, etc. brings about all kinds of new perspectives to method creation and usage. The resulting Method management system can be depicted as shown in figure 1.

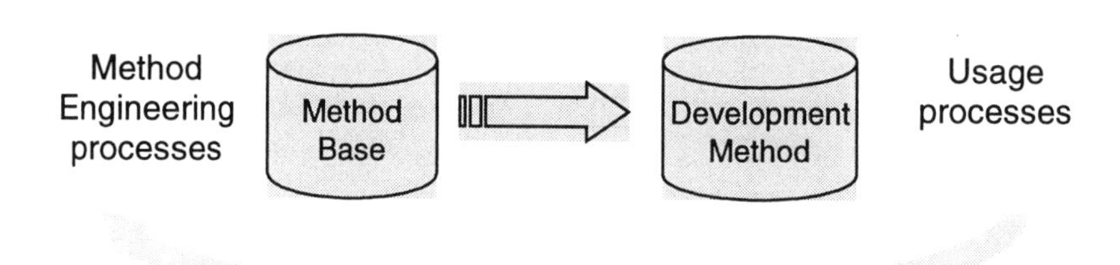

Figure 1. A Web-enabled Method Management System

We identified the following changes:

1. *Electronic availability*: The availability of the method on the intranet supported by the corporate network infrastructure obliterates the need for a paper method. Each worker can be granted access to the method, and distribution of the method is reduced to the notification of the location of the method web-site. Costly and time-consuming distribution of paper methods is history.
2. *Instantaneous deployment*: The improvement cycle of methods is reduced from years to hours, as extensions can be updated on-line and made available

to the users after propagation to all replicas of the development method site. This reduces the communication burden in large or geographical dispersed organisations significantly.

3. *Multi-media methods*: All media supported in web technology can be utilised in the method. This applies especially for training presentations in slide or video format, audio, and animations. Complete examples of deliverables, software process simulations, screen cams of tool usage are other potentials.
4. *Efficient access*: Whereas a paper method is bound to a linear presentation format, complemented with subject indices, a web-enabled method can support any access path through the method knowledge base through the proper placement of hyperlinks. Various entrances to the method base are created based on the categorised requirements for methodical assistance. We will illustrate this in the next section.
5. *Active user participation*: Integration with mail-to links and discussion news groups allows method users to interact with the method developers in a simple manner. Questions and answering for clarification, or suggestions for improvement can be communicated between the method engineers and the users very efficiently.

In the next chapter we will present an illustration of the method management system of the Baan Company for the world-wide development of its ERP software products.

3. Case: the Baan Development Method

Baan Company develops and sells so-called Enterprise Resource Planning (ERP) software all over the world. The Baan Research and Development departments consist of about 1500 employees in 17 offices distributed over the world. The development approach of Baan is called the Baan Development Method (BDM).

3.1 The BDM web-site

After several releases of an internal, paper based software development method, Baan decided in September 1997 to transform the next release of BDM to a web-enabled method. The resulting BDM web-site was launched in February 1998, making Baan one of the first companies world-wide to have a web-enabled development method on its intranet. The distributed organisational structure of Baan R&D, as well as an intensive Software Process Improvement program were among the essential stimulating factors for this effort.

The home page of the BDM web-site is shown in figure 2. On the left are links to generic information, such as the background of BDM, some explanations, a terminology list, a listing of the method engineering organisation (under process areas), and a search engine. Novices to BDM usually read these pages only once to get familiar with the overall structure of BDM. Experienced users check just the "What's New" button.

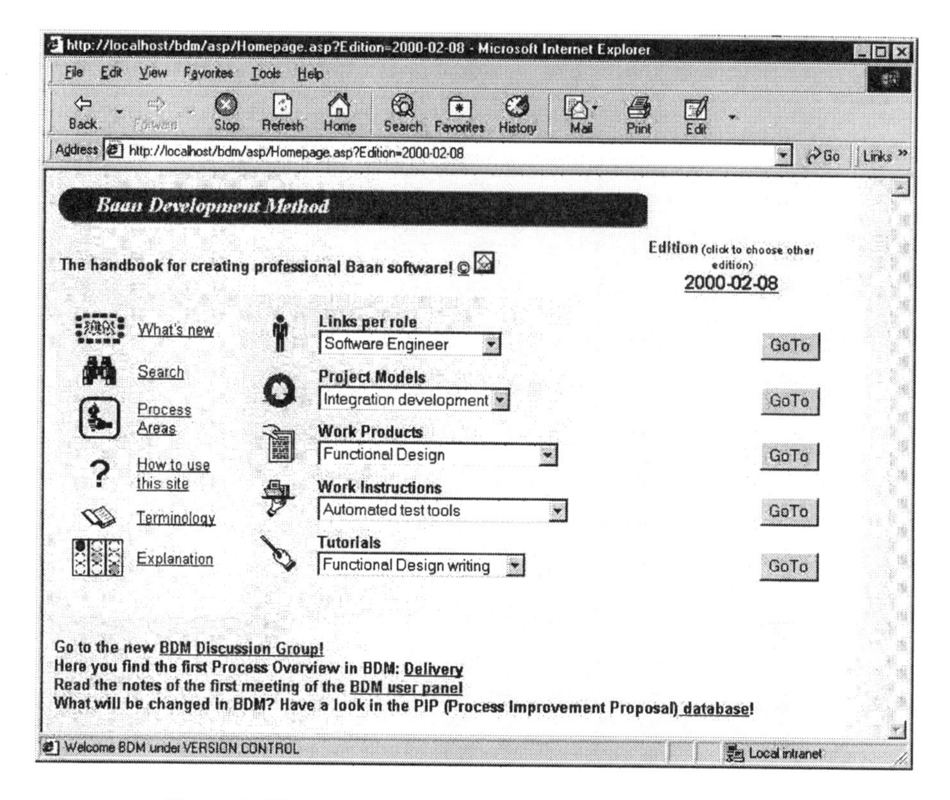

Figure 2. The home page of the Baan Development Method

The right side provides the main entrance to the 5 categories of method fragments: We will discuss these in the next section. At the right-upper corner a listing of earlier editions of BDM can be accessed. on The previous editions of BDM remain accessible for the projects that have been based on these editions. At the bottom the BDM news group can be directly invoked, as well as the database containing data on the ongoing process improvement initiatives can be queried.

In figure 3 an example of a BDM page is displayed. It shows a work instruction for the Software Language Edit, a tool for the translators of the textual labels of the Baan user-interface. Detailed instructions in a step-by-step and screen-by-screen presentation format aid the translators to perform their work in an orderly and optimal supported manner.

The maintenance and implementation of BDM resides in the Software Engineering Process Group (SEPG) of Baan. Seven full time employees perform the method engineering tasks, where each focuses on a particular area: Requirements Management, Design, Realisation, Testing, Configuration Management, Project Management, and one for the overall BDM structure. At the moment BDM consist of about 140 web pages, totalling to more than 5000 A4 pages, mainly contributed by the numerous examples, document templates and deliverable instructions.

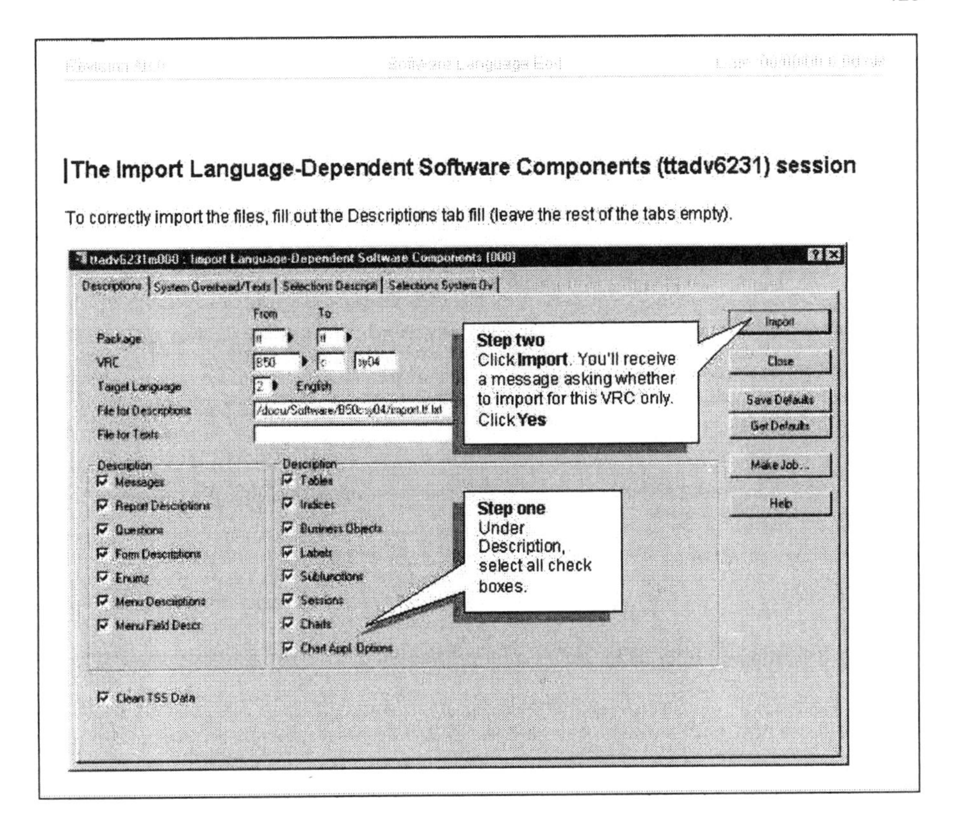

Figure 3. A Work instruction for Software Language Edit

3.2 The structure of a web enabled method

Where the description of IS engineering methods on paper media leads to method fragments such as phases, specification techniques, or basic concepts, the web-enabling of method engineering created a whole new classification of method fragment types. The users, i.e. developers, project managers, consultants, have various needs for methodical support ranging from obtaining a global impression to a very detailed tool instruction.

The BDM web-enabled method is highly deliverable oriented, i.e. focussing on the result, not the way it has been achieved as this left to the discretion of the developers. This implies that the balance on the usual dichotomy between the product and process perspectives, goes over to the product side. Process support is mainly superficial and just for situations of knowledge deficiencies, such as the training of new employees, or the handling of rare cases. The description of documentation standards for the software development products is very elaborate, e.g. formats for data models, definitions of application software, specification of tool integrations, system architectures, etc.

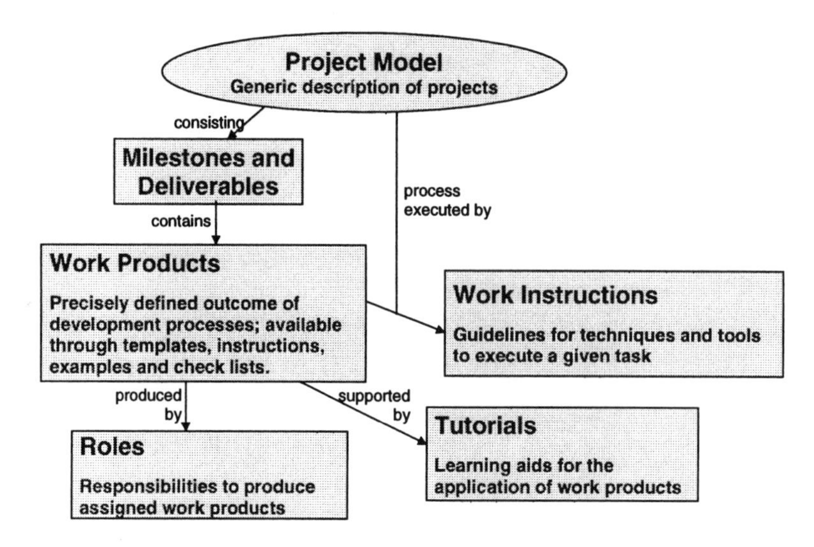

Figure 4.The global structure of BDM

The aim to provide support within three clicks for all BDM users has led to the following categories of method fragments in BDM:

- *Project Model*: Project Models describe the listings of all deliverables needed to complete a project milestone. Project models are available for Release delivery, Software development. Knowledge transfer, Integration test, and System test.
- *Milestone and Deliverables*: Project milestones are groups of deliverables, expressed in terms of Work Products. A Work Product may be reused in several milestones. For example the Work Product for a project plan appears in all project models.
- *Work Products*: Work Products are standards for documentation of deliverables of BDM. Instances are, for example, project plan, version definition, definition study, software unit, test design. For most Work Product some intermediate states are defined to identify the progress of development: draft (work just started), preliminary (not yet reviewed), and actual (completed and valid for current release). Document templates, instructions for the writing of a Work Product, and complete examples are available. Check-lists for Peer Reviews and Fagan Inspections are given. To support some variety of development situations, variants of Work Products can be selected from. There exists a Functional Design for applications, one for integrations, as well as a Functional Design for development tools.
- *Work Instructions*: Work instructions are detailed procedures for the completion of well-defined smaller tasks, such as performing a software unit test, or a project audit, or risk management. Work instructions consist of a purpose description, a flow chart, and a detailed description of the steps in the chart. In most work instructions the internal conventions for documenting development artefacts are listed.
- *Role support*: Currently the following roles are distinguished: Moderator (of Fagan Inspections), Project Leader, Project Leader Testing, Software

Engineering, Test Co-ordinator, and Test Engineer. This entrance summarises all links that are relevant for employees in that role.

- *Tutorials*: All training materials for the various tasks in BDM are accessible through this link. Developers can read training presentation to prepare for a course or to refresh their knowledge.

Method engineering is performed at the Software Engineering Process Group. The method engineers are responsible for all kinds of method improvement activities in their respective domains. Method improvements can be triggered in various ways: known method deficiencies, suggestions of the user communities, departmental improvement plans started as a follow-up to a CMM assessment, or by a so-called Process Action Teams (PAT). PATs are short living teams that work on the extension of BDM in a well defined small methodical domain, e.g. handling customer requirements, or definition of unit tests. Internal quality procedures assure the optimal reuse of method fragments and the adherence to the structural conventions in BDM. Needless to say, is that these method engineering procedures are also offered by BDM.

4. Findings and Research Themes

Web technology is *the* way to go for methods. Various surveys within Baan R&D show that the adoption of BDM is extremely high, especially in contrast to the former paper based situation. Most developers frequently visit the BDM site for checking the updates and extensions. Furthermore, many method improvements are based on contributions from the user community.

The optimal usage of a corporate method management system, such as BDM, has some critical success factors:

- Adequate infrastructure where every workstation linked to the corporate network (WAN) has direct access to the method management system.
- Standard technology used for the creation and deployment of method pages, documentation standards, and procedure descriptions.
- Active support group, which continuously reviews and improves the contents of the method
- Mandatory usage in projects as users should concentrate on developing applications and not spend their time building their own methodical deviations.

The transfer from the paper to the web gives rise to several new research questions, which we wish to put forward for the international ME community. First, the current Method Management System is still based on plain HTML pages, although they share a standard structure derived from the method fragment categories explained in section 3.2. This page based structure requires to perform the completeness and consistency checking manually. One could envision a method base that enforces the structures and content of the method fragment categories and from which the method pages are generated. Secondly, as the user friendliness of the Method Management System is essential for its adoption, the optimisation of knowledge access paths from user needs should be investigated. Storage of individual or role based usage patterns can reveal frequent and missing methodical support. Analysis of the usage patterns should then be input to various improvement of the user interface of the Method Management System. Currently, an extensive usability study is being performed at Baan, of which we hope to report shortly.

As explained the current BDM contains all previous editions. However, due to the limited technology these editions are just complete copies of the then existing BDM, without any sharing of pages from earlier editions. A Configuration Management System for the BDM is lacking completely. Such a CMS might also improve the cumbersome insertion of new or updated method pages.

Finally, as stated the deployment of new method fragments is very fast in a web-enabled method. However, that does not mean that every user impacted by this extension of methodical knowledge is automatically aware and immediate an active user. There exist various strategies for corporate knowledge management, i.e. the collection and exploitation of organisational expertise [9]. Suitable implementation strategies for web-based method introduction need further empirical investigation.

Acknowledgements

We wish to thank Kuldeep Kumar, Motoshi Saeki, Matti Rossi, Karel Lemmen, Kees van Slooten, Frank Harmsen for the numerous discussions on Method Engineering over the years. We also gratefully acknowledge the colleagues of Baan, especially those activily involved in the creation and extension of the Baan Development method and website: Judith Cornelisse-Vermaat, Cor van Dijk, Jaap Meeuse, Ranjana Narawane, Gilberto Onodera, and Lutzen Wijngaard.

References

1 Brinkkemper, S., Method Engineering: Engineering of Information Systems Development Methods and Tools. *Journal of Information and Software Technology*, Vol. 38, Nr. 4, pp. 275-280, 1996.

2 Brinkkemper, S., and S.M.M. Joosten (Eds.), Method Engineering and Meta-Modelling. Special Issue. *Information and Software Technology*, vol. 38, nr. 2, pp. 259-305, 1996.

3 Booch, G., I. Jacobsen, and J. Rumbaugh, The Unified Modeling Language User Guide, Addison-Wesley, 1998.

4 Brinkkemper, S., K. Lyytinen and R.J. Welke (Eds.), *Method Engineering: Principles of Method Construction and Tool Support*. Chapman and Hall. 1996.

5 Brinkkemper, Sjaak, Motoshi Saeki, and Frank Harmsen, Meta-Modelling Based Assembly Techniques for Situational Method Engineering. *Information Systems*, vol. 24, No. 3, pp. 209-228, 1999.

6 Harmsen, F., *Situational Method Engineering*, Moret, Ernst&Young, January 1997.

7 Harmsen, Frank, Sjaak Brinkkemper, and Han Oei, Situational Method Engineering for Information System Project Approaches. In: *Methods and Associated Tools for the Information Systems Life Cycle.* IFIP Transactions A-55, North-Holland, 1994, pp. 169-194.

8 Jarke, M., R. Gallersdörfer, M.A. Jeusfeld, M. Staudt, and S. Eherer, ConceptBase: A deductive object base for meta data management. *Journal for intelligent information systems*, 4(2), pp. 167-192, 1995.

9 Klooster, Marnix, Sjaak Brinkkemper, Frank Harmsen, Gerard Wijers, *Intranet Facilitated Knowledge Management: A Theory and Tool for Defining Situational Methods.* In: Proceedings of the 9th International Conference on Advanced Information Systems Engineering, Lecture Notes in Computer Science 1250, pp. 303-317, Springer Verlag, 1997.

10 Kelly, S., K. Lyytinen, and M. Rossi, *MetaEdit+: A Fully Configurable Multi-user and Multi-Tool CASE and CAME Environment.* In: Proceedings of the 8th Conference on Advanced Information Systems Engineering. Lecture Notes in Computer Sciences 1080, pp. 1-21, Springer Verlag, 1996.

11 Kumar, K. and R.J. Welke, Methodology Engineering: A Proposal for Situation-Specific Methodology Construction. In: W.W. Cotterman, J.A. Senn (Eds.), *Challenges and Strategies for Research in Systems Development*, Wiley, 1992.

12 Lemmen, Karel, Fred Mulder and Sjaak Brinkkemper, An empirical study on the educational effects of a course in Method Engineering for Information Systems. To appear in: *Education and Information Technology*. 2000.

13 Olle, T.W., H.G. Sol and A.A. Verrijn Stuart (Eds.), *Information System Design Methodologies - Improving the Practice*, Proceedings of CRIS-86 conference, North Holland Publ. Co., 1986.

14 Rossi, M. and S. Brinkkemper, Complexity Metrics for Systems Development Methods and Techniques. *Information Systems*, vol.21, nr.2, pp.209-227, 1996.

15 Rolland, C., C. Souveyet, and M. Moreno, An approach for defining ways-of-working. *Information Systems*, 20(4), pp. 337-359, 1995.

16 Slooten, C. van, and S. Brinkkemper, A Method Engineering Approach to Information Systems Development. In: *Information Systems Development Process*. Elsevier Science Publishers (A-30), pp. 167-186, September 1993.

17 Saeki, M., and K. Wen-yin, *Specifying Software Specification and Design Methods.* In: Proceedings of the 6th International Conference on Advanced Information Systems Engineering (CAiSE'94), Lecture Notes in Computer Science 811, Springer Verlag, pp. 353-366, Berlin, 1994.

Developing Data Warehouses with Quality in Mind

Yannis Vassiliou
Department of Computer and Electrical Engineering
National Technical University of Athens
Athens, Greece

Abstract

Data Warehouses provide large-scale caches of historic data. They lie between information sources gained externally or through online transaction processing systems (OLTP), and decision support or data mining queries following the vision of online analytic processing (OLAP). In developing and operating Data Warehouses, one can distinguish between different processes, each of which raises quality considerations. Using the framework of a general formal architecture developed in the DWQ project, this paper discusses some of the research conclusions regarding development of Data Warehouses with quality factors on mind. Basic processes and the related quality dimensions are considered. The quality factors, metrics and measurement methods are also presented.

1 Introduction

Data Warehouses were first defined by Inmon [1] as: "subject-oriented, integrated, time variant, non-volatile collections of data in support of the organization's decision making process." In essence, Data Warehouses provide large-scale caches of historic data. They sit between information sources gained externally or through online transaction processing systems (OLTP), and decision support or data mining queries following the vision of online analytic processing (OLAP).

Quality considerations have been integral in data warehouse research from the beginning. A large body of literature has evolved in addressing the problems introduced by the Data Warehouse approach, such as the trade-off between freshness of DW data and disturbance of OLTP work during data extraction; the minimization of data transfer through incremental view maintenance; and a theory of computation with multi-dimensional data models. However, we content that we are far from a systematic understanding and usage of the interplay between quality factors and design options in data warehousing. The European Esprit project DWQ* [2] set as a goal to address these issues by developing, prototyping and evaluating comprehensive Foundations for Data Warehouse Quality, delivered through enriched meta data management facilities in which specific analysis and optimization techniques are embedded. Major topics include the definition of an extended meta model for data warehouse architecture and quality [3], inference techniques for improving source integration [4], working with multidimensional data models [5, 6, 7], systematic design of refreshment policies for data warehouses [8], and optimization concerning the choice of materialized views [9].

In this paper, after introducing the basic notions and the DWQ formal framework to capture a Data Warehouse environment, we discuss some of the conclusions regarding development of Data Warehouses with quality factors on mind. References to the main body of research performed in the DWQ project are made throughout the paper. A complete report of the DWQ results, which is used as a basis for this paper, is [10].

2 Framework

Architecture, processes, and quality are three viewpoints for a Data Warehouse, which are not independent from each other (Fig. 1). The data warehouse processes operate, use and affect the objects of the architecture. Quality is important for DW architecture as well as processes. This provides the basis of a methodology for the quality-oriented design of a data warehouse and gives a guideline for how the quality of a data warehouse can be measured.

Fig. 1 Three viewpoints for data warehouses

* DWQ is a European project involving three universities (NTUA Athens / Greece, RWTH Aachen / Germany, Roma La Sapienza / Italy) and three research centers (DFKI / Germany, INRIA/France, IRST / Italy).

The next three sections summarize these three viewpoints in turn, as they were formulated in the DWQ project.

The real value of such a framework comes when all data warehouse components, processes and data are tracked and administered from a *metadata repository*. The metadata repository serves as an aid both to the administrator and the designer of a data warehouse. The metadata repository serves as a roadmap that provides a trace of all design choices and a history of changes performed on its architecture and components. For example, the latest version of the Microsoft Repository [11] and the Metadata Interchange Specification (MDIS) [12] provide different models and application programming interfaces to control and manage metadata for OLAP databases.

3 Data Warehouse Architecture Model

A Data Warehouse system is a collection of technologies aimed at enabling the knowledge worker (executive, manager, analyst, etc) to make better and faster decisions. The traditional data warehouse generic architecture (Fig. 2) exhibits various layers of data in which data from one layer are derived from data of the lower layer. Data sources form the lowest layer. They may consist of structured data stored in database systems and legacy systems, or unstructured or semi-structured data stored in files.

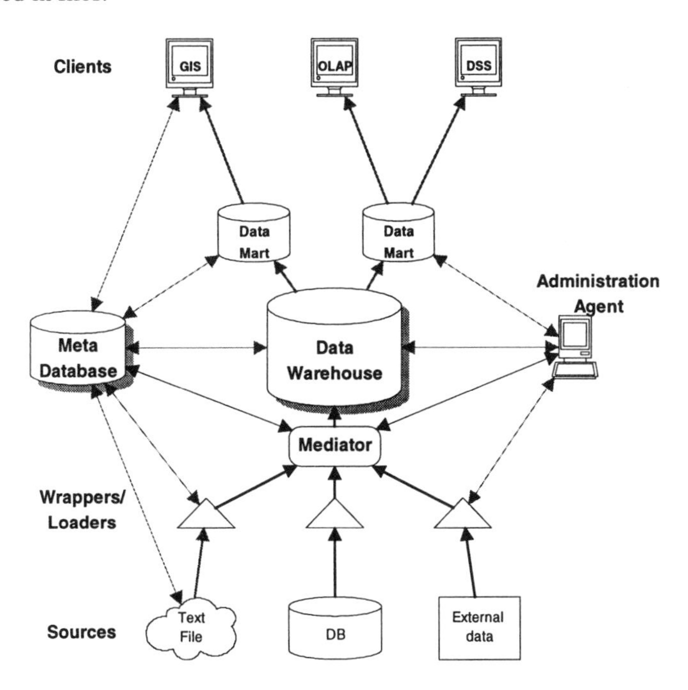

Fig. 2 A generic, traditional data warehouse architecture [2]

The central layer of the architecture is the primary Data Warehouse, which keeps a historical record of data that result from the transformation, integration, and aggregation of detailed data found in the data sources (through wrappers / loaders / extractors or mediators). Usually, an operational data store of volatile, low granularity data is used for the integration of data from the various sources. The operational data store, serves also as a buffer for data transformation and cleaning so that the data warehouse is populated with clean and homogeneous data. The next layer of views are the local, or client warehouses, which contain highly aggregated data, directly derived from the global warehouse. There are various kinds of local warehouses, such as the data marts or the OLAP databases, which may use relational database systems or specific multidimensional data structures.

Although many data warehouses have already been built and follow this architecture, a major observation is that there is no common methodology, which supports database system administrators in designing and evolving a data warehouse. In reality, the architecture in Figure 2 cannot express, let alone support, important quality problems and management issues, such as:

- How do the components exchange (meta-) data?
- How can the quality of a data warehouse be evaluated and designed?

Formally, the purpose of an **architecture model** is to provide an expressive, semantically defined and computationally understood meta modeling language, based on observing existing approaches in practice and research.

The first major development in DWQ was the formulation of a three-perspective meta modeling architecture for Data Warehouses (Fig. 3). All the important objects of data warehouse architecture are presented grouped with respect to their nature (conceptual, logical or physical).

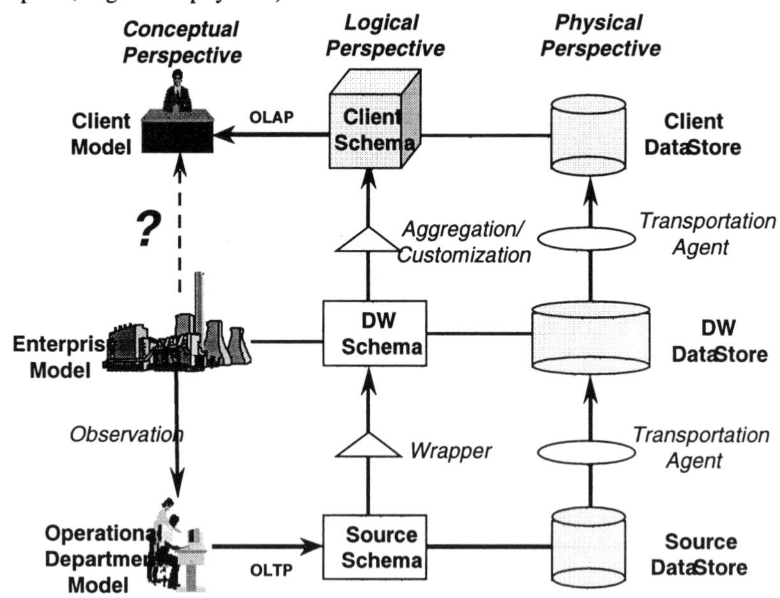

Fig. 3 Data Warehouse Metadata Framework [3]

By introducing an explicit business / conceptual perspective (left side of Fig. 3), the wrapping and aggregation transformations performed in the traditional data warehouse literature (logical perspective in Fig. 3) can thus all be checked for interpretability, consistency or completeness with respect to the enterprise model. At the same time, the logical transformations need to be implemented safely and efficiently by physical data storage and transportation – the third perspective in our approach. These three perspectives, and their interrelationships, are *orthogonal* to the three traditional layers of data warehousing, namely sources, data warehouse, and clients.

This framework can be instantiated (Fig. 4) by information models (conceptual, logical, and physical schemas) of particular data warehousing strategies, which can then be used to design and administer the instances of these data warehouses.

The extended meta-model architecture resulting from this approach was implemented in a repository. The meta database was used to store an abstract representation of data warehouse applications in terms of the three-perspective scheme. The architecture, as well as the other models, are represented in Telos [13], an extensible meta modeling language with a graphical syntax and a frame syntax, which are mapped to an underlying formal semantics based on standard deductive databases. Using this formal semantics, the Telos implementation in the ConceptBase system [14] provides query facilities, and definition of constraints and deductive rules.

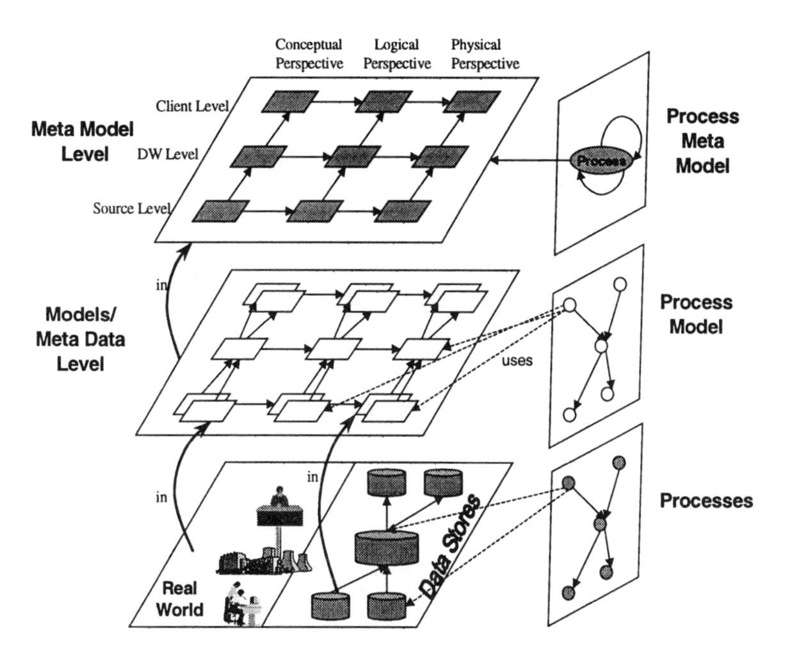

Fig. 4 Repository Structure for Capturing Product and Process of Data Warehousing [3]

To summarize, the purpose of the architecture model is to provide an expressive, semantically well-defined and computationally understood meta modeling language, based on observing existing approaches in practice and research. Expressiveness and services of the metadata schema are crucial for data warehouse quality

4 The Process Model

The *static* description of the architecture parts of the data warehouse can be complemented, with a meta-model of the *dynamic* parts of the data warehouse, i.e. the data warehouse processes (right part in Fig. 4). Providing a process model for data warehouses practically captures the behavior of the system and the interdependencies of the involved stakeholders. A three level instantiation is followed: a *Process Meta-model* deals with generic entities involved in all DW processes (operating on entities found at the DW meta-model level), the *Process Model* covers specific processes of a specific DW by employing instances of the meta-model entities and the *Process Traces* are capturing the execution of the actual DW processes happening in the real world. DWQ's process and workflow modeling work has been influenced by ideas on dependency and rationale modeling stemming from [15, 16, 17] and the Workflow Reference Model, presented in [18].

The process meta-model describes all the functionalities and service provided by the data warehouse system. Processes and dynamic relationships between these processes are described at this level, using different formalisms and diagrams. Each process is considered as an object, which has a conceptual definition, logical specification and physical implementation. This object can operate in one or several perspectives and in one or several levels. Figure 5 offers an intuitive view for the categorization of the entities in the process meta-model. The model has three different perspectives covering distinct aspects of a process: the *conceptual, logical* and *physical* perspective. A detailed description and use of the process model in DWQ can be found in [10].

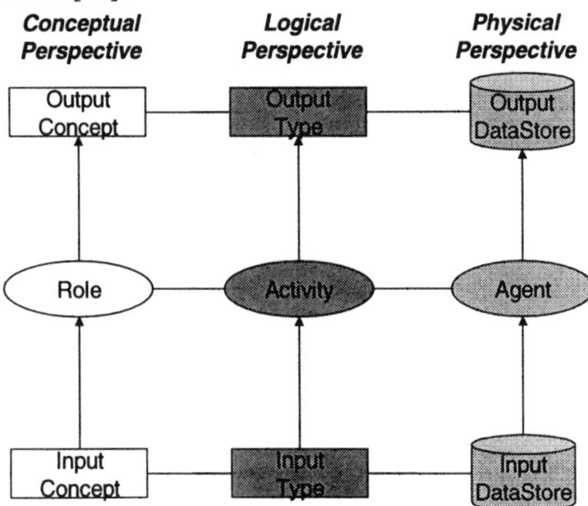

Fig 5 The three perspectives of the process meta-model

5 Data Warehouse Quality Model

Quality is defined and quantified in [19] as performance over expectation. As buffers in the data propagation path between operational databases and end users, Data Warehouses can bridge the gap between subjective user requirements for information quality and objective detection and measurements of data deficiencies. The data warehouse, being a collective *off-line*, multi-layered system, can serve as a "data cleaner" for the information presented to the user. Recording the quality of the data warehouse data and processes in a metadata repository is providing, the data warehouse stakeholders with sufficient power to evaluate, understand and possibly react against the structure and content of the warehouse.

From the viewpoint of information systems research, the problem of how to design, construct and administer data warehouses that satisfy specific, well-defined quality criteria is apparent.

This of course does not imply that there is no substantial activity on the commercial side. As the data warehouse market is rapidly evolving in the last few years, all major database companies have already created tools and products in order to support data warehouse solutions. A number of smaller companies have managed to develop and market specialized tools for data warehouses. Most (if not all) of those tools affect in some way the quality of the resulting data warehouse, but only few of them explicitly deal with data quality. The quality of data in a Data Warehouse is obviously affected by three factors:

- Data warehouse schema design;
- Quality of the data inserted in the data warehouse;
- Manipulation of data in the data warehouse.

Each of those factors is dependent on the set of tools, which are used for a particular data warehouse.

The design of the data warehouse schema is responsible for the (semantically) correct, complete and meaningful integration of the sources. If the design process fails to include all the required information in the data warehouse schema then the data may be ambiguous or even incomplete. If the semantics of the source data is misinterpreted or if the various sources are not properly integrated then the data warehouse will contain incorrect data. Also if the design process does not identify the required integrity constrains the data warehouse may store meaningless or incorrect information. All the quality dimensions defined in [20]: complete, unambiguous, meaningful and correct are affected by the design process. The design of the data warehouse schema is a complicated process involving the analysis of requirements, analysis of the available data, schema extraction and integration (of the sources) as well as other general database design steps. The
standard tools that may assist in this are: CASE tools, Data Modeling, Database Design, Schema Integration, Metadata Management, and Data Reverse Engineering.

Obviously the data stored in the data warehouse depends on the quality of data used to load / update the data warehouse. Incorrect information stored at the data sources may be propagated in the data warehouse. Still, the data is inserted in the data warehouse through a load/update process, which may (or may not) affect the quality of the inserted data. The process must correctly integrate the data sources and filter out all data that violate the constraints defined in the data warehouse. The process may also be used to further check the correctness of source data and improve their quality. In general, the most common examples of dirty data are: format differences, information hidden in free-form text, violation of integrity rules (e.g., in the case of redundancy), missing values or schema differences. The tools that may be used to extract/transform/clean the source data or to measure/control the quality of the inserted data can be grouped in the following categories [21]: Data Extraction, Data Transformation, Data Migration, Data Cleaning and Scrubbing, and Data Quality Analysis. As an example, SQL*Loader module of Oracle [22] can extract, transform and filter data from various data sources (including flat files and relational DBMSs). Another example is Integrity from Vality, which can be used for the application of rules that govern data cleaning, typical integration tasks (esp. postal address integration), etc.

The data in a data warehouse is usually handled by a Database Management System (DBMS) and cannot be updated by users. The most common manipulations are aggregations and multidimensional data reorganization, which are carried out by the DBMS. This means that the quality of data is generally preserved inside the data warehouse and it is hardly affected by the manipulation processes. The major database/software vendors (IBM, Oracle, Informix, Sybase, Red Brick Systems, Software AG, Microsoft, Tandem) provide quality-oriented tools that belong to nearly all the previously mentioned categories. Each vendor provides a set of tools that can be used to design and implement a complete data warehouse

Despite the variety of tools employed for data warehouse design, data refreshment and cleaning, the task of investigating and understanding explicitly the quality of data is still very hard. The DWQ project introduced formal models for Data Warehouse quality and, in order to make them useful, developed measurement techniques to populate them as well as a methodology to ensure quality during both the design and operation / usage of a Data Warehouse. As this paper can only sketch the above, the reader is again referred to [10] for a comprehensive coverage.

The formal requirements for the quality meta-model can be summarized as follows: The categorizations of the meta-data framework that has been adopted should be respected. Thus, all three perspectives and layers of instantiation should be clearly present in the quality meta-model. This would increase both the usability of the approach and the re-usability of any achieved solutions. Moreover, the exploitation of quality should be done through the use of a repository, enabling in this way the potential measurement of the involved quality factors through the use of well-established automated techniques and algorithms.

A large body of research on software and data quality exists, as well as in quality engineering and management. The concepts from the Goal-Question-Metric (GQM) approach [23] have influenced the work on quality in DWQ. The quality meta-model that was developed supplements coherently the architecture and process models (Fig. 6). The three layers of instantiation are consistently adopted. At the meta-model level, a generic framework that follows GQM is given, extended with associations applicable to any data warehouse environment. At the meta-data level, specific quality goals concerning each particular data warehouse are given. Finally, concrete values are the traces of measurement in the real world.

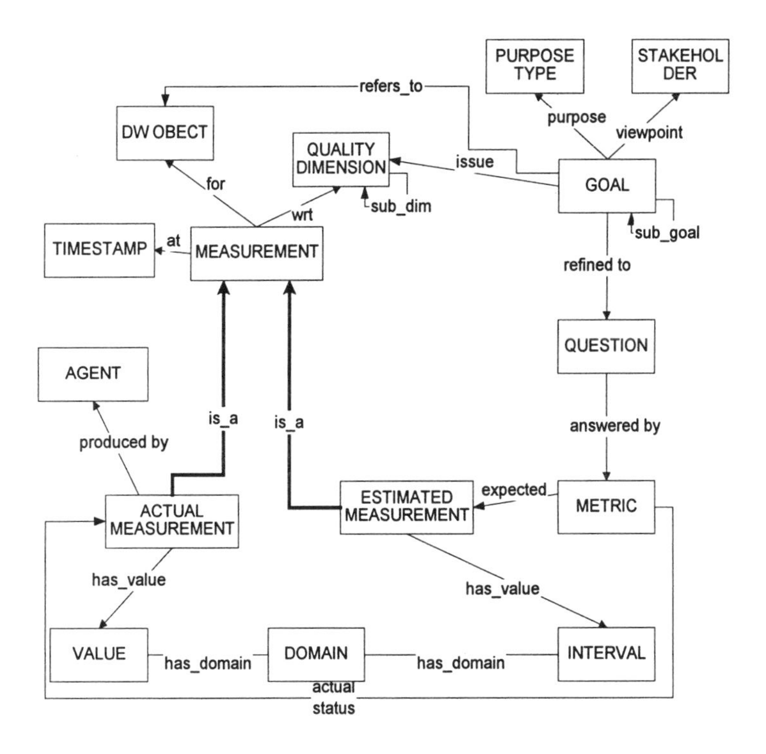

Fig 6 The quality meta-model

A (quality) *goal* is a project where a stakeholder has to manage (e.g., evaluate or improve) the quality of the data warehouse, or a part of it. This roughly expresses natural language requirements like 'improve the availability of source s1 until the end of the month in the viewpoint of the DW administrator'. Quality dimensions (e.g. 'availability') are used as the vocabulary to define abstractly different aspects of quality, as the stakeholder perceives it. Of course, each stakeholder might have a different vocabulary and different preferences in the quality dimensions. Moreover, a quality goal is operationally defined by a set of (quality) *questions*. Each question is dispatched to concrete (quality) *metrics* (or quality factors) which are groupings for concrete measurements of quality. A metric, thus, is defined over a specific data warehouse object and incorporates expected / acceptable values, actually measured values (*measurements*), methodology employed, timestamps etc.

Again the value of the above is that the quality meta-model is part of the meta-data repository of the data warehouse, in order to be consistent with the overall approach of data warehouse management.

6 Employing the Framework - Conclusions

For each process in the operation of a Data Warehouse (design, software implementation / evaluation, data loading and data usage) quality metrics and their relevance to quality measurements and dimensions have been defined in DWQ. As an illustration, we present here the ones, which relate to the *design* process.

Schema quality refers to the ability of a schema or model to represent adequately and efficiently the information. The *correctness* factor is concerned with the proper comprehension of the entities of the real-world entities, the schemata of the sources (models) as well as with the user needs. The *completeness* factor is concerned with the preservation of all the crucial knowledge in the data warehouse schema (model). The *minimality* factor describes the degree up to which undesired redundancy (due to cycles of relationships, transitivity of inclusion dependencies, or other forms of integrity constraints) is avoided during the source integration process. The *traceability* factor is concerned with the fact that all kinds of requirements of users, designers, administrators and managers should be traceable to the data warehouse schema. The *interpretability* factor ensures that all components of the data warehouse are described for easier administration. The *metadata_evolution* is concerned with the way the schema evolves during the data warehouse operation.

As it is obvious from the above, the design and administration quality factors are related to models and schemata. Consequently they are linked to the respective objects (Logical Object and Physical Object) in the architecture model.

The design major quality factors are schema quality and metadata evolution, which are in turn analyzed to other factors. The correctness, completeness, minimality and traceability factors should be measured after the source integration process in order to guarantee the absence of errors. The interpretability factor should be fairly documented in order to help the administrator know the system and data architecture, the relationship models, schemata and physical storage of the data and the processes of the data warehouse. The metadata evolution factor is used to track down the way the schema and models of the data warehouse evolve. All these are summarized in Fig. 7.

Factor	Methods of measurement	Metrics
Schema quality		
Correctness	final inspection of data warehouse schema for each entity and its corresponding ones in the sources	number of errors in the mapping of the entities
Completeness	final inspection of data warehouse schema for useful entities in the sources, not represented in the data warehouse schema	number of useful entities, not present in the data warehouse
Minimality	final inspection of data warehouse schema for undesired redundant information	number of undesired entities in the data warehouse
traceability	final inspection of data warehouse schema for inability to cover user requirements	number of requirements not covered
interpretability		
	physical part of the architecture (e.g. location of machines and software in the data warehouse)	number of undocumented machines/pieces of software
	logical part of the architecture (e.g. data layout for legacy systems and external data, table description for relational databases, primary and foreign keys, aliases, defaults, domains, explanation of coded values, etc.)	number of pieces of information not fully described
	conceptual part of the architecture (e.g. ER diagram)	number of undocumented pieces of information
	mapping of conceptual to logical and from logical to physical entities	number of undocumented mappings between conceptual, logical and physical entities
metadata evolution	metadata evolution (versioning/timestamping of the metadata)	number of metadata evolutions not documented

Fig 7 Design Factors – Measurement – Metrics [25]

Figure 8 shows how the traditional data warehouse architecture is extended by the DWQ approach.. All models (architecture, process, quality) are represented in the repository, which is based on ConceptBase, which provides a computational engine. Quality data (i.e., values of measurements) are entered into the ConceptBase system by external measurement agents, which are specialized analysis and optimization tools. In the DWQ project, four such tools were developed. ConceptBase can trigger these agents based on the timestamp associated to them in the repository.

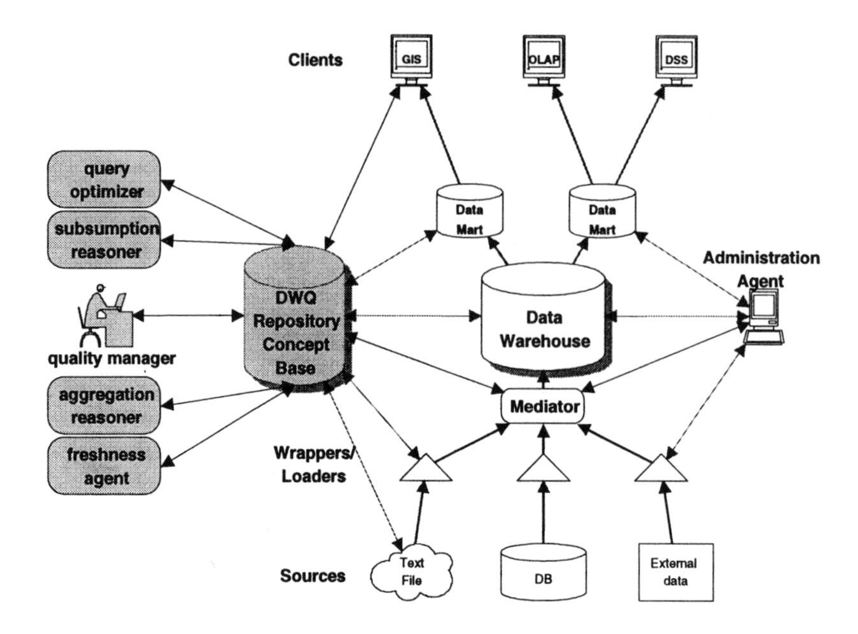

Fig 8 Mapping the DWQ repository approach to the traditional Data Warehouse [10]

The basic principle is that a meta-database is used as a repository for meta-information on DW components. Having goals operationally defined over a set of questions, the Data Warehouse repository becomes dynamic and can be exploited in a systematic manner. The DWQ methodology for quality management [10] is composed of three main phases: (a) the design phase which elaborates a quality goal by defining its purpose, the set of questions to solve it and the set of quality factors which answer to these questions; (b) an evaluation phase which deals with the computation of quality factors; (c) an analysis / improvement phase which gives an interpretation to the quality goal evaluation and suggests a set of improving actions.

Acknowledgments

The paper is solely based on work performed by the participants of the DWQ project, a project which I had the privilege to co-ordinate. This has a great value to me since it deals with an area – information systems development – in which I have been strongly influenced and inspired by the valuable contributions of my friend and colleague, Janis Bubenko.

References

1. W. H. Inmon, Building the Data Warehouse, John Wiley & Sons, second edition, 1996.
2. M. Jarke, Y. Vassiliou, Foundations of data warehouse quality – a review of the DWQ project. In: Proceedings 2nd Intl. Conference Information Quality (IQ-97), Cambridge, Mass. 1997.

3. M. Jarke, M.A.Jeusfeld, C. Quix, P. Vassiliadis, Architecture and quality in data warehouses: An extended repository approach. Information Systems, 1999; 24(3):229-253. (a previous version appeared in CAiSE '98, Pisa, Italy, 1998)
4. D. Calvanese, G. De Giacomo, M. Lenzerini, D. Nardi, and R. Rosati, Information integration: Conceptual modeling and reasoning support, In: Proceedings of the 6th International Conference on Cooperative Information Systems (CoopIS-98), 1998, pp 280-291.
5. F. Baader and U. Sattler, Description Logics with Concrete Domains and Aggregation, In: Proceedings of the 13th European Conference on Artificial Intelligence (ECAI-98), 1998, pp 336-340.
6. M. Gebhardt, M Jarke and S. Jacobs, A toolkit for negotiation support on multi-dimensional data, In: Proceedings of ACM SIGMOD International Conference on Management of Data. Tucson, Arizona, 1997.
7. P. Vassiliadis. Modeling Multidimensional Databases, Cubes and Cube Operations. In: Proceedings of 10th SSDBM Conference, Capri, Italy, 1998, pp 53-62.
8. M. Bouzeghoub, F. Fabret, F. Llirbat, M. Matulovic and E. Simon, "Designing data warehouse refreshment system", DWQ Technical report, 1997.
9. D. Theodoratos, T. Sellis. Data Warehouse Configuration. In: Proceedings 23rd International Conference on Very Large Data Bases (VLDB-97), Morgan Kaufmann,, 1997, pp 126-135.
10. C. Quix, P. Vassiliadis, M.Bouzeghoub, M. Jarke et al., Quality-Oriented Data Warehouse Design, DWQ Technical report, 1999.
11. P.A. Bernstein, Th. Bergstraesser, J. Carlson, S. Pal, P. Sanders, D. Shutt, Microsoft Repository Version 2 and the Open Information Model. Information Systems, 1999; 24(2):78-93.
12. Metadata Coalition, Meta Data Interchange Specification, (MDIS Version 1.1), August 1997, available at http://www.he.net/~metadata/standards/
13. J. Mylopoulos, A. Borgida, M. Jarke, M. Koubarakis, Telos: a language for representing knowledge about information systems, ACM Transactions on Information Systems, 1990; 8(4).
14. M. Jarke, R. Gallersdörfer, M.A. Jeusfeld, M. Staudt, S. Eherer. ConceptBase - a deductive objectbase for meta data management. J. Intelligent Information Systems, 1995; 4(2):167-192.
15. M. Jarke, M.A. Jeusfeld, T. Rose: A software process data model for knowledge engineering in information systems. Information Systems, 1990; 15(1): 85-116.
16. B. Ramesh, V. Dhar: Supporting systems development by capturing deliberations during requirements engineering. IEEE Trans. Software Engineering, 1992; 18(6).
17. D. Georgakopoulos, M. Rusinkiewicz. Workflow management: From business process automation to inter-organizational collaboration. In: Tutorials of the 23rd International Conference on Very Large Data Bases, Athens, Greece, 1997.
18. D. Hollinsworth. The workflow reference model. Technical Report TC00-1003, Workflow Management Coalition, 1995.
19. D. H. Besterfield, C. Besterfield-Michna, G. Besterfield and M. Besterfield-Sacre, Total Quality Management, Prentice Hall, 1995.
20. Y. Wand, R.Y. Wang. Anchoring Data Quality Dimensions in Ontological Foundations. Communications of the ACM, 1996; 39(11): 86-95.
21. R.J. Orli , Data Extraction, Transformation, and Migration Tools, Kismet Analytic Corp, 1997.
22. Oracle7 Server Utilities User's Guide, SQL*Loader Concepts, 1996.
23. M. Oivo, V. Basili. Representing software engineering models: the TAME goal-oriented approach. IEEE Trans. Software Engineering, 1992; 18(10):886-

User Software Engineering: a Retrospective

Anthony I. Wasserman
Software Methods and Tools
San Francisco, USA
tonyw@methods-tools.com

Abstract

The User Software Engineering methodology was developed two decades ago to support the development of interactive information systems. This paper describes the key concepts of User Software Engineering and how they apply to modern client-server and web application development.

1 Historical Background

The User Software Engineering methodology was conceived during the period 1975-1980. This period predated the widespread availability of personal computers. Networks were extremely limited, and most applications ran on a central computer, possibly reading and writing from a terminal. Even though it had been possible to write interactive applications since the mid-1960's using such a terminal, there was little attention given to user interface design or principles of computers and human interaction.

This period was marked by a transformation in thinking about software development. The failure of numerous large-scale software projects during the 1960's led researchers to seek out systematic ways to develop software systems. The term "software engineering" was coined in 1967, and was the theme for a pair of NATO-sponsored workshops. Two key ideas that emerged from these workshops were new approaches to management of software development and the concept of "structured programming", presented by Edsger Dijkstra. Dijkstra, Hoare, Wirth, and others viewed structured programming as a form of hierarchical problem decomposition, where the end result was an algorithm that could be implemented as a solution to the problem. In practice, though, the concept of structured programming devolved into a set of rules for writing programs, including (often emphasizing) the avoidance of unlimited transfers of control (the *go to* statement) as a way to enhance program readability. The programming language Pascal became a highly popular tool for teaching programming and for writing well-structured programs.

For many researchers, though, structured programming failed to address the problem of how one came to understand the requirements for a system and how to specify its intended behavior. Much of the theoretical work related to structured programming was demonstrated for small systems that could be clearly described in a few sentences at most. The vast majority of these approaches did not hold up for complex real world systems, such as those used for banking, airline reservations, or operating systems.

An important exception to this statement is Parnas' notion of modularity [1], which has stood the test of time, remaining one of the most important ways to organize systems, both large and small. His concept of information hiding is at the heart of many advances in software design and development, and was a strong influence on the development of object-oriented concepts in the 1980's.

Beyond that, though, were numerous projects, both in industry and in the research community, aimed at those phases of the software development process that precede programming. Stevens, Myers, and Constantine [2] developed structured design, as later popularized by Yourdon [3], as a way to decompose a system into a hierarchical modular architecture, including metrics for evaluating the quality of the design. By the mid-1970's, there were many different groups working on the problems of requirements definition and system specification. The various projects could be partitioned along several different dimensions.

In many respects, the most important of these partitions was between data-oriented and function-oriented analysis. In the latter case, projects followed the ideas of structured programming and structured design, creating models that were compatible with those ideas. One such approach was Structured Analysis, as developed by DeMarco [4]. Structured analysis produced data flow diagrams, a hierarchical decomposition of system functions connected by flow of information between functions and by "data stores". It was often straightforward to transform a set of data flow diagrams into the structure charts used in structured design, and hence into a program architecture ready for implementation.

While function-oriented analysis was quite natural for those who had worked in scientific programming, data-oriented analysis was much more natural for people working in information systems with files and databases. Data-oriented analysis focused on the items that have real world meaning for the application. Peter Chen's Entity-Relationship modeling [5] was among the first methods to emphasize a data-oriented view of a system.

This data-oriented view is the basis for object-oriented development, where an object is an instance of a class, where a class is defined to have a set of externally visible behaviors. The data itself is encapsulated (hidden) inside the class, so that only well-defined operations can be performed on the data. These operations could be defined formally, as with a first-order predicate calculus [6], or in terms of a programming language.

2 Concepts of User Software Engineering

User Software Engineering (USE) added another dimension to this partition: that of the user at an interactive device, typically an alphanumeric display. Systems were seen to have three major aspects: data, operations on the data, and events that caused those operations to be performed. In the case of USE, the events were often initiated by a human user, typically entering data, often in response to an output from the program.

At that time, systems were frequently imposed on users, and users rarely participated in definition and design of a system. There was (and is) a broad communication gap between system designers and non-technical users which made it difficult to explain system concepts to a user and for a user (or even a user surrogate) to describe tasks and thought processes in a way that could influence system design and use.

The work in structured programming exemplifies the problem. Structured programming is grounded in the concept of top-down decomposition, where the "top" is the function to be performed by the system. Successive refinement of the function leads to "smaller" functions, but this decomposition is often orthogonal to the user view of the system, and fails to capture some real world aspects of a system, such as user errors that can lead to exceptional conditions.

What was needed for such interactive systems was not top-down design, but rather outside-in design, where "outside" represents the user's perception of the system. *Outside-in modeling* is one of the most important concepts of the USE methodology. By emphasizing the external view of a system, it became much easier to communicate with users. The key observation is that users have little interest in the structure of the system; their only concern is whether the system makes it easier for them to get their job done.

Each user interaction with the system was viewed as an event that could trigger an activity and/or a response. For example, the system could display a menu of choices to the user, with the user's input determining the program action, eventually leading to either program termination or another request for user input.

The user interaction and the system behavior was modeled as a hierarchical set of transition diagrams. This approach had two major benefits. First, even on paper, it was possible to walk through a dialogue with a potential user of a system, validating the overall scenario. Second, and more significant, transition diagrams are a formal, executable model, making it possible to build an executable version of the emerging system.

The executable nature of transition diagrams led to the most important innovation of the USE methodology: *rapid prototyping of user interfaces.* The state transition diagrams, including specification of the user inputs and system outputs, were encoded in a transition diagram language TDL. This language also included the ability to specify executable program units. A tool, RAPID/USE, was built to interpret the TDL and execute the associated program units. In this way, RAPID/USE could be used both for prototyping the user interface and for running a complete program.

In this way, users could begin to work with the emerging system at a very early stage of development, to the extent that they could actively contribute to the definition of the system and the style of the user interface. This notion of *user involvement in the software development process* is, in many respects, the central idea of the USE methodology. This user participation was extremely valuable for improving the quality of systems, identifying problem areas at a phase of the development process when they could still be addressed. The RAPID/USE system also gathered metrics on user behavior, making it possible to track error conditions, task completion times, and other measures of system usability. As workstations became available, the toolset was extended to include a graphical transition diagram editor to draw the transition diagrams and annotate them. These diagrams were compiled into TDL and executed much as before.

In general, developers would start by designing part of the user interface, implementing it, and adding functions or pseudo-functions as place holders, as well as beginning design of the relational database model. These functions could be specified formally, with a behavioral specification, or informally with natural language or a programming language, either with or without a formal specification. Because both transition diagrams and relational databases have formal underpinnings, it was possible to use *formal methods* as part of the USE methodology if so desired, another aspect of the USE methodology

The next key notion of the USE methodology was support for *an incremental approach to application development*, a sharp contrast to the waterfall approach in widespread use at the time. The traditional waterfall approach was poorly suited to the development of interactive systems, since it could not accommodate user participation in the early stages of the development process, particularly the user interface prototypes used to support requirements gathering and validation.

The ability to separate the user interface component from the program operations led to another significant concept of the USE methodology: *a three-tier architecture*. Figure 1, from [7], shows the logical structure of interactive information systems:

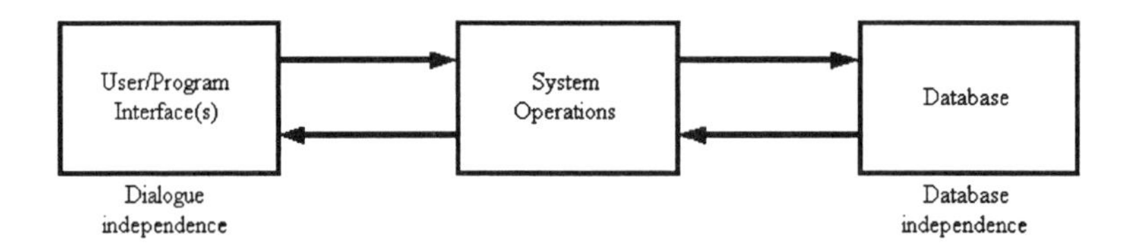

Figure 1 – The three-tier architecture of the USE methodology

This architecture shows the separation between the various components of the information system, anticipating client-server systems of the early 1990's and modern n-tier architectures. In addition, it shows the possibility of associating multiple user interfaces with a set of system operations. Such an approach permitted separate interface designs for novice and expert users, as well as an application programming interface that could be used to drive test cases or to integrate the system with another system.

Finally, the last important aspect of the USE methodology was an *integrated tool environment* that provided direct support for the methodology. The USE tools included:

1) The RAPID/USE rapid prototyping system with the Transition Diagram Interpreter and Transition Diagram Editor;
2) a relational database management system that could be used both as database for the application and a repository for system development data;
3) the PLAIN programming language, a Pascal-like language that supported data abstraction, exception handling, database integration, and other features needed for the development of interactive information systems, and;
4) the Module Control System, a version control and configuration management system that stored information about system components and versions in the relational database.

In summary, then, the User Software Engineering methodology was built around six concepts:

1) outside-in modeling, in contrast to top down refinement;
2) rapid prototyping of user interfaces;
3) effective user involvement in the software development process;
4) ability to use both formal and informal specification methods;
5) separation of implementation concerns through a three-tier architecture, and;
6) integrated development environment to support use of the methodology.

3 The Revolution in Computing

Concepts of computing and software development have changed drastically since the late 1970's. At that time, programs ran on a single computer with a single stream of input (typically from a file or a terminal) and a single stream of output. Distributed computing, networks, and client/server systems existed only in certain segments of the research community. The first personal computers had just been created. Graphical user interfaces were not generally available until the release of the Apollo and Sun workstations in 1982 and the Apple Macintosh™ in 1984. It was another 10 years before tools such as Visual Basic for Windows made it easy to design GUI's. In the late 1970's, the first object-oriented languages were also in the research community, long before their popularization. OO analysis and design methods didn't appear until the late 1980's. And, of course, the World Wide Web (Web) wasn't invented until 1990, and didn't see much use until the development of the graphical Mosaic browser several years later.

In short, computing has undergone revolutionary change in the last twenty years. Many of today's most widely-used applications running in a client/server or n-tier environment, often across a heterogeneous network. It is very common to see desktop machines running Microsoft Windows™ serving as clients for Unix™ servers. Sophisticated e-commerce applications use multiple tiers, with a browser client, a web server (or "server farm") with server side CGI programs and/or Java servlets, middleware processing in an application server (e.g., BEA WebLogic™), and a database server. A typical architecture is shown in Figure 2.

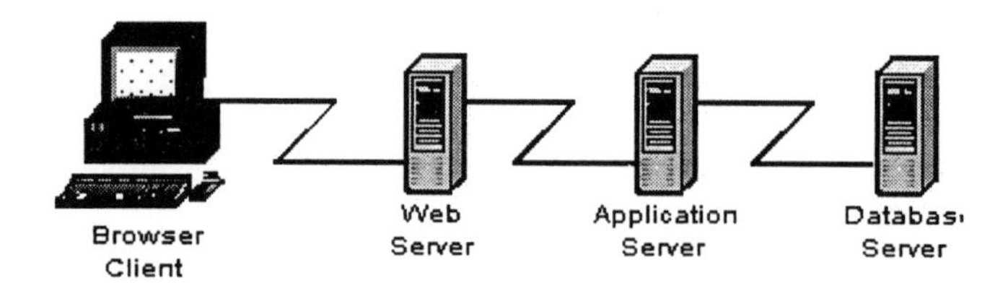

Figure 2 - Multi-tier architecture for Web application

Data modeling is of the few concepts that has not undergone drastic change in all those years. The ideas pioneered by Chen, Bubenko, and the Smiths [8] not only remain in widespread use, but have also been incorporated into numerous other modeling notations, such as the Unified Modeling Language (UML) [9].

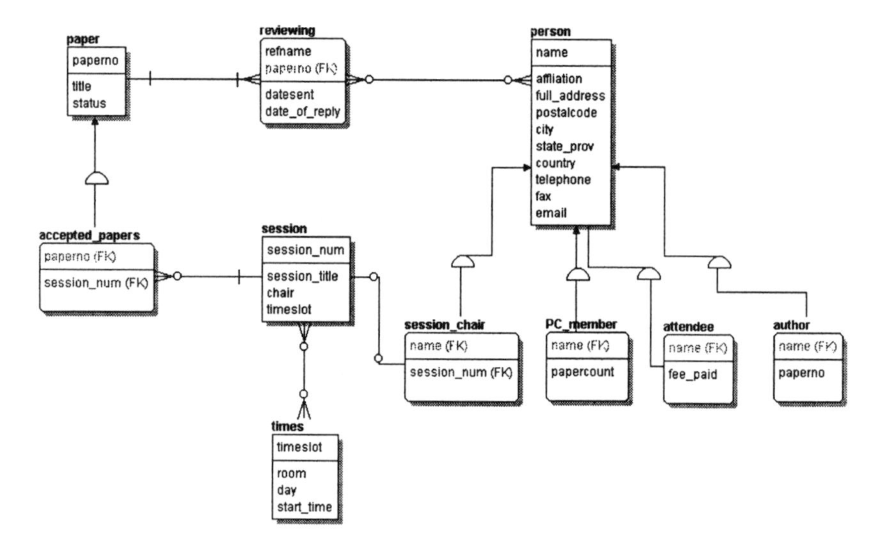

Figure 3 – Extended entity-relationship model of technical conference database

Graphical modeling tools, such as Software through Pictures™ [10], were developed in the early 1980's, and supported graphical editing and presentation of process and data models, as well as generation of database schemas and code skeletons. A partial model of a database for the technical conference problem, described in [11,12], is shown in Figure 3. (Named relationships are omitted.)

This model can serve as the basis for a UML class model, to which operations are added to the name and attribute information. Figure 4 shows a partial class model for the same technical conference problem, which includes some of the operations needed for the system. UML is a highly complex notation that includes other diagram types for describing, both formally and informally, use cases, activities, state transitions, and other system phenomena.

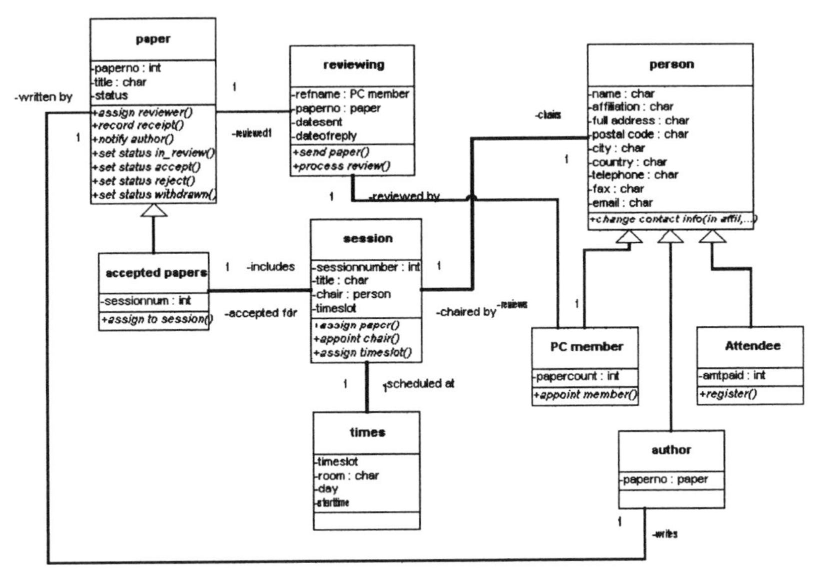

Figure 4 - Preliminary class model for conference management system

4 User Software Engineering today

Even though many aspects of computing have changed drastically, many of the concepts of User Software Engineering remain valid and highly applicable to today's systems. In particular, there is a greater need than ever for effective user involvement in the software development process and for an integrated suite of tools to support the process.

Many of today's applications are routinely used by many thousands of average consumers. The most popular sites on the Web, such as Yahoo!, Microsoft, AOL, and CNN, receive millions of visitors each month. If a site is difficult to navigate or takes too long to display, users will go elsewhere. Reliability and robustness of Web-based applications is also critical, particularly with E-commerce and other financial transaction-based sites.

Usability of applications is now receiving considerable attention. Jakob Nielsen has written several books on usability, including one specifically on Web applications [13]. The best way to determine usability is through prototyping of the site design, combined with testing of the user experience.

There are numerous commercial tools that can be used to create a prototype of one or more pages of a web site. Even though the final version of a site may use hand-coded HTML, the prototype can be created with a WYSIWYG editor, such as Macromedia Dreamweaver™ or Adobe GoLive™, combined with suitable tools, e.g., Adobe Photoshop™, for creating images. These pages can be loaded onto a machine with a web server, e.g., Apache, where users may view them and use them. Since web servers can do extensive logging of HTTP requests, the prototyping process can be used to obtain not only the subjective experience of the user, but also quantitative data on the download time, the user response time, and the pages presented. Log analysis programs, such as Analog from Cambridge University, presents the log information as an HTML page. In this way, it is easy to see the number of times that a user made an error in completing a form or requested help.

The advent of modern GUI building tools, such as Visual Basic™, along with the Web design tools, revolutionizes this prototyping process. One of the key benefits of these tools is that they are interpretive, making it easy to make changes. In this way, it's easy to compare alternative interface designs and to make iterative enhancements to the designs.

For example, Figure 5 shows a preliminary design of a Web-based interface to the conference management system. This design uses frames, giving the user the ability to click a button in the left frame, select a hyperlink in the right frame, or click a button at the bottom of the right frame. This design is intentionally not very good, since the text offers 6 hyperlinks, and there is no easy way to see how those 6 hyperlinks are related to the 3 different button types. There are no help or exit options, either. It would take only a couple minutes of effort, though, to add these options and to improve consistency between the buttons and the hyperlinks.

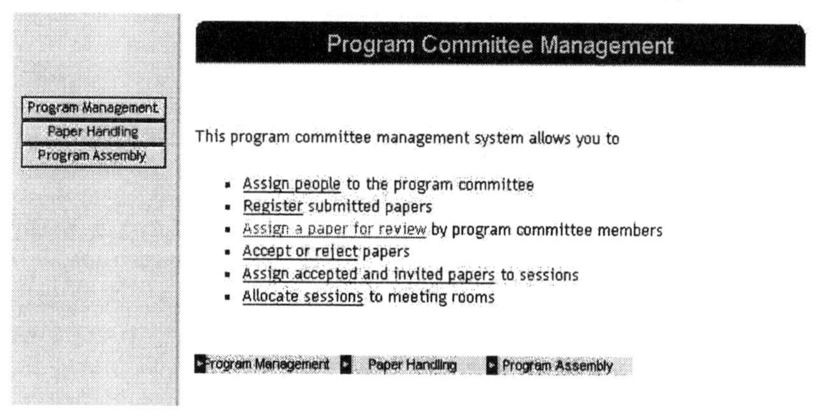

Figure 5 - Web browser home page for conference management system

Figure 6 is a prototype of a page for assigning a submitted page to a reviewer. Note that Help and Quit buttons have been added based on experience with the previous page, and that the form is a straightforward layout of the reviewer selection process.

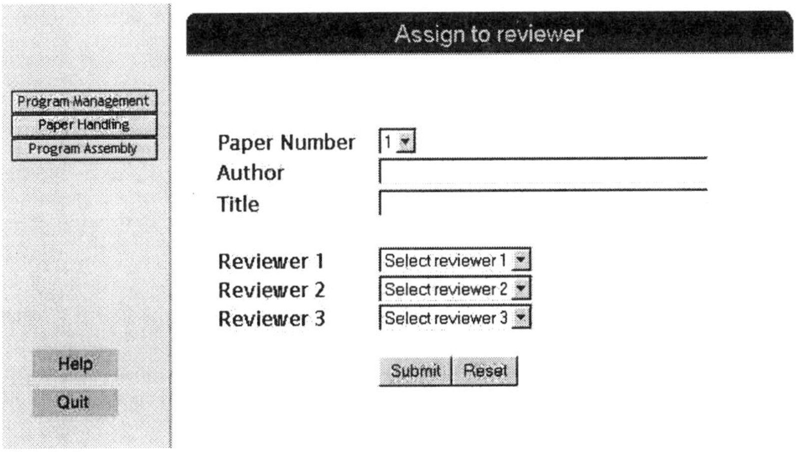

Figure 6 - Web form for reviewer selection
in conference management system

Figure 6 shows the page at a very early stage of the prototyping process. As requirements gathering proceeds, it is possible to create a database schema from the database model shown in Figure 3, and then to connect fields in the web form to information in the database. For prototyping purposes, one can use any of a wide number of databases, taking advantage of ODBC and SQL to create a prototype of the production database.

With that next step, one can envision populating the database through a form for entering information about new reviewers and another form for entering information about submitted papers. Thus, selecting a paper number can then display a paper title and author name, and the reviewer selection fields can display a list of reviewers who have not yet received their maximum number of papers to review.

In other words, the key concepts of User Software Engineering not only remain valid today, but are more clearly needed than they were at the time these concepts were first developed. There are now numerous widely available commercially supported tools for following this methodology. This approach works both for web-based applications, as described here, and for client-server applications, where Visual Basic might serve as the prototyping tool, as well as the final implementation tool.

The separation of user interface design from database design and program function works very well with these commercial tools, since the user interface component can be easily connected to the database. Numerous scripting languages, as well as Visual Basic, Java, and C++ can be used to build the operations.

While UML is of no direct help in design of the user interface, the class diagram helps with the database design, and also helps to identify the operations that must be supported in the user interface. If user interface prototyping is done in parallel to UML modeling, they can combine with one another to develop a complete and consistent class model. Today's interpretive user-interface design tools make it possible to build and modify user interfaces very quickly. The more sophisticated Web design tools provide for easy design and modification of page templates, supporting creation of consistent user interfaces across all pages of a site.

In summary, User Software Engineering remains a viable approach to analyzing and developing software systems, despite the vast changes to computing technology, both hardware and software, over the past two decades.

Acknowledgments

The ideas for the User Software Engineering methodology came from many sources. I would particularly like to acknowledge my colleagues in IFIP Working Group 8.1 (Design and Evaluation of Information Systems), where preliminary discussions of these notions was of great help in formulating the ideas. I am also grateful to Peter Pircher, David Shewmake, and Martin Kersten for their participation in various phases of the work and for their development of the initial set of tools that supported the methodology.

References

1. Parnas, DL. On the Criteria to be used in Decomposing Systems into Modules. Communications of the ACM, 1972; 15:1053-1058.
2. Stevens, W, Myers, GJ, Constantine, LL. Structured Design. IBM Systems Journal, 1974; 13: 115-139.
3. Yourdon, E, Constantine, L. Structured Design. Prentice-Hall, Englewood Cliffs, NJ, 1979.
4. DeMarco, T. Structured Analysis and System Specification. Prentice-Hall, Englewood Cliffs, NJ, 1979.
5. Chen, Peter P-S. The Entity-Relationship Model – Toward a Unified View of Data. ACM Transactions on Database Systems, 1976; 1: 9-36.
6. Bubenko, JA, Jr. Information Modeling in the Context of System Development. In: Proceedings 1980 IFIP Congress. North Holland, Amsterdam, 1980.
7. Wasserman, AI, Pircher, P, Shewmake, D, et al. Developing Interactive Information Systems with the User Software Engineering Methodology. IEEE Transactions on Software Engineering, 1986; 12: 326-345.
8. Smith, JM, Smith, DCP. Database Abstraction: Aggregation and Generalization. ACM Trans. Database Systems 1977; 2: 105-133.
9. Rumbaugh, J, Jacobson, I, Booch,G . The Unified Modeling Language Reference Manual. Addison-Wesley, Reading, MA, 1999.
10. Wasserman, AI, Pircher, P. A Graphical, Extensible Integrated Environment for Software Development. In: ACM SIGPLAN Notices, 22 (1), 131-142.
11. Olle, TW, Sol, HG, Verrijn-Stuart, AA (eds.) Information Systems Design Methodologies: a Comparative Review. North-Holland, Amsterdam, 1982.
12. Wasserman, AI, "The User Software Engineering Methodology: an Overview," In: [11], 591-628.
13. Nielsen, J. Designing Web Usability. New Riders, Indianapolis, 1999.

Application and Process Integration – Concepts, Issues, and Research Directions[1]

P. Johannesson, P. Jayaweera[2]
Department of Computer and Systems Science, Stockholm University/KTH
Stockholm, Sweden, email: pajo@dsv.su.se, prasad@dsv.su.se

B. Wangler
Department of Computer Science, University of Skövde
Skövde, Sweden, email: benkt@dsv.su.se

Abstract

The need for integrating applications is growing as a consequence of organisational demands and enabling technologies, in particular the Web and enterprise software packages. In this Chapter, we introduce the basic concepts of application integration, discuss a number of the most important issues in the area and outline promising research directions: process libraries, methodologies for process integration, adaptive and flexible process enactment, and moving business logic from systems to processes.

1. Introduction

As organisations are becoming increasingly dependent on information technology, the need for integrating applications is growing. As an answer to this need, technologies in Enterprise Application Integration (EAI) have been proposed. EAI can be defined as "the unrestricted sharing of data and business processes among *any* connected applications and data sources in the enterprise", [1]. The demand for EAI is driven by many forces, where one of the most important is the move to process orientation in many organisations. Traditionally, organisations have been functionally divided, i.e. companies have been separated into departments such as marketing, sales, procurement, production, and service. However, such a functional organisation has been shown to have a number of weaknesses. In particular, it requires a huge administration to handle issues crossing functional borders. In order to overcome the problems of a functional organisation, companies have been concentrating on business processes, that is the connected activities that create value

---

[1] This work was sponsored by NUTEK (Swedish National Board for Industrial and Technical Development) and SIDA (Swedish International Development Cooperation Agency)
[2] On leave from Institute of Computer Science, University of Ruhuna, Sri Lanka

for the customers. These processes cross the internal borders of an organisation as well as the external borders to other organisations, thereby supporting supplier and customer relationship management, virtual enterprises, and extended supply chain management.

Supporting cross-functional and inter-organisational processes puts new demands on IT systems and applications. Traditionally, the applications have been built up around departments or functions in the companies. The result has been a "stovepipe like" relation between the functions and the applications, where every function in the company is supported by its own system or application (see Fig. 1), but where the applications work as "islands of automation" with limited communication among each other. This architecture is not satisfactory for process oriented organisations; to support the business processes in full the applications need to be integrated.

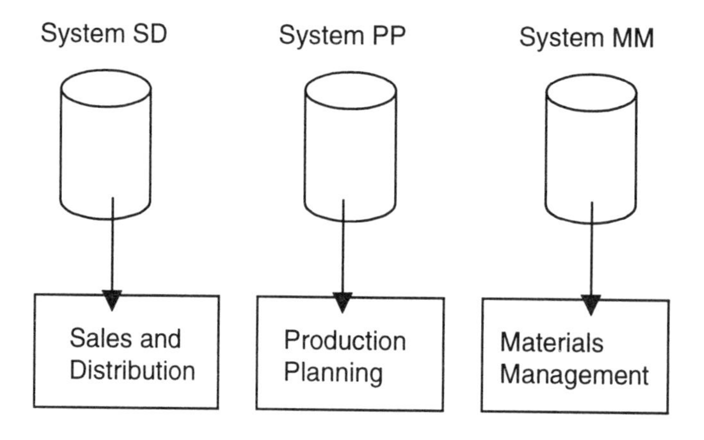

Fig. 1 IT systems supporting single functions

In addition to organisational forces, a number of enabling technologies drive the demand for EAI, in particular Internet and enterprise software packages. Internet provides an environment that can link a company's customers, suppliers, partners, and its internal users. Enterprise software packages offer an integrated environment for supporting business processes across the functional divisions in organisations. Some packages, like enterprise resource planning (ERP), for example SAP R/3 or Oracle Applications, manage back-office requirements, while other packages provide front-office capabilities, e.g. customer services. Common to Web applications as well as enterprise software packages is the need for application integration. Application integration is required to connect front office systems with back office systems, to transfer business processes to the Web, and to create extended supply chains involving customers, partners, and suppliers. Application integration is also needed for wrapping legacy systems and for migrating to new environments.

The purpose of this Chapter is to introduce the basic concepts in EAI, to discuss a number of the most important issues in the area, and to point out promising research directions.

2. Technologies for EAI

Enterprise Application Integration can take place on different levels, which is a fact that has been recognised since a long time, [2] [3]. In [1], D. Linthicum identifies the following four levels:

Data level EAI. In data level integration, the integration takes place between data stores. More concretely, data is extracted from one database and used to update another database, possibly after appropriate modifications. A major advantage of this approach is that it does not require any modifications of the existing applications. Furthermore, the approach relies on inexpensive and established technology, i.e. database oriented middleware such as ODBC and JDBC.

Application interface level EAI. An application interface is an interface that gives access to services provided by a custom application or a standard package. It is possible to distinguish between three types of services, [1]: business, data, and object services. A business service provides access to some business logic, e.g. calculating prices or updating customer information. A data service provides a route to the logical or physical database and is in this similar to data-level access tools. An object service is the combination of business and data services packaged as an object. An advantage of objects is that integrity constraints cannot be violated as updates to data are always carried out by the appropriate methods of an object.

Method level EAI. In method level EAI, applications are integrated by being able to share a set of common methods. These methods can be stored on a central server, or they can reside as distributed objects in a network. Method level EAI is closely related to reuse. By introducing a number of common methods, different applications can reuse these methods.

User interface level EAI. User interface level EAI (also known as "screen scraping") is the most primitive form of EAI. Applications are integrated through the user interfaces, i.e. information are accessed from user screens by programmatic mechanisms. This approach may seem quite unattractive, but in many cases it is the only one available for integration. Furthermore, it has the advantage of not requiring any changes to the applications to be integrated.

Integration of applications can be supported by many different architectures. One architecture for integrating applications is the point-to-point solution where every application is directly connected to every other application, see Fig. 2. This solution could work for a small number of applications, but as the number of applications increases, the number of connections quickly becomes overwhelming. The Message

Broker technology reduces this complexity, see Fig. 3. The main idea is to reduce the number of interfaces by introducing a central Message Broker and thereby make it easier to support the interfaces. If one of the applications changes format, only one connection has to be changed: the one to the Message Broker. The Process Broker, see Fig. 4, is an extension of the Message Broker. In addition to handling format conversions, the Process Broker also encapsulates the process logic for connecting applications. When all process logic resides in one place, it becomes possible to study, analyse, and change the processes using a graphical interface. This visualisation reduces the complexity and enables different categories of people to take part in the process design.

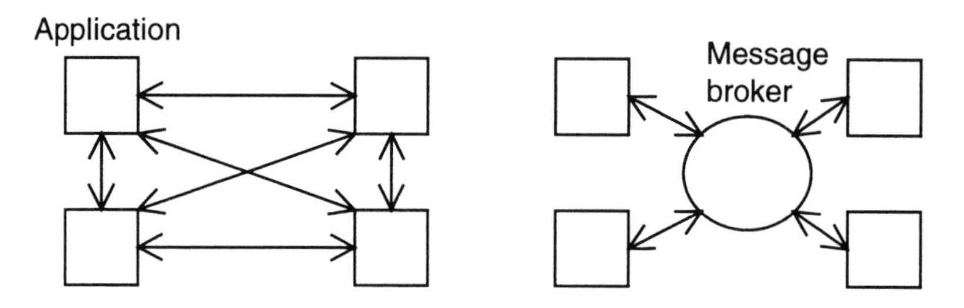

Fig. 2 Point to point integration

Fig. 3 Integration through a message broker

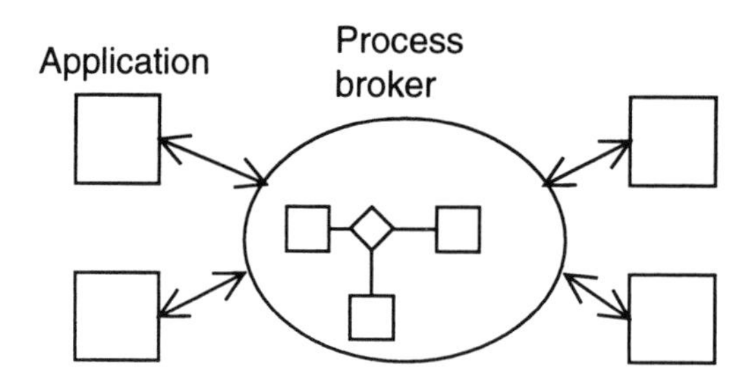

Fig.4 Integration through a process broker

The Process Broker technology can be seen as a continuation of workflow management systems. It is possible to categorise workflow systems into four generations starting from application specific, hard coded workflow capabilities realised as proprietary, closed systems. In the second generation, workflow capabilities factored out from the application domains to separate workflow applications with limited 3rd party application integration and tailorable process definitions via scripting languages. The current trends categorised into third generation can be distinguished as tailorable workflow services accessible to other applications through APIs with open standards based architecture. Today proprietary workflow interfaces and interchange formats has the full interaction capability to 3rd party applications and tailorability to workflow definition via user friendly GUIs. In the next generation, workflow services will be fully embedded with other middleware services and towards standardised interfaces and interchange formats for workflow enabled applications where workflow capabilities are ubiquitous but invisible.

3. Research Directions

In this section, we identify a number of research directions within the area of application and process integration. Many of the problems and solutions are relevant also for other areas, but they take on an added significance when considered in the context of application integration.

3. 1 Process Libraries

A process library contains a large number of processes as well as support for navigation and customisation. A typical usage scenario is a business designer constructing a new process for sales. She could navigate the process library to identify a variety of alternative sales processes, e.g. sales by retail, mail order, or Internet. After having compared the alternatives, she can choose one or more processes, combine parts of them, and restructure the resulting process. Using a process broker, she can finally generate software for supporting the process and integrate with requested applications. Process libraries already exist as parts of commercial products, e.g. the Process Reference Model in SAP R/3 and the ARIS tool set, [4]. A research prototype of a large process library is the MIT Process Handbook, [5]. We believe that the most important research problems for process libraries are mechanisms for structuring the libraries in such a way that they become easy to browse and search. Below, we outline some promising structuring mechanisms.

Decomposition and Specialisation. A basic mechanism for structuring processes is decomposition. A process consists of subprocesses, which can be decomposed into other subprocesses and eventually into tasks. An example is shown in Fig. 5, where the "Sell product" process contains two subprocesses "Presales" and "Postsales", which are decomposed into tasks such as "Identify customer" and "Receive payment". In addition to being structured by decomposition, processes can also be organised in a specialisation hierarchy with the most general processes at the top and the most specialised processes at the bottom. As pointed out in [5], such a hierarchy will be similar to an object inheritance hierarchy in conventional object-oriented analysis and design. An object inheritance hierarchy consists of increasingly specialised objects that are associated with actions (i.e. methods). A process hierarchy, on the other hand, consists of increasingly specialised actions (i.e. processes and tasks) that are associated with objects. An example of a specialisation of the process in Fig. 5 is given in Fig. 6. Process hierarchies provide two main benefits for a process library. First, they enable concise representations. When a new process is to be added, it can inherit large parts of its description from a more general process and only the differences to that process have to be explicitly stated.

Secondly, hierarchies facilitate navigation in a process library: users can traverse the hierarchy upward in order to identify more general versions of a process and they can move across the hierarchy to find related processes.

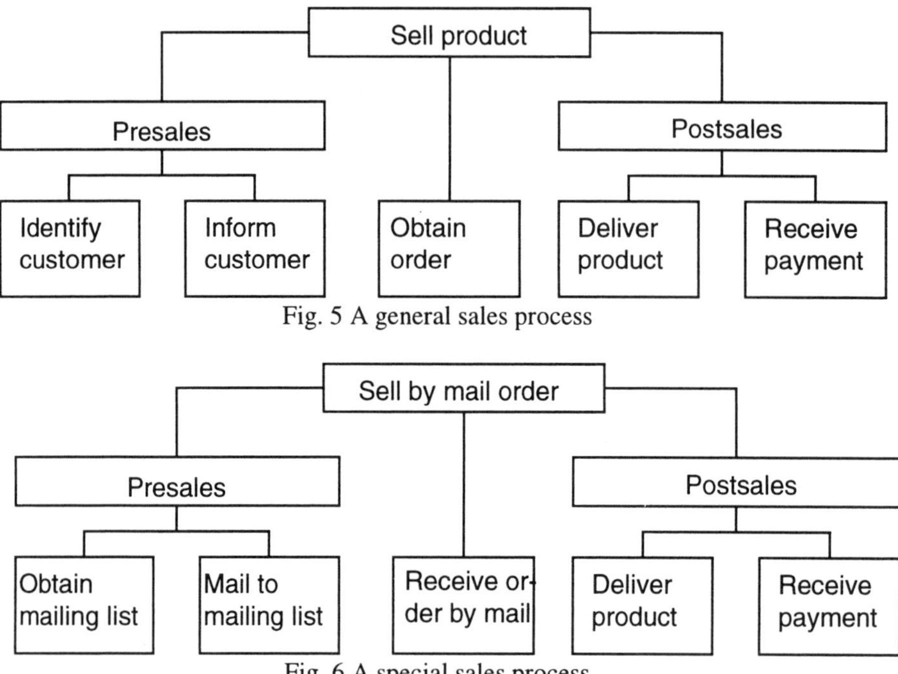

Fig. 5 A general sales process

Fig. 6 A special sales process

Co-ordination. Another mechanism for structuring process libraries is based on co-ordination theory, [6]. Co-ordination is seen as the management of dependencies among activities, and three basic kinds of dependencies are identified: flow, sharing, and fit. *Flow dependencies* arise whenever one activity produces a resource that is used by another activity. *Sharing dependencies* occur when multiple activities all use the same resource. *Fit dependencies* arise when multiple activities together produce a single resource. Each dependency can be managed by different co-ordination processes. For example, a sharing dependency can be handled by co-ordination mechanisms such as first come/first serve, human decisions, bidding procedures, and priority orders. A flow dependency could be managed by, for example, a make to order or a make to inventory process. Dependencies and their possible co-ordination mechanisms can help structure a process library by allowing designers different views of the same process. A designer can switch back and forth between a dependency and the co-ordination mechanism by which it is handled. Furthermore, a designer can specify a particular dependency and browse through all the possible co-ordination mechanisms for that dependency.

Communication. Another structuring mechanism is based on a communicative language/action approach. This approach provides a way to structure the constituents of processes, see Fig. 7, [7]. At the bottom level, we find instrumental acts (e.g.

producing goods or delivering products) as well as communicative acts (also called "speech acts"). The latter are used to make requests, commitments, and declarations through which obligations, permissions, and authorisations are established. The next layer contains the business transactions. A business transaction is the smallest sequence of acts that results in a new deontic state, e.g. a state in which an obligation to carry out some action has been created, or a state where an authorisation for certain actions has been established. A single speech act is in most cases not sufficient to achieve a deontic effect. In general, at least two messages are required. For example, a customer requesting a supplier to deliver a product and the supplier promising to do so; these two acts together constitute a business transaction. The top layer in Fig. 7 is the process level, where a process is built up by a number of business transactions. A language/action approach can help structuring a process library by providing a basis for decomposing processes into their constituents.

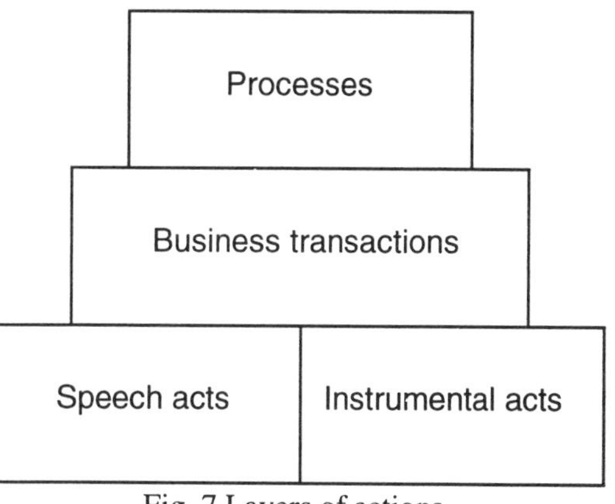

Fig. 7 Layers of actions

3. 2 Methodology

Effective EAI requires a comprehensive methodology including identification of data, processes, and application interfaces, performance considerations, process design, maintenance and much more. Three of the most important activities that would benefit from better methodological support are understanding the enterprise data, understanding the enterprise processes, and designing processes.

Understanding the enterprise data. Understanding data is obviously required for EAI at the data level, but an understanding of databases is also needed for EAI at the other levels. Understanding data starts by identifying the data by cataloguing which databases exist, who owns these databases, their physical location, data formats, etc. When data has been identified, an enterprise model of the data is to be built. This model will contain not only traditional data dictionary information, but also additional information such as security information, connected processes, communication mechanisms, and integrity issues, [1]. Research on reverse engineering of databases will be relevant for this activity, especially when extended to the environment of the databases and not only their schemas.

Understanding the enterprise processes. Understanding processes also begins by identifying the processes in the enterprise. When the processes have been identified, a business model is constructed that describes the processes, their owners, the databases they interact with, and the technologies in which they are implemented. General process libraries, as introduced above, can be used as a means for organising the processes of a particular enterprise. Their navigation mechanisms can also be used for searching among the processes.

Designing processes. Application integration often results in highly unstructured and complex process models. One reason for this is that exception handling makes up a large part of an application integration specification and thereby easily obscures the main business logic. Furthermore, there is often extensive communication between a process broker and different applications, which also tends to conceal the business logic. Another characteristic of a process broker is that it does not maintain control over external applications, which means that these applications can be updated without the process broker being notified. As a consequence, it is often desirable to maintain redundant information that duplicates parts of the information in the external applications. This duplication of information requires some mechanisms for handling possible inconsistencies, which makes the process model even more complex. In order to overcome these problems of process modelling in the context of EAI, adequate design support is needed. One approach, as suggested in [8], is to model a process through a series of views starting with a customer oriented view on the business level, which models the interactions between the process broker and the customer. The succeeding views add more and more details moving from a business perspective to a technical perspective by adding interactions with external applications, exception handling, etc. Each view is an extension of the previous one, either through adding subprocesses or through introducing new components into the existing diagrams. This approach gives the designer increased control by allowing her to focus on different aspects of a process in different views.

3.3 Adaptive Process Management

Current software systems for supporting process management, such as process brokers and workflow management systems, are effective for predictable and repetitive processes. However, they are typically unable to adapt to a dynamic environment where unexpected situations have to be managed. It is possible to distinguish among four different types of exceptions that can occur during the execution of a process, [9]: *basic failures*, which are failures of the software system or its environment; *application failures*, which are failures of the applications that are invoked by the software system; *expected exceptions*, which are predictable deviations from the normal flow of the process; *unexpected exceptions*, which are mismatches between the actual execution of a process and its definition in the software system.

Basic failures and application failures are usually handled by the underlying layers of the process management system, e.g. by a database management system that supports recovery. Expected exceptions and unexpected exceptions, on the other hand, are special for process management software and require their own techniques and methodologies, [10]. An expected exception is directly related to the process domain and can therefore be modelled as a part of the process, even if it represents a deviation from the normal, or desired, course of events. Expected exceptions can be caused by temporal events (e.g. the expiration of a deadline) or by actions by external agents (e.g. a customer cancels her reservation). Typically, these events take place asynchronously with the tasks in the process, which means that they are not easily modelled by a graph of tasks, which is the most common way of specifying processes. Instead, a more promising approach for representing expected exceptions is to use ECA (Event-Condition-Action) rules, as suggested by [10]. The event part in the rules specifies the occurrence of a possibly exceptional situation, the condition part checks that an exception has really taken place, and the event part specifies the response to the exception. An unexpected exception is caused by a change in the process that could not have been anticipated at design time, e.g. a new government regulation. Unexpected exceptions can be handled in two ways. First, a running process instance that encounters the exception can be modified. This means that the process definition is left unchanged, and other future process instances will execute according to that definition. Secondly, the process definition can be modified so that also future process instances will execute according to the new definition. The latter alternative gives rise to research issues in determining when a running instance can be migrated to a new process definition.

3.4 Moving Application Logic Out of Systems and into Processes

Integration between applications concerns not just the transfer of data, but also the business logic that controls the sequencing and ensures the integrity of the business transactions. Each step may have to complete before the next can commence. Data entered by people and systems communicating with the business process have to fulfil correctness criteria, usually expressed through (business) rules.

In object-oriented analysis and design, e.g. UML, rules are often captured during the definition of use cases, [11]. Later they may be represented e.g. as cardinality constraints or generalisation relationships of class diagrams, or as triggering event rules or guard conditions of state transition diagrams.

It is not obvious where the business rules that together determine the business logic should be coded. Are they part of a new application, part of existing applications, or should they be kept in a separate integration layer? In object orientation, it is sometimes claimed that rules belong to the objects they control. However, many rules concern several objects. Hence, it is not obvious to which objects these rules should be assigned. In TEMPORA [12], a project aiming to develop a systems development platform based on the explicit representation of business rules, this was found to be a serious methodological problem.

If a business rule is embedded in the code of an application, irrespective of whether it is existing or new, any time that rule changes, the application must be carefully modified, tested and then re-deployed. In business process oriented integration, it is often found that the functional behaviour of the systems is required to be integrated, not just the data.. Therefore business logic should, as much as possible, be kept outside of applications. The logic that is specific to a certain subtask is kept within the component that implements that task and the activity is associated with an application definition, in which it only needs to specify the work to be done and the resources required by a client application. Service combinations are implemented through logic that is specific for each combination and hence kept in the service (combination) definition. When the application changes, or is replaced, the only alteration necessary is to change the definition to point to another application.

Moreover, customers should be able themselves to specify which service combination they want. This calls for the ability to augment the service definition with guard conditions that hinder the use of services that customers are not authorised to use or that are not possible to combine with other services selected. For example, in the ordering process it should be possible to specify which combination of services as regards e.g. delivery and payment methods one wishes. Constraints that guard the gate (guard conditions) to task components should be put in the process definition.

The fact that many companies move to using standardized enterprise resource planning systems and other COTS products tend to make them more alike each other, hence removing individual differences that may represent competitive advantage. Moving logic out of the applications and into the process definitions is then becoming a means for companies to retain their uniqueness.

4 Concluding Remarks

In this Chapter, we have discussed the rationale for application and process integration as well as a number of possible research directions in the area. In the marketplace, a number of middleware vendors are adding process modelling and simulation capabilities to their products, thereby moving into the Process Broker market. Some of the major products in this market are: Viewlocity's Business Integration Modeler and Manager [13], Extricity Software's AllianceSeries [14], Vitria Technology's BusinessWare [15], and HP's Changeengine [16]. We believe that research as outlined in this Chapter will benefit the future development and use of products in this category. Within the NUTEK (Swedish National Board for Industrial and Technical Development) sponsored project ProcessBroker [http://www.dsv.su.se/~pajo/arrange/index.html], we will pursue the research directions discussed here.

Acknowledgements

The authors are grateful to other participants of this project, especially Birger Andersson, S.J. Paheerathan, Erik Perjons, and Nasrin Shakeri at KTH and Christer Wåhlander at Viewlocity for contributing ideas and for commenting on earlier versions of this paper.

References

[1] Linthicum D. Enterprise Application Integration. Addison-Wesley, 2000

[2] Bubenko J A Jr, Wangler B. Research Directions in Conceptual Specification Development. Conceptual Modeling, Databases and CASE: An Integrated View of Information Systems Development. Eds. Loucopoulos P, Zicari R. John Wiley & Sons, Ltd. 1992

[3] Ahlsén M, Bubenko J A Jr. Interoperability in Federated Information Systems. Second International Workshop on Intelligent and Cooperative Information Systems: Core Technology For Next Generation Information Systems, Ed. Brodie M L, Ceri S. Villa Olmo, Como, Italy, Dipartimento di Ellettronica, Politecnico di Milano, 1991

[4] Curran T, Ladd A. SAP R/3 Business Blueprint. Prentice Hall. 2000

[5] Bernstein A, Klein M, Malone, T. The Process Recombinator: A Tool for Generating New Business Process Ideas. In: International Conference on Information Systems, 1999.

[6] Malone T, Crowston K. The Interdisciplinary Study of Coordination. ACM Computing Surveys 1994; 26

[7] Papazoglou M. Distributed, Interoperable Workflow Support for Electronic Commerce. Lecture Notes in Computer Science, volume 1402, 1998

[8] Johannesson P, Perjons E. Design Principles for Application Integration. In: International Conference on Advanced Information Systems. Springer. 2000

[9] Eder J, Liebhart W. The Workflow Activity Model WAMO. In: 3rd International Conference on Cooperative Information Systems. University of Toronto Press. 1995, pp. 87-98

[10] Casati F, Fugini M, Mirbel I. An Environment for Designing Exceptions in Workflows. Information Systems 1999; 24: 255-273

[11] Allen P, Frost S. Component-Based Development for Enterprise Systems. Addison-Wesley, 1998

[12] Loucopoulos P, McBrien P, Schumacker F, Theodoulidis B, Kopanas V, Wangler B. Integrating database technology, rule-based systems and temporal reasoning for effective information systems: the TEMPORA paradigm. Journal of Information Systems 1991, 129-152.

[13] Wåhlander C, Nilsson M, Skoog A. Introduction to Business Integration Model (BIM). Copyright Viewlocity, 1998

[14] Extricty AllianceSeries, Extricity Software, http://www.extricity.com/products/alli_series_over.html, 2000-02-21

[15] Atwood R. Bringing Process Automation to Bear on the Task of Business Integration. Vitria Technology 1999 http://www.vitria.com/products/white papers/seyboldwp.html, 1999-11-25

[16] HP Changengine Overview, Hewlett Packard Company, http://www.ice.hp.com /cyc/af/00/101-0110.dir/aovm.pdf, 1999-10-04

Transactions, Processes, and Modularization

A. T. Berztiss
Department of Computer Science, University of Pittsburgh
Pittsburgh, PA 15260, USA
(e-mail: alpha@cs.pitt.edu)
and
SYSLAB, University of Stockholm

Abstract

The term information system has lately acquired two meanings. Under one meaning the concern is with representation techniques, and infrastructure, such as animation, workflow or data base management, internet access, and mobile computing. We call this the technology aspect. Under the second meaning the concern is with the modeling of domains and enterprises, and the development of systems based on such models. This we call the conceptual aspect, and our contribution relates to the conceptual aspect. A conceptual model can be regarded as consisting of an information model and a process model. Here we consider modularization principles for the process model, and their effect on modification of the models in the contexts of product-line development and maintenance.

1 Introduction

Nowadays, in discussing information systems, it has to be made very clear what one is talking about. There are numerous conferences on web-based computing, mobile computing, e-commerce, data warehousing and data mining, graphical interfaces, and the like. This all relates to representational issues, infrastructure, and knowledge acquisition. We shall refer to such issues as the technology aspect. More or less independent of the technology aspect is the formulation of domain and enterprise models, and the development of systems based on these models. We call this the conceptual aspect. At some point the two aspects come together — decision making under an enterprise model may need information from the World Wide Web or knowledge mined from a data warehouse.

Despite the dependence of an enterprise software system on technology, enterprise models can be developed without having to consider the technological aspect. The primary concern is how information is to assist in furthering the goals of the enterprise, not where the information originates.

In [1] we developed a software process model for business reengineering based on an enterprise model. Parts of a sixteen-step plan for business reengineering [2] were combined with an enterprise model [3, 4] to arrive at a model for the development of software systems in the context of business reengineering. This model is well suited for the development of one-of-a-kind systems. But increasingly we need to consider software product-lines, i.e., products with similar purposes that have some context-dependent components. For such software it is particularly important that it be well modularized. The modularization presents several challenges.

First, the software contains what we call procedural and transactional components, and they have to be treated differently. In Section 2 we discuss the differences between these two kinds of software. Second, a software system typically implements a dynamic process. The combining of transactions into processes is considered in Section 3. This has relevance to both reuse and maintenance. In Section 4 we study a particular system, and identify components of the system that could be used in different contexts, but would then differ in detail. Section 5 is a summary. It also contains a number of questions in need of answers.

2 Procedural and Transactional Computation

For some time it has been realized that there are two major classes of computation. Harel [6] distinguishes between reactive and transactional systems, where reactive systems are for the most part driven by external events. Wegner [6] analyzes interactive computing, which he shows to be more powerful than algorithmic computation. Interactive computing has been studied in some depth by Kurki-Suonio and Mikkonen [7]. We have grouped computations into transactional and procedural classes [8], with a software system likely to contain components of both classes.

The distinguishing characteristic under our classification is whether or not a system undergoes visible state changes, such as changes in a persistent data base. If it does, then it is transactional. The classification is consistent with those of Harel and Wegner — reactions or interactions imply a long-lived system, and the evolution of such a system is reflected as changes in its state.

We model all information sources as collections of functions. A function is a set of pairs $<x, f(x)>$, where both x and $f(x)$ may be composite data objects, such as sets, or tuples, or even tuples of sets. In an immutable function, e.g., the cosine, $cos(x)$ is always the same for a given x; in a mutable function $f(x)$ may change, e.g., the salary of an employee x changes over time. A procedure is a device F that transforms an input x into an output $f(x)$. Queries are procedures: the query is x and the answer is $f(x)$. Transactions change mutable functions.

Wegner contends that proofs of correctness of interactive systems are impossible [6]. Let us consider this in some detail. The proof of a procedure may be difficult, but in principle it is straightforward: a specification describes the form of $f(x)$, and relates $f(x)$ to x; the proof consists of showing that, for all x in the input domain, F generates $f(x)$ that satisfy the specification. Such an approach is too simple for interactive systems because now there is a sequence of inputs, the inputs may be determined by intermediate results of computation, and the inputs may determine how the system proceeds. Thus, there is no simple way of relating the ultimate outcome of the computation to its inputs.

Nevertheless, transactional computations should still have specifications, but now they would define valid system states and valid state transitions. Such specifications can be written in terms of, for example, Z schemas [9], and can take two forms. One consists of global invariants, i.e., predicates that the system has to satisfy. Another relates to individual transactions, and consists of preconditions that are to ensure that a transaction will not put the system into an invalid state or induce an invalid state transition.

Mature engineering disciplines are characterized by the ability of designers to determine at design stage what properties the engineered product based on the design will have. To reach maturity, software engineering needs to put more emphasis on methods for establishing the properties of software systems before they have been built. This implies the use of proofs in software development, the word *proof* suggests mathematics, and methods based on mathematics are known as *formal* [10].

There are three levels of proof. In a formal logical proof each step has to be justified by reference to a rule of inference. Engineers in established disciplines do not use logical proofs, but logical proofs in the software context can in principle be automated, and this is their appeal. For example, a proof can show that a procedure is consistent with its specification. Most proofs in mathematics are informal. They use generally accepted lines of reasoning. We shall call them conventional proofs. The third level of proof is statistical, and it establishes expected values with associated confidence bounds for parameters that characterize an engineering product.

We have shown that conventional proofs can be quite rigorous when applied to programs [11]. Programs are formal representations in that they are written in languages with well defined syntax, but conventional proofs can be applied even to representations that are not formal. This means that it is not necessary to have formal specification of transactions.

Let us determine the application areas of transactional computing. The main characteristic of a transactional system is that it undergoes state changes. Such state changes are caused by events in the environment of the system, and the purpose of the software system is to react to the events. In the case of embedded control systems the events are sensor readings when they reach critical values. The system reacts by performing appropriate control actions. There is some activity in the data base, but it is fairly minor. For example, a history of past control actions is maintained so that fine tuning of the host system can be undertaken. The other main application domain is business systems. Now the events are, for example, messages from customers and suppliers, and the system reacts to these inputs. Of course, based on the current state of the data base, the system may itself initiate a transaction, such as reordering of items that are in short supply. Data bases of business systems are much more extensive than those of control systems.

3 Transactions and Processes

A transactional system is composed of transactions and a communication and coordination mechanism that combines transactions into processes. The transactions of an information system operate on a data base; those of a control system operate on activators. In both cases this results in state changes, but there is a major difference. A data base is itself software, so that representations of transactions can describe state changes in detail. State changes in the host of a control system occur in hardware, i.e., outside the area of competence of software developers.

The developers of control systems are not expected to define how activators affect host systems. The question arises whether the developers of information systems should be concerned with the organization of data bases or even their type. Our view is that an abstract conceptual model of a data base in terms of functions provides the appropriate level of detail. The users of the information system can formulate queries in terms of the functional model, without having to know how the data base is implemented. On the other hand, the user of the information system has complete control over changes in the data base. This satisfies Davenport's recommendation that the "owner" of a business process is to "own" the data associated with the process [12]. Translation of the functional schema into appropriate implementation structures, and of a functional query into a retrieval procedure can be automated. A functional representation can support complex queries, but here we are not going to demonstrate how such queries are constructed.

The distinction between information systems and control systems is not all that clear. At one end of the scale we have decision support systems based on non-volatile data warehouses. At the other end are embedded control systems. But in the middle we find systems in which the information or control aspect dominates to a greater or lesser degree, as in, for example, workflow systems. We have found that the specification of these middle-ground systems is best started with a definition of the transactions that are needed. The functions of the definition of the data base evolve in parallel with the transaction definitions: transactions need information and they deposit items of information in the data base. A data base schema in the form of functions is thus built up in a natural way, and no special methodology has to be established for defining an information model. This is different for data warehouses. Their lack of volatility implies that a model based on state transitions is inappropriate. Since data mining or other utilization of a data warehouse is strictly procedural, there are no transactions, and the information model is everything. When emphasis is on information rather than process, the organization of the information base is an ontological problem – for an overview of ontologies see [13].

By means of student projects we have thoroughly studied rental processes [14, 15]. They include car rentals, video rentals, "rental" of a place on a flight, even "rental" of a place in a university course. In general terms, a rental is the allocation, on demand, of a resource for a limited time, generally against payment. The first stage of our approach was the definition of a generic rental process in terms of 15 natural-language capsule descriptions of 75-250 words each. A capsule describes the activities of a transaction without much attention to how the transaction is to interact with other transactions. In the next stage the generic capsules are adapted for a specific application. This version is still in natural language, but now a prescribed format with three main components is to be followed. The components tell what the transaction does, how it is initiated, and what transactions it may in turn initiate. Linkages that define a process are thus being introduced. The students were able to develop C++ code directly from the formatted descriptions, but a formal specification phase may be interposed in case a specification is needed for formal analysis, e.g., of mission-critical systems.

This is an example of a top-down approach to product line development. The generic rental process has been used to define several specific systems. We have developed capsules for numerous smaller processes. Two capsules that relate to the assembling of items are given in full, Specializations include setting up a meeting in an office as soon as all participants are free, the assembly of a purchase order, and the assembly of parts from suppliers.

Assemblage request. A set of items S is checked against a set of available items AV. If S is a subset of AV, then S becomes an assembly, and it is subtracted from AV and added to a set *assembled*; otherwise S is added to *requests*. A special case arises when multiple copies of the items of S are to enter an assembly.

Availability. Whenever the supply of available items in AV increases, elements of *requests* are tested against AV. How this supply increases is determined by the specific application. If there is now a set S in *requests* such that S is a subset of AV, then S is subtracted from AV and added to *assembled*.

In constructing a new application from the generic model one should examine the applications that have already been constructed from this model. It is likely that some formatted specializations can be used unchanged in the new application, allowing full code reuse. In other cases the formatted definitions may need to be changed minimally so that substantial code reuse becomes feasible.

In summary, a new application will differ from other applications based on the same generic model, but the differences need not be large. This is closely related to maintenance. Basili [16] points out that maintenance can be regarded as the construction of a new system under extensive reuse of components of the old system. Maintenance consists of correction, prevention, adaptation, and enhancement. Corrective maintenance removes faults, preventive maintenance improves software structure to make maintenance easier in the future, and adaptive maintenance reacts to changes in the environment. Enhancement is the addition of new features to a system, or the introduction of an improvement in the way the system performs its tasks. However, we do not consider the addition of new features as maintenance, on the basis of the analogy that adding new rooms to a house is not regarded as maintenance of the house.

Experience shows that our approach keeps corrective maintenance to a minimum, and the structure of a system developed from our capsule descriptions needs no preventive maintenance. Adaptive maintenance is unlikely to affect the generic capsules, and the partition of a process into well-defined transactions restricts change to a few easily identified locations. Enhancement would affect the generic capsules, and this can be a major undertaking. It is therefore advisable to forestall it by getting the system right initially, but, if management sees a need for enhancement, modularization based on transactions should aid it.

It has been conjectured that success in the development of product lines depends primarily on a focus on simplification, minimization, and clarification, and on a software architecture that can be easily adapted to future customer needs [17]. We hope to have succeeded in showing that the capsule-based approach satisfies both these success factors. It is also consistent with the basic principle of business reengineering, namely that an organization should be defined in terms of processes [12, 2]. Our modules implement processes, and our transactions are submodules of the process modules.

4 Case Study: Banking Machines

The approach of the preceding section is top-down: the specification of a complete generic process is refined into a specialized application. But some of the capsules belonging to one generic process could be used in some totally different context. Under a bottom-up approach a system is constructed from capsules that can derive from various sources. This is one reason why interconnection of transactions is not emphasized at the generic capsule level. Here we shall study a particular application, a banking machine system, and identify in it components that have high reuse potential under a bottom-up approach.

We consider a case study used in our software design course, which is loosely based on a study in [18]. A system of automatic teller machines (ATMs) is operated by a consortium of banks. There is a central computer belonging to the consortium. An ATM reads a user's bank card and transaction details, communicates these transaction data to the central computer, receives transaction authorization from the central computer, and dispenses cash and a hard copy statement of account balance. All ATM transactions are withdrawals. From the bank card the ATM reads a code identifying the bank that issued the card, and an account identifier. The user supplies a password, and the amount to be withdrawn. If the ATM cannot complete a transaction, an appropriate message is generated. Each bank has its own computer. The central computer communicates ATM transaction data to the appropriate bank computer, and receives transaction authorization from the bank computer. An ATM becomes inoperative when its cash reserve or paper supply for the balance statements are exhausted, or there is a communication breakdown that prevents transaction authorization.

The first step was to identify the normal sequence of transactions:

a. Customer inserts bank card, and ATM determines that the card is valid.
b. ATM sends bank card data to the consortium, which issues authorization for acceptance of the card in the form of a transaction identifier.
c. ATM asks for password and customer enters it.
d. ATM sends transaction identifier and password to the consortium, which accepts it.

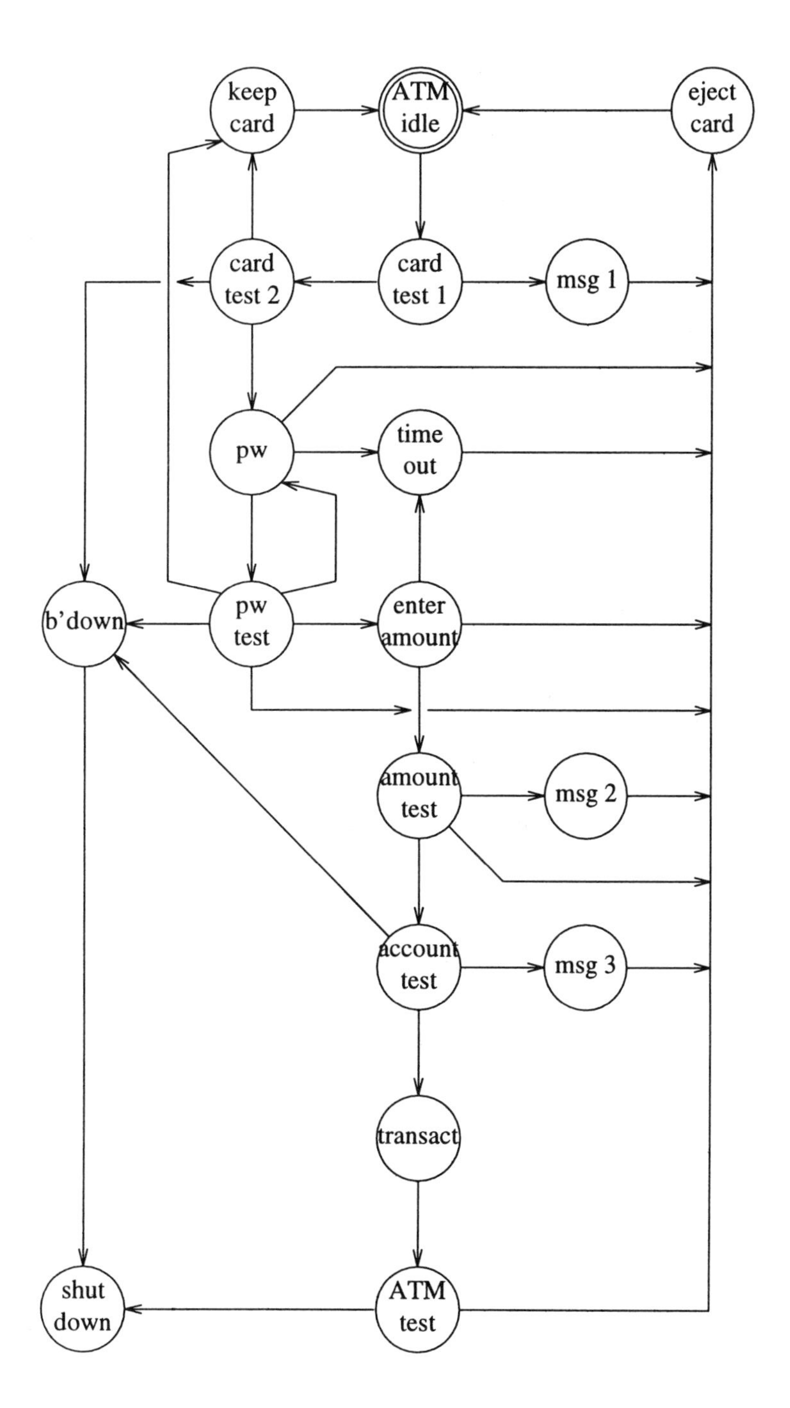

Fig. 1 – State transitions of an ATM

e. ATM asks for the withdrawal amount, which the customer enters.

f. ATM determines that the amount is within limits for the machine, and that its cash reserve is adequate.

g. ATM sends transaction identifier and amount to the consortium, which passes the information to the appropriate member bank. The latter authorizes payment, adjusts the balance, and sends authorization and new balance to the ATM via the consortium.

h. ATM adjusts its cash reserve total, dispenses cash and balance statement, returns bank card.

Next we identify abnormal situations, i.e., exceptional conditions that cause deviation from the normal sequence:

i. Invalid card inserted – bank does not belong to consortium or card has expired.

ii. Consortium recognizes card as reported stolen, or there has been excessive usage of the card.

iii. Wrong password entered. A special case arises when wrong passwords are tried too many times.

iv. Withdrawal amount exceeds limit for the machine.

v. Cash reserve does not cover the amount requested.

vi. No paper for printing new balance.

vii. Bank does not authorize payment because of insufficient funds.

viii. Customer cancels transaction.

ix. Customer takes too long over an input.

x. There is a communication breakdown between the ATM and the consortium, or the consortium and the bank holding the account.

The operation of the ATM is illustrated by Fig. 1, which was provided with a node dictionary. We show the entry in the node dictionary for *account test*. Students were able to start coding directly from the node directory.

NODE: *account test.*

Reached from: *amount test*, when this test is passed.

Action: test via network that the amount to be withdrawn is covered, and, if it is, cause customer balance to be adjusted and sent to ATM.

Outgoing options:

(a) *msg 3*, taken if the test fails;

(b) *transact*, taken if test is passed;

(c) *b'down*, taken if network communication is broken.

Our purpose here is to identify in Fig. 1 tasks that could be used in other contexts. We can distinguish five generic activities:

1. *Entry protocol*, comprising nodes *card test 1, card test 2, keep card, pw, pw test, msg1.*

2. *Release of product*, represented by nodes *enter amount, amount test, msg2, account test, msg3, transact.*

3. *Clean-up*, comprising nodes *ATM test, shut down, b'down, eject card.*

4. *Time out*, arising from excessive time taken with inputs, and represented by the single node *time out*.
5. *Cancellation by customer*, available while the transaction sequence *pw, pw test, enter amount, amount test* is in progress. This is a customer-initiated exceptional condition, represented by node *eject card*. Hence it has a slight overlap with *Clean-up*.

We note a shift in architecture. Instead of a two-way partition of the process into a normal sequence of operations and abnormal situations or exceptions, we now have five subprocesses, each with its own complement of exceptions. Each of the five can be given a capsule description, which can be quite elaborate (*Entry protocol*).

As an example, specializations of *Release of product* include, besides the ATM operation, dispensing of goods against payment from a vending machine, and release of supplies from a stockroom against an authorization. In both cases it has to be made sure that a sufficient quantity of a product is available (*amount test*). In the stockroom application *account test* is replaced by validation of the authorization; nothing corresponding to *account test* is needed for the vending machines.

5 Summary and Agenda for the Future

We have outlined an approach to the design of transactional systems based on capsule descriptions of tasks. The tasks are assembled into processes, which are the major modules of a complete system. The tasks are submodules. One example of our approach is the rental process, which we did not consider in detail because we have done so elsewhere. This process consists of 15 tasks, and the 15 tasks were identified by means of a top-down development process. For our second example we took a system that had been developed earlier in an ad hoc manner, and identified five tasks in the system. Once a repository of such tasks has been built up, systems can be developed bottom-up from components in the repository. Figure 1 contains 18 nodes, and 17 of them are distributed over the five tasks. These 17 nodes represent transactions, which are our smallest modularization units.

A significant advantage of our approach is that it is based on a thorough analysis of each task. By avoiding consideration in early design stages of how a task interacts with other tasks the analysis is focused on the task itself, and does not get sidetracked into issues only marginally related to the task.

A notable feature of task design is that exceptional conditions are considered at the level of tasks rather than that of entire processes. A capsule description includes a checklist of exceptional conditions, and this is to be consulted when a generic capsule is refined for a specific application. Designers of applications are expected to add to the checklists, and in time they become a valuable resource.

Because of the special attention given to exceptions, the task modules are robust, which is an important concern for reuse. Not only does our approach permit design reuse, there is also potential for code reuse. In [19] we discuss the relationship of code reuse to design reuse in some detail. Moreover, the formatted specializations of the generic capsules, referred to in Section 3 and described in [14], can be fairly easily developed into formal specifications and, provided the specification language is designed for this, a code generator can produce code from such a specification automatically. Whatever properties of a task definition are established by (conventional) proofs, such properties will also be possessed by the code.

There remain several problems that need to be addressed. The most important is the determination of the level of generality that the designer of capsules should aim at. Our capsules are closer to the building patterns of Christopher Alexander [20] than to the patterns described in, say, [21, 22]. The latter can be excellent for teaching general design principles, but – for the most part – we find them to be at too high a level for actual design.

With our approach, too, finding the right level of detail can become difficult. For example, in an analysis of the general problem of resource allocation we began with a formulation that could relate to the matching of a job description against available personnel, of the dimensions of a proposed building against available sites, or of the requirements for a textbook against available textbooks. Moreover, the match could be exact, as for building sites, or approximate, as for personnel and textbooks. This was much too broad. Our first restriction was to consider exact matches alone.

Resource availability A. In order to find all resources of a given type that satisfy a set of conditions, generate a requirement configuration, and, for each available resource, an availability configuration, and find the resources whose availability configurations can accommodate the requirement configuration.

This solution is still too broad. We found that we had to narrow it further by giving a specific interpretation to "configuration". The specific context in which the problem arose made us consider a variant in which the configuration relates to times.

Resource availability B. Define a timeline and on the timeline indicate all the times at which the resource is required. Also, for each applicable resource, define a timeline indicating all times when the resource is not available. There is to be no overlap of the times marked on the two timelines.

A second problem is the determination why some problems are easier to deal with than others. The easy answer is that some problems are inherently more complex than others, but we are interested in determining why this is so. A problem can in general be addressed in different ways. The simplest solution is usually purely transactional, from which we can advance to solutions in which there is an increasing dependence on procedures. It seems that too often an attempt is made to define a task in terms of transactions when procedures are more suitable. We illustrate this by a simple example: a date and place are to be selected for a meeting, and participants are to be registered for the meeting. At the lowest level a steering committee selects a date and place for the meeting, and an information system merely performs registration transactions. At a higher level the steering committee selects a date, but now considers several places for the meeting. A place is to be selected for which the travel costs of the participants are minimized, and this should done by a procedure rather than a transaction.

Perhaps the most important task facing us is the development of an ontology of the tasks that the capsules describe. According to [13], ontologies are theories about the sorts of objects, properties of objects, and relations between objects that are possible in a specified domain of knowledge. Our domain is that of generic tasks and processes, with capsules defining the properties of tasks. In the interests of flexibility we have deliberately refrained from relating generic tasks to each other, but they stand in a part-of relation to processes.

References

1. Berztiss AT, Bubenko JA. A software process model for business reengineering. In: Solvberg A, Krogstie J, Seltveit AH (eds) Information systems development for decentralized organizations. Chapman & Hall, 1995, pp 184-200
2. Berztiss A. Software methods for business reengineering. Springer, 1996
3. Bubenko JA. Extending the scope of information modelling. In: Olive A (ed) Proc. fourth international workshop on the deductive approach to information systems and databases. Departament de Llenguatges i Sistemes Informatics of the Universitat Politecnica de Catalunya, Barcelona, 1993, pp 73-97
4. Bubenko JA, Rolland C, Loucopoulos P, DeAntonellis V. Facilitating "fuzzy to formal" requirements modelling. In: Proc IEEE Internat Conf on Requirements Engineering, 1994
5. Harel D. Statecharts: a visual formalism for complex systems. Science of Computer Programming 1987; 8: 231-274
6. Wegner P. Why interaction is more powerful than algorithms. Comm. ACM May 1997; 40(5): 80-91
7. Kurki-Suonio R, Mikkonen T. Harnessing the power of interaction. In: Jaakkola H, Kangassalo H, Kawaguchi E (eds) Information Modelling and Knowledge Bases X. IOS Press, 1999, pp 1-11
8. Berztiss AT. Transactional computation. In: Bench-Capon T, Soda G, Tjoa AM (eds) Database and Expert Systems Applications (DEXA'99). Springer, 1999, pp 521-530 (Lecture notes in computer science no.1677)
9. Spivey JM. Understanding Z. Cambridge University Press, 1988
10. Hall A. Seven myths of formal methods. IEEE Software Sept 1990; 7(5): 11-19
11. Berztiss A. Programming with generators: an introduction. Ellis Horwood, 1990
12. Davenport TH. Process innovation: reengineering work through information technology. Harvard Business School Press, 1993

13. Chandrasekaran B, Josephson JR, Benjamins VR. What are ontologies, and why do we need them? IEEE Intelligent Systems Jan/Feb 1999; 14(1): 20-26

14. Berztiss AT. Domains and patterns in conceptual modeling. In: Information Modelling and Knowledge Bases VIII. IOS Press, 1997, pp 213-223

15. Berztiss AT. Failproof team projects in software engineering courses. In: Proc. 27th Frontiers in Education Conference. IEEE CS Press, 1997, pp 1015-1019

16. Basili VR. Viewing maintenance as reuse-oriented software development. IEEE Software Jan 1990; 7(1): 19-35

17. Dikel D, Kane D, Ornburn S, Loftus W, Wilson J. Applying software product-line architecture. Computer Aug 1997; 30(8): 49-55

18. Rumbaugh J, Blaha M, Premerlani W, Eddy F, Lorensen W. Object-Oriented Modeling and Design. Prentice-Hall, 1991

19. Berztiss AT. Reverse engineering, reengineering, and concurrent engineering of software. International Journal Software Engineering Knowledge Engineering 1995; 5: 299-324

20. Alexander C, Ishikawa S, Silverstein M, et al. A pattern language. Oxford University Press, 1977

21. Gamma E, Helm R, Johnson R, Vlissides J. Design patterns – elements of reusable object-oriented software. Addison-Wesley, 1995

22. Coplien JO, Schmidt DC (eds). Pattern languages of program design. Addison-Wesley, 1995

METADATA: The Future of Information Systems

Keith G Jeffery
CLRC-Rutherford Appleton Laboratory
Chilton, UK

Abstract

Information systems are increasing in complexity. There are greater volumes of data, users, processes and transactions. There are greater interdependencies between components. The range of available storage, user interface and computing devices is increasing so producing heterogeneity at the physical system level. The utilisation of multiple information sources to solve a problem (or create an opportunity) creates a need for homogeneous access over heterogeneous information sources. The optimal utilisation of multiple computing resources demands the creation of a uniform computing landscape. The key to homogeneous access to heterogeneous resources (not only information) lies with metadata. The future of advanced information systems depends on metadata. Metadata is the core of the emerging UK GRIDs project.

1 Introduction

The title makes an assertion that metadata is the future of information systems. The purpose of this paper is to support that assertion. The premise is that metadata is an essential, and the most important, component in advanced information systems engineering. The topic of Metadata has recently found the limelight, largely due to a sudden realisation of its necessity in making the WWW (World Wide Web) usable effectively. Metadata (data about data) is essential for WWW to scale, for finding information of relevance and for integrating together data and information from heterogeneous sources. Metadata is essential for refining queries so that they select that which the user intends. Metadata is essential for understanding the structure of information, its quality and its relevance. Metadata is essential in explaining answers from ever more complex information systems. Metadata assists in distilling knowledge from information and data. Metadata assists in multilinguality and in multimedia representations. The engineering of systems from components (data, processes, software, events, subsystems) is assisted by metadata descriptions of those components.

Metadata has been used in information systems engineering for many years – but usually in a specialist, one-off and uncoordinated way. Commonly the metadata has been human-readable but not specified sufficiently formally, nor accepted sufficiently widely, to be interpreted unambiguously by IT (Information Technology) systems. The ubiquity of WWW, the increasing need for access to heterogeneous distributed information and the increased use of multilingual and multimedia sources all demand some common representation of, and understanding of, metadata.

Metadata is attached to data to aid in its interpretation. Metadata processing systems interpret the data using the attached metadata. In addition to information systems such as WWW (update, retrieval) and systems engineering as described above, metadata is essential for electronic business from advertising through catalogue information provision through initial enquiry to contract, purchase, delivery and subsequent guarantee or maintenance.

Metadata is like the Rosetta Stone – which provided the multiway translation key between Greek, Demotic and Hieroglyphics – or, with an associated processing system, like the Babel fish [1]. To quote from [2] "The Babel fish is small, yellow and leech-like, and probably the oddest thing in the universe. It feeds on brainwave energy received not from its own carrier but from those around it. It absorbs all unconscious mental frequencies from this brainwave energy to nourish itself with. It then excretes into the mind of its carrier a telepathic matrix formed by combining the conscious thought frequencies with nerve signals picked up from the speech centres of the brain which has supplied them. The practical upshot of all this is that if you stick a Babel fish in your ear you can instantly understand anything said to you in any form of language."

2 What is it?

2.1 Metadata

Metadata is data about data. Metadata can describe a data source, a particular collection of data (a file or a database or a table in a relational database or a class in an object-oriented database), an instance of data (tuple in a relational database table, object instance in a class within an object-oriented database) or data associated with the values of an attribute within a domain, or the particular value of an attribute in one instance. Metadata can describe data models.

Metadata can also be used to describe processes and software. It can describe an overall processing system environment, a processing system, a process, a component of a process. It can describe a suite of software, a program, a subroutine or program fragment, a specification. It can describe an event system, an individual event, a constraint system and an individual constraint. It can describe a process and /or event model.

Metadata can describe people and their roles in an IT system. It can describe an organisation, a department, individuals or individuals in a certain role.
The process of standardisation of metadata – models, semantics and syntax – is only just beginning, and then mainly in the data domain. Particular application domains have their own metadata standards to assist in data exchange e.g. engineering [3], healthcare [4], libraries [5]. An attempt at a more general exchange metadata for internet resources – the Dublin Core - has been proposed [6] but unfortunately it is insufficiently formal to be really useful [7]. A general metadata model, RDF (Resource Description Framework) has been proposed [8] with the implementation language XML (eXtended Markup Language) [9].

This paper concentrates on the traditional data / information / knowledge aspects of metadata; however, there are clear linkages to processing (including events) and people – especially from the object-oriented and logic-based viewpoints.

2.2 A Classification of Metadata

Metadata is used for several purposes;
(a) describing data for the purposes of data exchange;
(b) describing data for the purposes of global access from query (including update) to optimise recall and relevance;
(c) describing data for the purposes of query optimisation;
(d) describing data for the purposes of answer integration and explanation;
(e) describing data for the purposes of correct analytical processing or interpretation, representation or visualisation.
(f) describing the data to overcome multilinguality and multimedia heterogeneities

All of these purposes require that the data be described:
(1) such that the resource is constrained formally to ensure integrity;
(2) such that the resource is reachable by automated means;
(3) such that there is sufficient description for the purposes to utilise the resource.
This requirement leads to a classification orthogonal to purposes but serving all of them proposed in a tutorial in 1997 and published in [7] (Figure 1: Metadata Classification). It should be noted that, whereas this classification is suitable for data it can also be used - and is sufficiently general for - metadata about processes, events, organisations or people.

1.1.1 Schema Metadata

Schema metadata constrains the associated data. It defines the intension whereas instances of data are the extension. From the intension a theoretical universal extension can be created, constrained only by the intension. Conversely, any observed instance should be a subset of the theoretical extension and should obey the constraints defined in the intension (schema). One problem with existing schema metadata (e.g. schemas for relational DBMS) is that they lack certain intensional information that is required [10]. Systems for information retrieval based on, e.g. the SGML (Standard Generalised Markup Language) DTD (Document Type Definition) experience similar problems.

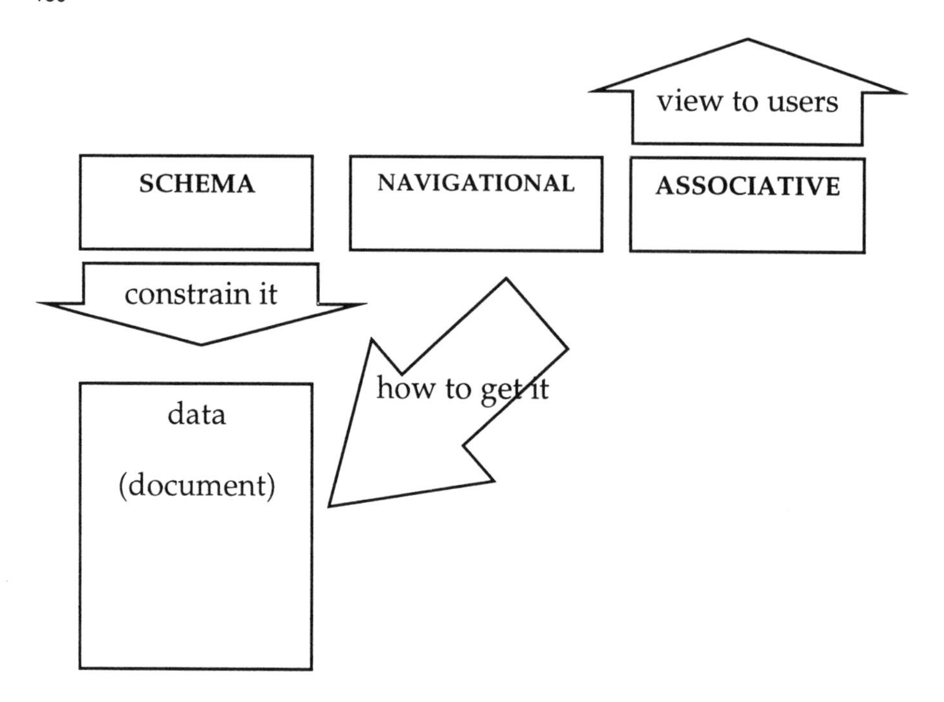

Figure 1: Metadata Classification

It is noticeable that many ad hoc systems for data exchange between systems send with the data instances a schema that is richer than that in conventional DBMS – to assist the software (and people) handling the exchange to utilise the exchanged data to best advantage.

1.1.2 Navigational Metadata

Navigational metadata provides the pathway or routing to the data described by the schema metadata or associative metadata. In the RDF model it is a URL (universal resource locator), or more accurately, a URI (Universal Resource Identifier). With increasing use of databases to store resources, the most common navigational metadata now is a URL with associated query parameters embedded in the string to be used by CGI (Common Gateway Interface) software or proprietary software for a particular DBMS product or DBMS-Webserver software pairing.

The navigational metadata describes only the physical access path. Naturally, associated with a particular URI are other properties such as:

(a) security and privacy (e.g. a password required to access the target of the URI);
(b) access rights and charges (e.g. does one have to pay to access the resource at the URI target);
(c) constraints over traversing the hyperlink mapped by the URI (e.g. the target of the URI is only available if previously a field on a form has been input with a value between 10 and 20). Another example would be the hypermedia equivalent of referential integrity in a relational database;

(*d*) semantics describing the hyperlink such as 'the target resource describes the son of the person described in the origin resource'

However, these properties are best described by associative metadata which then allows more convenient co-processing in context of metadata describing both resources and hyperlinks between them and – if appropriate - events.

1.1.3 Associative Metadata

In the data and information domain associative metadata can describe:

(a) a set of data (e.g. a database, a relation (table) or a collection of documents or a retrieved subset). An example would be a description of a dataset collected as part of a scientific mission;

(b) an individual instance (record, tuple, document). An example would be a library catalogue record describing a book ;

(c) an attribute (column in a table, field in a set of records, named element in a set of documents). An example would be the accuracy / precision of instances of the attribute in a particular scientific experiment ;

(d) domain information (e.g. value range) of an attribute. An example would be the range of acceptable values in a numeric field such as the capacity of a car engine or the list of valid values in an enumerated list such as the list of names of car manufacturers;

(e) a record / field intersection unique value (i.e. value of one attribute in one instance) This would be used to explain an apparently anomalous value.

In the relationship domain, associative metadata can describe relationships between sets of data e.g. hyperlinks. Associative metadata can – with more flexibility and expressivity than available in e.g. relational database technology or hypermedia document system technology – describe the semantics of a relationship, the constraints, the roles of the entities (objects) involved and additional constraints.

In the process domain, associative metadata can describe (among other things) the functionality of the process, its external interface characteristics, restrictions on utilisation of the process and its performance requirements / characteristics.

In the event domain, associative metadata can describe the event, the temporal constraints associated with it, the other constraints associated with it and actions arising from the event occurring.

Associative metadata can also be personalised: given clear relationships between them that can be resolved automatically and unambiguously, different metadata describing the same base data may be used by different users.

Taking an orthogonal view over these different kinds of information system objects to be described, associative metadata may be classified as follows:

(a) descriptive: provides additional information about the object to assist in understanding and using it;

(b) restrictive: provides additional information about the object to restrict access to authorised users and is related to security, privacy, access rights, copyright and IPR (Intellectual Property Rights);
(c) supportive: a separate and general information resource that can be cross-linked to an individual object to provide additional information e.g. translation to a different language, super- or sub-terms to improve a query – the kind of support provided by a thesaurus or domain ontology;

Most examples of metadata in use today include some components of most of these kinds but neither structured formally nor specified formally so that the metadata tends to be of limited use for automated operations – particularly interoperation – thus requiring additional human interpretation.

3 Why is it Important?

3.1 Complexity

It is observed that the number of available information sources increases, the number of users increases and the number of information requests increase in both number and complexity. The complexity arises because of several factors:
(a) the heterogeneity of the information sources, including character set, language, media, content quality (accuracy, precision), structure and semantics;
(b) the increased required expressivity of queries including more complex syntax and semantics, the use of graphical interfaces, query improvement or refinement to improve relevance and recall;
(c) the increased complexity of the logic of processes acting over the information sources where the query (or update) may include inline functions (e.g. the concept of 'inexpensive' requires a function involving price of the required object and person salary to be inline in the query);
(d) the increased complexity of integrating information from multiple sources, resolving different values or sets of values for the same required object and explaining the choices made to provide the answer, and the values in the answer itself.

The increasing number of information sources and increased number of users is due to the reduced cost of a person joining the world information society and the increased commercial and non-commercial opportunities for marketing information either for itself or as a step towards purchasing or obtaining traditional goods and services. The increased expressivity and complexity of queries is caused by increasingly educated end-users demanding more of the information systems than previously especially in relevance and precision of answers, structuring of answers, associated explanation and multimedia representation.

3.2 Utility

Metadata increasingly becomes essential to be used in optimising queries, explaining answers, mediating between information sources and between those sources and the querying client and in handling access rights and possible associated payments. Metadata, with associated processes to use it, becomes the glue that holds together the rich diversity of information, suppliers and consumers on the internet.

3.3 Problem

Unfortunately, this sudden realisation generally of its overwhelming importance comes too late; already there are multiple sectoral standards for metadata and attempts to find a commonly agreed set of standards have so far failed to be accepted widely. Even the RDF [8] recommendation from W3C (The World Wide Web Consortium) [11], which is a basic model for describing 'things' and 'connections between them' without semantics, has itself failed to obtain universal acceptance. Various proprietary models – some loosely related to RDF and commonly using XML [9] as the implementation language – have appeared, such as XMI [12].

4 What Exists Now?

4.1 Introduction

There are many good and usable metadata systems in operation every day. Usually, they are specific to a particular organisation (internal data exchange standards, internal IT System documentation standards), a pair of organisations (agreed data exchange standards) or organisations in a particular business sector where a common standard for data exchange or accessing each others' systems is agreed – for commercial benefit. Some of these latter metadata systems have reached international standard status, notably EDI [13] and STEP/EXPRESS [3].

Most of these systems are successful because they are implemented in a narrow domain where the syntax and structure of exchanged datasets have been agreed and where the semantics are well understood in that circumscribed community.

4.2 Some Specific Initiatives

More recently, the explosive growth of WWW has caused several interesting initiatives concerning metadata:

(a) PICS [14]: a method of tagging pages on WWW with content classification information such that compatible processing elements can prevent the pages being displayed. This system is targeted at privacy and parental protection of minors from accessing unsuitable material. This is a kind of associative restrictive metadata;

(b) DC (Dublin Core) [6]: an initial attempt to provide a general associative descriptive metadata element set for the description of content in a WWW page. The original 13 element set was extended to 15 by the Warwick Framework and subsquently there has been much discussion between those who wish to keep the DC simple and human-readable and those who wish to make it more formal and computer-readable;

(c) RDF [8]: The Resource Description Framework General Model for metadata proposed by W3C [14]. This proposal is based on a simple binary relational model such that it can be used universally as a descriptor. The problem is the potential diversity of content, structures and semantics placed upon this basic model – and such diversity is appearing already, especially since the implementation language is XML [9] which is very flexible, providing a syntax but no semantics unless declared externally;

(d) XMI (XML Metadata Interchange) [12] is a standard accepted by OMG (Object Management Group) [15] and brings together XML [9], UML (Unified Modeling Language) [16] and MOF (Meta Object Facility) [17] to provide a metadata facility for information exchange between information systems;

(e) XIF (XML Interchange Format) [18] which may be seen as a competitor to XMI from Microsoft with a consortium of independent repository vendors;

(f) A host of application domain or business domain initiatives such as: numeric and statistical data [19], geospatial information [20], music [21], works of art [22] (and, because in this cultural heritage area there are several standards; a useful crosswalk is provided at [23]), scientific metadata [24] [25], biosciences [26], healthcare [27], education [28] and a host of others. Digital library metadata has already been mentioned [5], [6];

(g) A major use of metadata is in electronic business: the UN (United Nations) EDI standard [13] is widely adopted and the XML/EDI initiative [29] [30] is gaining popularity. ICE (Information and Content Exchange) [31] is being implemented and utilises various security features based on W3C initiatives such as P3P [32] which is an example of associative restrictive metadata.

It is unclear exactly how these initiatives will develop and inter-relate. Some are proprietary, and there are parties with commercial or other interests in the groups defining open standards. Many of these application initiatives concern data exchange, but increasingly there are groups working on the underlying associative supportive metadata in the form of terminological thesauri or domain ontologies. The latter developments are particularly significant because such resources provide maximum flexibility for systems built using cooperating intelligent agents e.g. [33] and also provide greater support in both query refinement and answer integration and explanation e.g. [10].

It should be noted that the Information Systems Engineering community has utilised metadata for many years in attempts to improve systems construction management and systems maintenance. The major objective was to have well-understood communication between designers but also the metadata was used to drive tools which assisted in systems engineering. An early attempt was the extension of schema metadata with the IRDS (Information Resource Dictionary System) [34] followed by several attempts - such as Conceptbase [35] - to capture metadata for the purpose of describing systems.

4.3 Systems Using Metadata

Systems utilising metadata are similarly diverse. Basically, they may be classified into:

(a) systems with extensive human interaction to make choices based on metadata information (e.g. web browsing, use of web portals or query refinement systems accessing heterogeneous information sources [10])

(b) systems relying on profiles input by the client-user and the server(s) which then are used by mediating agents (e.g. electronic business systems utilising P3P for security [32], or CORBA-based systems accessing compliantly-wrapped information sources [33]);

(c) totally automated systems (e.g. automated sensor systems in scientific experiments or regular data exchange between earth observation devices).

5 Future

5.1 Metadata

Metadata has moved centre-stage as the most important component of the solution to the application requirements of the architecture and construction of modern information systems. Most modern systems are web-based, either within the organisation (Intranet) or public. In the latter case, especially, metadata is utilised to improve communication between heterogeneous information systems – for the purposes of obtaining and providing information, for communication between the user client workstation and the information servers and for electronic business between information systems.

The concept of separating the primary information resources from data and processes (metadata system) providing access to those resources is extremely important. This allows changes of access policy – such as changes in access restrictions for certain kinds of users in certain roles, changes in categorisation and classification and changes (additions) in descriptive metadata depending on viewpoints of different authorised users – without accessing the data resource itself.

5.2 Scale

The rapidly expanding internet community, and the ever increasing demand for services – largely WWW-based – demands that solutions must be scalable. Ever-increasing computer power, storage capacities and networking speeds only mitigate the problem – the expansion and consequential demand outstrip supply of the technological services. The technology, however, has predicted limitations varying from the need to develop a technology other than CMOS for processors through the need to develop faster and denser storage devices to the need for provision of inexpensive and faster communications technology than even that based on fibre.

Thus the solution must lie with better systems engineering – the 'brute force' methods will not provide the whole solution.

A major component of that systems engineering solution has to be intelligent utilisation of resources. This implies better refined queries, better constructed databases, better utilisation of distribution and parallelism for algorithms acting on data resources, and better concurrency. For all of these aspects quality metadata – accessed and used by intelligent agent technology, is the basis for the solutions.

5.3 GRIDs

5.3.1 Background

There has emerged through 1998-1999 in North America the concept of a Computation Grid [37] closely followed by the same concept in Europe [38]. In UK the concept of GRIDs was first articulated completely in the summer of 1999 (but was dependent on much internal work before that finding its roots in the Distributed Computing Systems programme of the late 1970s) and captured succinctly in [39] which described the 3-layered Computation / Data, Information and Knowledge grids architecture as proposed by the author. By September 1999 the North American community had also considered data access [40] and overall architecture [41] so moving from computation (linking supercomputers for compute-power) to the world of data. The 'Grid Bible' published in July 1998 gives some flavour of the challenges [37] although rooted in computation. However, the North American and European architectural view is less comprehensive than that in [39] which overviews underlying detailed considerations of access, security and rights strategies as well as uniform information access over heterogeneous sources and a uniform computation landscape.

The UK view of GRIDs has been driven by requirements in science, engineering and technology and is being promoted through the UK Government Office of Science and Technology with the label 'e-Science'. It is expected that this pull will lead to solutions later (but quickly) applicable to general commercial and business processes, especially e-Commerce. The author has coordinated a meeting of leading UK academics and industrial representatives who endorsed enthusiastically the architecture and who are now working with scientists in the application areas to refine specific requirements and implement component GRIDs systems. An early application will be the management of data streaming from the LHC (Large Hadron Collider) at CERN (Centre Europeen pour la Recherche Nucleaire) in Geneva where it is proposed that each member country will need to support a large data centre with data cascading to its scientists. In UK the GRIDs architecture will be used. Similarly, UK groups working on biosciences – and especially genomics, environmental systems, advanced materials science, engineering modelling and social science systems are active.

5.3.2 GRIDs and Metadata

The architecture envisaged by the UK community attempts to bring together the (upward) refinement of data to information and knowledge and the (downward) application of knowledge to information handling and data collection through feedback loop control (Figure2 : GRIDs Architecture). The computation / data grid has supercomputers, large servers, massive data storage facilities and specialised devices and facilities (e.g. for VR (Virtual Reality)). The main functions include compute load sharing / algorithm partitioning, resolution of data source addresses, security, replication and message rerouting. The information grid resolves homogeneous access to heterogeneous information sources. The knowledge grid utilises knowledge discovery in database technology to generate knowledge and also allows for representation of knowledge through scholarly works, peer-reviewed (publications) and grey literature, the latter especially hyperlinked to information and data to sustain the assertions in the knowledge [7].

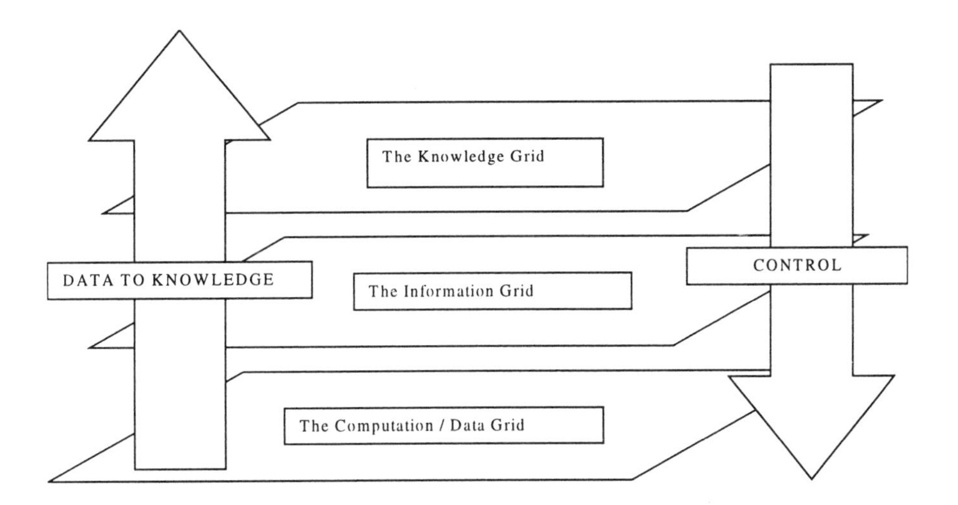

Figure 2: GRIDs Architecture

The concept is based on the idea of a uniform landscape within the GRIDs domain, and external appliances - ranging from supercomputers, storage access networks, data storage robots, specialised visualisation and VR systems, data sensors and detectors (e.g. on satellites) to user client devices such as workstations and WaP enabled Mobile phones. The connection between the external appliances and the GRIDs domain is through agents representing the appliance (and thus continuously available to the GRIDs systems). These representative agents handle credentials of the end-user in their current role, appliance characteristics and interaction preferences (for both user client appliances and service appliances), preference profiles and associated organisational information. These agents interact with other agents in the usual way to locate services and negotiate use. The key aspect is that all the agent interaction is based upon available metadata.

5.3.3 GRIDs , the CAiSE Conference Series and Janis Bubenko

Key functionalities in the GRIDs will be based on computer science results (some stretching back 30 years or more) in database, hypermedia, visualisation & VR, information retrieval, document systems, workflow-supported processes, knowledge management (including inferencing and dealing with uncertain and incomplete information), knowledge discovery in database (including data scrubbing, warehousing and mining) cooperative working and related topics. The building of the GRIDs will be a continuous, incremental systems development process based on results from the CAiSE conference series among others. The unique feature of the project is the size, complexity and open aspects of this systems development exercise, which has the problems also of facilitating interoperation of legacy systems. It is noteworthy that many of the technologies being used are based on the work of Janis Bubenko, especially in information systems modelling and systems development methods.

Acknowledgments

The metaphors of metadata as the Rosetta Stone and as a Babel fish come from a discussion with Chris Pound from BT IT Strategy Group not long before his tragic early death in late 1999. Its inclusion is a small tribute to his ability and humour.

I should like to acknowledge the ideas, stimulation and encouragement from key co-workers on major metadata projects with which I have been involved, including: Liz Gill on the environmental data integration projects 1973-1978, Ray Pollard and Duncan Collins on the Oceanographic projects 1974-1978, Pete Sutterlin and Liz Gill of the Filematch team (geological data exchange) from 1976-1977 and associated work with Gordon Williams, the IRAS (Infra-red astronomy satellite) team 1975-1980, John Hart from the High Energy Physics Book-keeping project of 1980-1984, Fulvio Naldi and Sam Zardan on the IDEAS and EXIRPTS (research documentation) projects 1983-1990, the European Commission CERIF 1991 standardisation group 1988-1991 especially Fulvio Naldi and Jostein Hauge, Carole Goble on the semantic constraints (healthcare) project 1992-1994, Jana Kohoutkova on the Hypermedata (medical systems interoperation) project 1995-1998, the European Commission ERGO project and CERIF revision project teams 1996-1999 especially Anne Asserson and Eric Zimmerman, the RDF Working Group of W3C 1996-1999, Anne Asserson and Hana Konupek from the Norwegian electronic thesis project (1998-) and Nigel Turner of BT (1999-).

Discussions with numerous colleagues at the CAiSE conference series initiated by Janis Bubenko and Arne Solvberg has provided much inspiration for which I am extremely grateful.

Colleagues in my 'home team', working on many aspects of metadata for many years, deserve much credit – especially Judy Lay on the electronic library project 1981-1983 and the IDEAS and EXIRPTS (research documentation) projects 1983-1990, Michael Wilson on the MIPS project 1992-1995, Jan van Maanen (STEP/EXPRESS projects over many years) and Brian Matthews (RDF/XML). Simon Dobson and Brian Matthews have worked closely with me on refinements of the detailed GRIDs architecture.

References

1. Pound C. Personal communication 1999

2. Adams D. The Hitch-Hikers Guide to the Galaxy. Pan Books, London, 1979
3. http://www.nist.gov/sc4/www/stepdocs.htm
4. http://www.chime.ucl.ac.uk/HealthI/GEHR/Deliverables.htm
5. http:/lcweb.loc.gov/marc/marc.html
6. http://purl.oclc.org/dc/
7. Jeffery, K G: ‘An Architecture for Grey Literature in a R&D Context’ Proceedings GL'99 (Grey Literature) Conference Washington DC October 1999 http://www.konbib.nl/greynet/frame4.htm
8. http://www.w3.org/TR/1999/REC-rdf-syntax-19990222
9. http://www.w3.org/TR/1998/REC-xml-19980210
10. Jeffery,K G; Hutchinson,E K; Kalmus,J R; Wilson,M D; Behrendt, W; Macnee, C A: 'A Model for Heterogeneous Distributed Databases' Proceedings BNCOD12 July 1994; LNCS 826 pp 221-234 Springer-Verlag 1994
11. http://www.w3.org/
12. http://www.oasis-open.org/cover/xmi.html
13. http://www.harbinger.com/resource/edifact/
14. http://www.w3.org/PICS/
15. http://www.omg.org/
16. http://www.omg.org/uml/
17. http://www.omg.org/cgi-bin/doc?ad/99-06-05
18. http://msdn.microsoft.com/repository/technical/xif.asp
19. http://www.icpsr.umich.edu/DDI/
20. http://www.opengis.org/
21. http://www.oasis-open.org/cover/smdlover.html (music)
22. http://www.getty.edu/gri/standard/cdwa/HOMEPAGE.HTM
23. http://www.getty.edu/gri/standard/intrometadata/crosswalk.htm
24. http://www.getty.edu/gri/standard/intrometadata/crosswalk.htm
25. http://www2.echo.lu/oii/en/science.html#science
26. http://www.nbii.gov/tools/non-spatial.html
27. http://www.mcis.duke.edu/standards/HL7/sigs/sgml/index.html
28. http://www.manta.ieee.org/p1484/
29. http://www.geocities.com/WallStreet/Floor/5815/index.html
30. http://www.geocities.com/WallStreet/Floor/5815/guide.htm
31. http://www.w3.org/TR/NOTE-ice
32. http://www.w3.org/TR/P3P/
33. http://img.cs.man.ac.uk/tambis/index.html
34. http://burks.bton.ac.uk/burks/foldoc/72/56.htm
35. http://www-i5.informatik.rwth-aachen.de/CBdoc/index.html
36. http://www.privaseek.com/
37. Foster I and Kesselman C (Eds). The Grid: Blueprint for a New Computing Infrastructure. Morgan-Kauffman 1998
38. http://www.egrid.org/
39. http://www.itd.clrc.ac.uk/Activity/GRIDs
40. http://www.sdsc.edu/GridForum/RemoteData/Papers/architecture.html
41. http://www.gridforum.org/building_the_grid.htm

Intention Driven Component Reuse

Colette Rolland
Université de Paris 1 Panthéon Sorbonne
Paris, France
rolland@univ-paris1.fr

Abstract

Despite the growing acceptance of component based systems, there is inadequate understanding of the requirements engineering process to support component selection and assembly. In contrast to the current approach which starts from components and finds the best fitting ones to meet requirements, we propose here to start from requirements and select/assemble components that meet these. We adopt the intention driven approach where customers express their requirements in terms of tasks that are to be supported and component suppliers are invited to express component features in terms of the organisational tasks they support. This paper proposes the notion of a map as a means for expressing both, customer requirements and component features, introduces the matching process, and exemplifies it by the acquisition of a COTS product in the electricity supply industry.

1.Introduction

In the hope of reducing risks and costs associated with software development, organisations that rely on software systems are increasingly shifting from bespoke development to component based systems. Components can be commercial off the shelf software (COTS) or complex ones (CCOTS) such as ERP systems which are developed for a whole market and therefore have to satisfy a wide range of requirements. They can be predefined solutions to generically identified problems and then, take the form of patterns. Indeed, although patterns have been applied for some time to the software engineering domain, and in particular to object technologies domain [5] [4] [12] [11], there is now a significant application in the organisational and business sphere [6] [7] [13].

In all cases, from a requirements point of view, the situation is difficult [10] [17][16]. First, to be reusable, components are generic. This raises the need to build variations in order to cope with the specificity of the problem at hand. Secondly, to understand which variation is required it is necessary to match the customer requirements to the features provided by components. Imagine a specific feature required by the customer, e.g. the system shall accept data x and as a result produce data y. Some component suppliers may provide the feature as a standard part of their product, others may provide it as a tailor-made extension, and still others may not have that feature but suggest that the customer use two other features to obtain essentially the same result. We see the problem in two parts : (a) how to express customer requirements and (b) how to describe component features in order to facilitate the matching process?

As shown in Figure 1, there are two broad approaches that can be followed in matching requirements and components. One is the component driven approach which starts from components and finds the best fitting ones to meet requirements. When doing this, it is necessary to align requirements with any assumptions about business processes, organisational structures, methodologies etc. made in these components. This is the prevalent approach. The alternative approach shown in the Figure is the requirements driven one. Here, customer requirements are repeatedly refined and components meeting these refined requirements are located. This approach has the potential to bring about a natural alignment between requirements and components and is therefore likely to find greater customer acceptance.

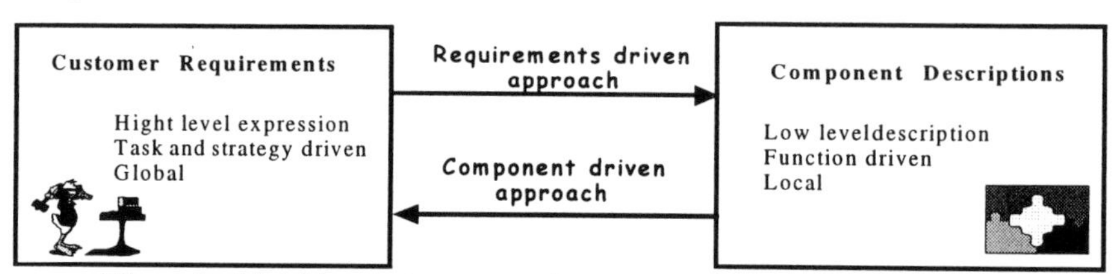

Figure 1 : Matching customer requirements and component features

We investigate the requirements driven approach in this paper. We propose to use the same uniform representation system for customer requirements description and component description. As a consequence, there is a smooth transition from the former to the latter.

Customers like to express high level requirements in terms of goals, intentions and strategies in opposition to component features described in terms of low level functionality such as screen pictures, input/output flows and data structures. Customers are interested by setting the context in which the suitable assembly of components will be used instead of focusing on product features. Besides they favour a global view of the context whereas a component description provides a local functional view. Our view is that it is difficult to understand the features of a product without understanding its usage context. Therefore, we choose a representation means suitable for high level description of requirements.

In essence this representation allows customers to specify their requirements in terms of tasks and strategies to perform them and invite component suppliers to describe components features in terms of how they are able to support these tasks and their associated strategies. We have chosen to describe both requirements and components as *maps*. The map uses two fundamental notions, *intention* and *strategy*. An intention captures the notion of a task whereas the strategy is the manner in which the intention can be achieved. The nodes of the map are intentions whereas its edges are labelled with strategies. The directed nature of the map identifies which intention can be done after a given one.

When used to express customer requirements, the map provides a strategic view of what is expected from the component based system in terms of organisational tasks to support and strategies to achieve them. When used to describe component features, the map provides a strategic view of tasks and their strategies that the component supports. The component map describes the tasks that individual components support as well as that when components are assembled together. The former is captured in the intentions of the map whereas the latter is captured in the flows from intention to intention via strategies. This allows the matching process to reason about the fit of requirements to each component considered separately and their assembly as well.

Clearly the map representation system takes a teleological view that favours the expression of requirements at the intentional level of tasks and strategies to achieve them and the description of component features in terms of how they support these tasks and their associated strategies. We believe that this approach shall lead to an increased understanding of both requirements and component features and a better coverage of customer demands.

Furthermore, maps can be defined recursively, a map can be refined in several other maps. This shall help mastering the matching process step by step. If a component meets the requirements at a level *i*, a refined view through a map at *level i+1,* shall help identifying finer grained matching or the absence of a complete match, and consequently, the required adjustments and adaptations of components and/or the development of missing components.

The remainder of the paper is organised in two main sections. The next section first introduces the notion of a map and then shows, with the help of an example, how it can be used to describe and assemble components. In section 3 we develop the process of matching maps as part of a larger RE process. This process is then illustrated by an example in the electricity supply domain. Conclusions and suggestions for further work are presented in section 4.

2. Maps and Components

In this section we introduce the notion of a map and apply it for describing components and their assembly. We also show that a map forms the basis of the process of aligning customer requirements to components features.

2.1 Introducing the Map

In general terms a map is a process model in which a non-deterministic ordering of *intentions* and *strategies* has been included. It is a labelled directed graph with intentions as nodes and strategies as edges between intentions. The directed nature of the graph shows which intentions can follow which one.

An *intention* is a goal that can be achieved by the performance of the process. It refers to a task that is part of the process and is expressed at the intentional level. Each map has two special intentions, *Start* and *Stop*, to start and end the process respectively.

A *strategy* is an approach, a manner to achieve an intention. The strategy S_{ij} characterises the flow from the source intention I_i to the target intention I_j and the way I_j can be achieved.

A *section* is the key element of a map. It is a triplet $<I_i, I_j, S_{ij}>$ and represents a way to achieve the target intention I_j from the source intention I_i following the strategy S_{ij}. Each section of the map captures the specific manner in which the task associated with the target intention can be performed. Sections of a map are *connected* to one another. This occurs (a) when a given task can be performed using different strategies. This is represented in the map by several sections

between a pair of intentions. Such a map topology is called a *multi-thread.* (b) when a task can be performed by several combinations of strategies. This is represented in the map by a pair of intentions connected by several sequences of sections. Such a topology is called a *multi-path.* In general, a map from its *Start* to its *Stop* intentions is a multi-path and may contain multi-threads.
It is possible to *refine* a section of a map at level *i* into an entire map at a lower level *i+1* to view a task together with its strategy as a complex graph of subtasks and their associated strategies. Refinement as defined here is an abstraction mechanism by which a complex assembly of sections at level i+1 is viewed as a unique section at level *i.* Since refinement results in a map, it produces multi-path/multi-thread structures at level *i+1.* As a result, a pair of subtasks can be connected through several strategies and different paths are provided to support different combinations of strategies and subtasks. Thus, a level *i* section is a more complex structure than a simple composition structure in that it provides (a) several alternative decompositions of the aggregate structure into its constituents *Ci* and (b) different alternatives to *Ci* constituents.

The key characteristics of the map are threefold
- Tasks are expressed intentionally, thus avoiding the bother of the details of task implementation.
- Strategies are made explicit, thus showing the different alternative ways of achieving a task.
- The graph includes a multiple assembly of tasks (through multi-path) with possibly multiple alternative ways of achieving tasks (through multi-thread) to reach the same result.

2.2 Maps As Component Descriptions

Our position is that the use of a map for describing complex assemblies of components facilitates the discovery of components matching customer requirements. This is achieved by simply relating each section in a map to a component. We illustrate this association in the following.
Consider *'Electricity Supply Management'* (**ESM**), a COTS product to support electricity distribution processes. Its map shown in Figure 2 is organised around two key intentions, *"Serve Customer Request"* and *"Sell Electricity".* In conformity with our approach, these two intentions have been selected to represent the organisational goals that can be achieved by using the product. Further, these intentions are generic in the sense that they exist in any electricity distribution process. Furthermore, the map indicates an ordering constraint: in order to sell electricity to a customer, his/her request for electricity provision has to be fulfilled first.
In the **ESM** map, it shall be noticed that there are two different strategies to achieve each of these two intentions. For example, the *"Advance Payment strategy"*, and the *"Credit strategy"*, are two alternative strategies to achieve the organisational goal " *Sell Electricity".*
On one hand, these map strategies identify two rather different business strategies to get the customer to pay for his electricity consumption. Indeed the *"Advance*

Payment strategy" refers to a solution based on the use of payment cards to energise the customer meter whereas the *"Credit strategy"* refers to the more conventional solution where the electricity company provides electricity to its customer and gets paid after consumption. On the other hand, these two map strategies identify two variations of the COTS product that can be selected depending on the situation at hand. Each variation is simply captured in a map section. The triplet <*Serve Customer Request, Sell Electricity, Credit stratgy*> is an example of section in the ESM map which corresponds to one COTS variation to "*Sell Electricity*".

More generally, every section in a map is associated with a component. Therefore, the map can be seen as a means to explain how a complex component is made of other components and in which way components co-operate to achieve collectively an organisational goal.

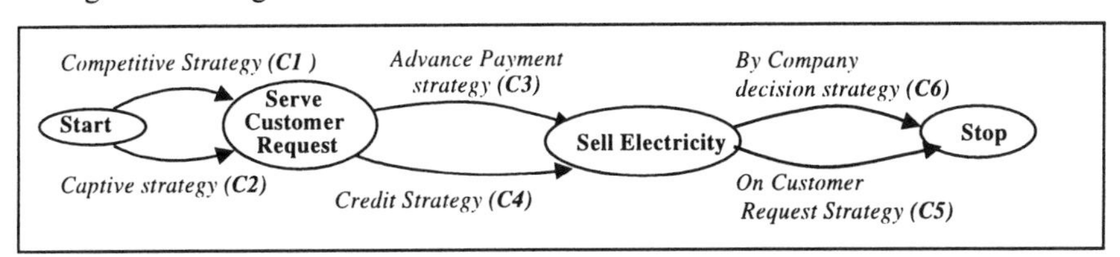

Figure 2 : The «Electricity Supply Management » Component Map

Table 1 briefly describes the six components of the ESM COTS product associated with the six sections of the map as indicated (with component's reference) in Figure 2. The description of the six components in Table 1 shows that each component has an *interface* and a *body*. The body is what the component really does whereas the interface is the visible part of the component. The body of the component C1 for example, provides IT support to install an intelligent front desk to deal with the various customer requests. Its interface is a couple <situation, intention> stating the precondition for the component to be used (the situation is that the current state of the company has been modelled in a so-called As-Is model) and the intention that can be fulfilled in that situation (to "Serve Customer Request") following a given strategy, namely the *"Competitive strategy"*.

Thus, there is a tight connection between the components and the map :

(a) each section in the map is associated to a component,
(b) the interface intention of the component is the target intention of the section completed by the name of the section strategy and,
(c) the interface situation refers to the state resulting from the fulfilment of the section source intention.

Therefore, the map provides a strategic view of what components can achieve both individually and when assembled together. The former is captured in a section of the map and the related component interface whereas the latter is captured in the flows from intention to intention via strategies.

Ref	Component Name	Component Interface	Component Body
C1	Customer servicing in a competitive environment	<(As-Is model), Serve Customer Request with Competitive strategy>	Provides IT support to -install an IFD to serve customer request -develop customer culture within the company and measure customer satisfaction -contract customers -market the company
C2	Customer servicing in a captive environment	<(As-Is model), Serve Customer Request with Captive strategy>	Provides IT support to -handle customer requests -keep track of customer complaints -manage customers and customer installations
C3	Electricity selling in an open market	<(Customer Id), Sell Electricity with Advance Payment strategy>	Provides IT support to install card based meters and keep track of their use
C4	Electricity selling in a conventional way	<(Customer meter Id), Sell Electricity with Credit strategy>	Provides IT support to manage the process chain of conventional meter reading, electricity consumption billing and payment collection (see Figure 3)
C5	Stopping electricity provision on customer request	<(Customer Id), Stop on Customer Request strategy>	Provides IT support for customer disconnection on customer request
C6	Stopping electricity provision on company decision	<(Customer Id), Stop by Company Decision strategy>	Provides IT support for customer disconnection by company decision

Table 1 : List of ESM components

The view adopted here is strategic in the sense that it abstracts from the details of intention achievement through task performance and emphasises the manner in which the task will be performed. As we will see later, this makes it easier to focus component based requirements engineering on intention matching (to understand if a component fits the organisational task it is aimed at supporting) and strategy matching (to understand if the strategy to assemble components in order to meet a global objective is valid or not). We believe that this is the most effective way to align customer requirements to component selection and subsequent assembly and is in conformity with usage driven requirements engineering as found in goal driven RE [2] [3] [21] [9] [1] and scenario based approaches [8] , [15] [19][22].
Strategic reasoning at different levels of abstraction is supported by section refinement introduced in 2.1. Through this mechanism, a section in a map i.e. a component, might be itself deployed as a map. This is illustrated in Figure 3 with the **ESM** section *<Serve Customer Request, Sell Electricity, Credit strategy>*, i.e. the C4 component.

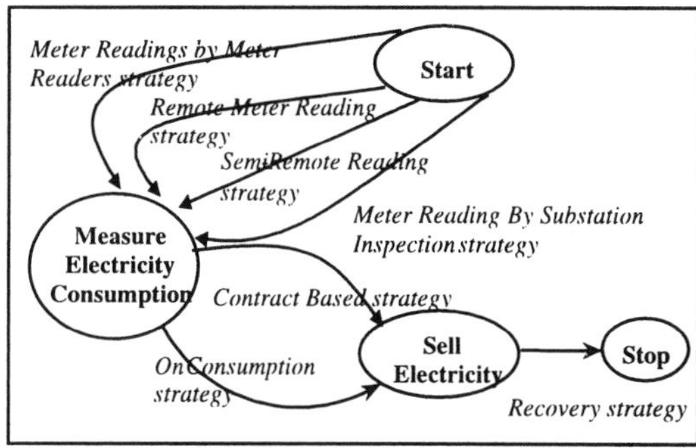

Figure 3 : Refinement of the C4 Component

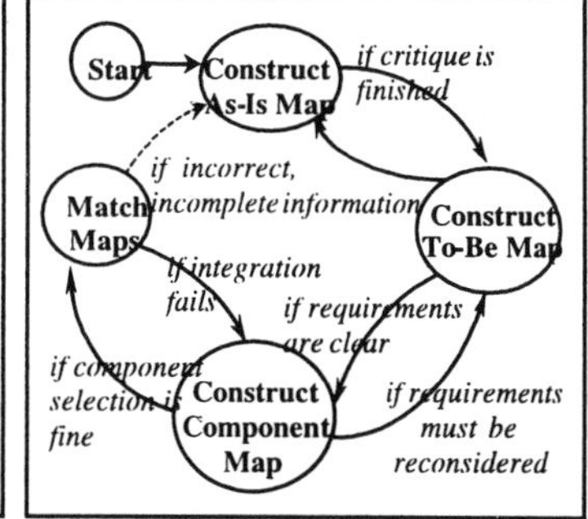

Figure 4 : The component based RE process

Section refinement helps to select the right customisation of the C4 component. At the level of abstraction in Figure 3, C4 is a multi-thread map presenting several strategies to achieve "*Measure Electricity Consumption*" and "*Sell Electricity*" respectively. Customisation could be to use classical "*Meter Reading by Meter Readers*" or a completely innovative one based on "*Remote Meter Reading*". The former is based on manual readings taken by meter readers visiting customer sites whereas the latter is automatically done but requires special meters to be installed. Customisation of "*Sell Electricity*" could similarly be driven by a choice of "*Contract Based strategy*" or "*Consumption Based strategy*".

When viewed as a whole, the map of Figure 3, is a multi-path and comprises several paths from *Start* to *Stop*. Each of these is a combination of two strategies, one to "*Measure Electricity Consumption*" and the other to "*Sell Electricity*". The complete customisation therefore involves the selection of one out of several strategy combinations.

The foregoing shows that customisation in our approach is strategy driven and consists of selection of a strategy in every multi-thread and a combination of strategies within a multi-path.

To sum up

(a) The map provides a uniform description of components at any level of complexity.
(b) The map establishes a relationship between component features and the tasks to be supported in the organisation.
(c) The map, through alternative strategies to achieve a given task, makes. component variants explicit.
(d) Map recursion allows customisation at different levels of abstraction.
(e) Component selection is strategy driven.
(f) Component assembly is combination-of-strategy driven.

3. Maps for matching

In this section we show how maps can help in matching component features to user requirements. In the approach user requirements and component features are both expressed as maps. Therefore the component based RE process performs the matching of these maps. The process is outlined in 3.1 and illustrated in 3.2 with the Electricity Supply Management COTS product.

3.1 Overview of the component based RE process

The process of matching maps is part of the larger view of component RE process as a change handling one [14]. The change handling process creates a movement from an existing situation captured in the As-Is model to a new one reflected in the a set of user requirements for the future captured in the To-Be model. While accepting this position, our intentional reasoning approach leads us to express these models as maps. This is reflected in Figure 4 by (a) *Construct As-Is Map* and (b) *Construct To-Be Map*. We extend this basic position with our matching process for matching component features with customer requirements. This is reflected in Figure 4 by (c) *Construct Component Map* and (d) *Match Maps*. Notice that processes (a) and (b) are in conformity with change handling, (c) and (d) have been specifically introduced to handle the map matching process. However, all these four processes use maps uniformly as the means of knowledge representation.
The *As-Is map* resulting from (a) abstracts from the organisation current practice to describe the currently achieved goals/requirements. It serves as a support for critiquing the current situation and thereby identifying customer requirements for the *To-Be map*. Additionally, it also serves as a reference to evaluate the new solution against current practice. The *To-Be* map reflects the customer requirements that the organisation would like to satisfy by acquiring and assembling a suitable set of components. These requirements are expressed in terms of tasks and their strategies that are to be supported by the assembled set of components. The *Component map* specifies component features in terms of tasks and their strategies that the component supports. Finally, *Match map* produces a map that meets user requirements by integrating together component maps. It may be the case that off-the-shelf components are not able to fully perform the required tasks according to the required strategies. In such .a case, *Match map* helps in the identification of any new components that may have to be built to fill in the missing gaps. The Figure shows the high amount of iteration in the process. Iteration may be because more information about requirements is needed or because *Match map* produces a map that does not exactly meet user requirements.

3.2 An Example

This example deals with the acquisition of a COTS product to support electricity supply management in an anticipated deregulated environment. Figure 5 shows the constructed As-Is map which represents the way in which customers are currently serviced. This is achieved by a sequence of four intentions. In order to "*Provide Connection*" customers visit the companies offices. This leads to long customer queues. It is also found that providing a connection itself takes too long. Once a connection is provided, "*Billing Customer Consumption*" of electricity is based on "*Reading Electric Meter*". The latter is done by meter readers according to a pre-defined plan of meter reading. It is found that there are too many errors in the bill and bills are inordinately delayed. In order to "*Collect Payment*" the company offers to its customers a number of strategies. Three of these are according to the conventional cash payment mode. Payments made to Post offices and Private offices are too expensive and not efficient. Finally, electricity supply can be "*Stopped*" either on customer request or by company's decision.

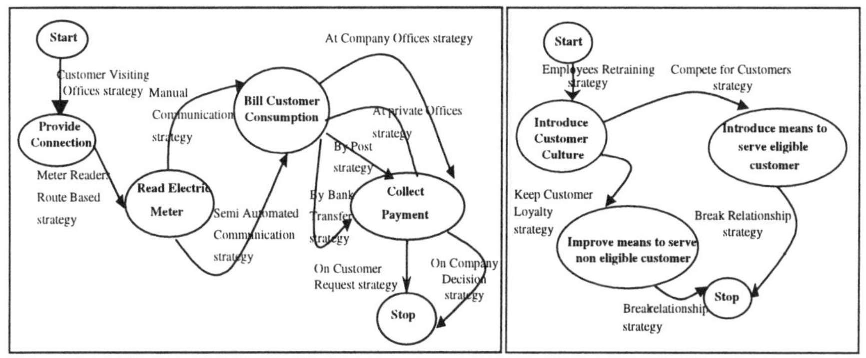

Figure 5 : The As-Is map **Figure 6 : The To-Be map**

Under the anticipated deregulation, the company needs to become more customer-friendly. This is reflected in the To-Be map shown in Figure 6 by the intention *"Introduce Customer Oriented Culture"* achieved through the *"Retraining Employees strategy"*. The deregulation introduces two kinds of customers, those who have the choice of their electricity supplier (called eligible customers) and those who do not (the non-eligible customers). The company must re-orient its processes to win over as many eligible customers as possible. This is reflected in the To-Be map by the intention *"Introduce Means To Serve Eligible Customer"* under the "*Compete For Customer strategy*". To protect its customer base for the future the company has to serve its non-eligible customers better. This leads to the intention "*Improve Means to Serve Non-eligible Customers*" under the "*Keep Customer Loyalty strategy*".

The **ESM** COTS product introduced in Figure 2 was selected as a good candidate to meet the requirements. This selection was based on its description in Table 1. The matched map resulting from the *Match Maps* process is shown in Figure 7. It is obtained by a customisation and integration of the To-Be and **ESM** maps.

Starting from the To-Be map, the **ESM** map is examined and it is found that the two sections from *Start* to "*Serve Customer Request*" are driven by strategies which are very close to the ones in the To-Be map for serving eligible and non-eligible customers. This leads to the acceptance of "*Serve Customer Request*" and its associated strategies. A detailed examination of the strategies now leads to the rejection of the "*Captive strategy*" and also shows that the "*Competitive strategy*" is a bundle of "*Marketing strategy*", "*Change Culture strategy*", "*Contracting strategy*", and "*Intelligent Front Desk (IFD) strategy*".

The "*Change Culture strategy*" to "*Serve Customer Request*" can obviously be related to the intention "Introduce Customer Oriented Culture" in the To-Be map. This confirms the need of the section *<Start, Serve Customer Request, Change Culture strategy>* in Figure 7. The "*Marketing strategy*" and the "*Contracting strategy*" to "*Serve Customer Request*" are not explicitly expressed in the To-Be map. Rather they are suggested by the COTS map. Evidently "*Introduce Customer Oriented Culture*" requires both marketing and new forms of contracting. In the former case the company is marketed to potential customers whereas in the latter contracts are developed to match the requirements of specific customer-profiles and are offered to them. This leads to the sections *<Start, Serve Customer Request, Marketing strategy>* and*<Start, Serve Customer Request, Contracting strategy>* in Figure 7. The adoption of the IFD strategy suggests special treatment for some important customers leading to *"IFD at Customer Premises strategy"* complementary to *"IFD at Company Premises strategy"*. Then, the two sections *<Start, Serve Customer Request, IFD at Company Premises strategy >* and *<Start, Serve Customer Request, IFD at Customer Premises strategy >* are included in Figure 7. Whereas the other strategies promote customer-orientation, the IFD strategies directly contribute to reducing connection delays and customer queues at offices. These were some of the difficulties of the current practice discussed earlier. The rejection of "*Captive strategy*" is motivated by the anticipation that non-eligible customers will not stay captive very long. Consequently, it is best to treat them in the same way as eligible customers.

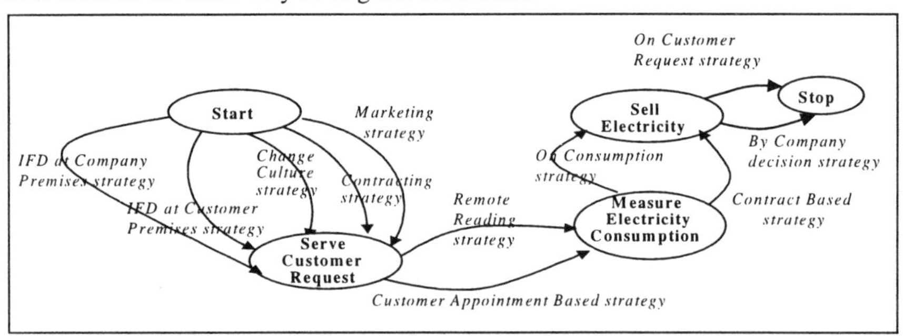

Figure 7 : The matched map

The **ESM** COTS product contains the "*Sell Electricity*" intention as well. An examination of this intention shows that it is dealing with electricity consumption, billing and payments. These are integral parts of customer servicing as can be confirmed from the As-Is map. In order for the To-Be map to be compliant with the As-Is map, it is clear that the intention "*Sell Electricity*" must be retained in the solution. The COTS offers two strategies for this intention. The "*Advance Payment strategy*" is found to be difficult to implement due to customer resistance to this change and high cost of replacing old meters with new ones. Therefore, the "*Credit strategy*" is accepted.

The matching process proceeds with the examination of the refinement of the section *<Serve Customer Request, Sell Electricity, Credit strategy>*. The relevant map is shown in Figure 3. Figure 7 contains the two intentions "*Measure Electricity Consumption*" and "*Sell Electricity*" found in Figure 3 as well as the multi-thread between these two. This leads to the two sections *<Measure Electricity Consumption, Sell Electricity, Contract Based strategy>* and *<Measure Electricity Consumption, Sell Electricity, On consumption strategy>*. Out of the various strategies for "*Measure Electricity Consumption*" in Figure 3, only *the "Remote Reading strategy*" is selected. The strategy is innovative and implies a large investment as all current electric meters have to be replaced by new ones. However the company believes that this investment is worse doing as it will give her the possibility to sell new services such as remote heating or cooling to its

customers; therefore providing a competitive advantage to its competitors. Besides, remote meter reading will avoid the mistakes in billing as well as the delay in billing which are critiques emerging from the current practice. On examination, it was found that remote reading would not work in all cases and a completely new *strategy "Customer Appointment Based strategy"* is introduced. This leads to the two sections *<Serve Customer Request, Measure Electricity Consumption, Remote Reading strategy>*, *<Serve Customer Request, Measure Electricity Consumption, Customer Appointment Based strategy >* in Figure 7.
Lastly, it can be noticed that in Figure 7 the strategies to *Stop* are those that are contained in the ESM map since these are found to be compliant with the As-Is map.
The foregoing example brings out that :

- The As-Is is useful since it provides domain knowledge which helps in understanding differences, finding incompleteness, reusing solutions and evaluating the future solution against critiques of the current practice.
- New strategies can be introduced in the matched map directly. These are not supported by any component and have to be developed in-house. Though not brought out by the example, it is possible for this to happen for intentions as well.
- Initial requirements in the To-Be map are refined and new ones discovered while examining functionality offered by components.
- The matching process might end up suggesting several alternative matched maps. This might occur when the decision for component selection and/or assembly requires further evaluation such as cost/benefit analysis, risk analysis etc.

4. Conclusion

In this paper we presented the concept of a map as a means to uniformly describe customer requirements as well as off-the-shell component features. The aim was to replace the prevalent functional requirements with intention and strategy graphs, so that customers can express their needs in terms of intentions and strategies to achieve them and also component suppliers could describe how assemblies of components support these intentions and their strategies. The map serves as a basis for the requirements engineering process to match both individual components and their assembly against requirements thus avoiding the selection of individual components that, when assembled together, do not meet the global need of the customer. The requirements engineering process is viewed as a change process where the As-Is model is also a map, therefore making possible to evaluate the component based final solution to the current practice.
The next step of our work is the development of process guidance (a) to construct a map and check its correctness and (b) to support the matching process. We are defining a set of operators to formalise the integration of maps [18] that were found useful in initial experiments. A set of situated rules [20] which will be fired to 'situate' guidance are being specified. A future step is also, in a longer term to abstract from experience a collection of matching patterns that could be made publicly available.

5. References

1. A.I. Anton, *Goal based requirements analysis*. Proceedings of the 2nd International Conference on Requirements Engineering ICRE'96, pp. 136-144, 1996.
2. J. Bubenko, C. Rolland, P. Loucopoulos, V De Antonellis *Facilitating 'Fuzzy to Formal' requirements modelling*. IEEE 1st Conference on "Requirements Enginering " (ICRE'94), pp. 154-158, 1994.
3. J. Bubenko Jr., Marite Kirikova. *Worlds' in Requirements Acquisition an Modelling*, 4th European - Japanese Seminar on Information Modelling and Knowledge Bases, Kista, Sweden, Kangassalo and Wangler (Eds.), IOS (pub),1994.
4. J. Coplien, D. Schmidt (eds.) *Pattern Languages of Program Design*, Addison Wesley, Reading, MA, 1995.
5. P. Coad *Object-Oriented Patterns*, Communications of the ACM, Vol. 35, No. 9 (Sep. 1992), pp. 152-159.
6. J. Coplien *A Development Process Generative Pattern Language*, AT&T Bell Laboratories, WWW Publication, http://www.bell-labs.com/people/cope/Patterns/Process/index.html, 1995.
7. J. Coplien *A generative Development - Process Pattern Language*, in 'Pattern Languages of Program Design', J. O. Coplien and D. O. Schmidt (ed.), Addison-Wesley, pp. 183-237.
8. A. Cockburn, *Structuring use cases with goals*,http://members.aol.com/acocburn/papers/usecases.
9. A. Dardenne, A. van Lamsweerde, S. Fickas, *Goal directed requirements acquisition*. Science of Computer Programming, 20(1-2), pp.3-50, April 1993.
10. A. Finkelstein., G. Spanoudakis, M..Ryan *Software package requirements and procurement*. In : Proc. of the 8th International workshop on Software Specification and Design, IEEE Computer Society Press, Washington, DC, 1996, pp 141-145.
11. M. Fowler *Analysis Patterns: Reusable Object Models*, Addison-Wesley, 1997.
12. E. Gamma, R. Helm, R. Johnson, J. Vlissides, *Design Patterns: Elements of Reusable Object-Oriented Software*, Addison Wesley, Reading, MA, 1995.
13. G. Grosz, P. Loucopoulos, C. Rolland, S. Nurcan *A, Framework for Generic Patterns Dedicated to the Management of Change in the Electricity Supply Industry*, Proceedings of the 9th International Conference and Workshop on Database and Expert Systems Applications, DEXA '98.
14. M. Jackson. Software *Requirements and Specifications - A lexicon of Practice, Principles and Prejudices*. Addison Wesley Press, 1995.
15. I. Jacobson, *The use case construct in object-oriented software Engineering*. In ''Scenario-based design: envisioning work and technology in system development'', John M. Carroll (ed.), John Wiley and Sons, 309-336, 1995.
16. S. Lauesen, M. Mathiassen *Use Cases in a COTS Tender*, REFSQ'99, Int. Workshop on Requirements For Systems Quality, Hiedelberg, 1999.
17. C. Ncube, N. Maiden., *Guiding parallel requirements acquisition and COTS software selection*, Int. IEEE Conference on Requirements Engineering, Limerick, Ireland, 1999.
18. J. Ralyté, C. Rolland; V. *Plihon Method Enhancement by Scenario Based Techniques*, in Proceedings of the 11th Conference on Advanced Information Systems Engineering, Springer Verlag, Heidelberg, Germany, June 14-18, 1999.
19. C. Rolland, C. Souveyet, C. Ben Achour. *Guiding Goal Modelling using Scenarios*, IEEE Transactions on Software Engineering, Special Issue on Scenario Management, Vol. 24, No. 12, 1055- 1071, Dec. 1998.
20. S. Si-Said, C. Rolland, G. Grosz, MENTOR : *A Computer Aided Requirements Engineering Environment*. Proceedings of CAISE' 96, Springer Verlag, Crete, Greece, May 1996.
21. A. G. Sutcliffe, M. Jarke, C. Rolland, J. Bubenko, P. Constantopoulos, *Defining visions in context models, process and tools for Requirements Engineering*. Information Systems Journal, Volume 21, no:6, pp. 515-547, 1997.
22. A.G. Sutcliffe, N.A.M. Maiden, S. Minocha, D. Manuel, *Supporting Scenario-based Requirements Engineering*, Transaction of Software Engineering, Special Issue on Scenario Management, Vol. 24, No. 12, 10- 10, Dec.1998.

Conceptual Modeling and Software Components Reuse: Towards the Unification

Barbara Pernici
Politecnico di Milano
Milano, Italy

Massimo Mecella
Università "La Sapienza" Dipartimento di Informatica e Sistemistica
Roma, Italy

Carlo Batini
Autorità per l'Informatica nella Pubblica Amministrazione
Roma, Italy

Abstract

Accurate and complete conceptual modeling is an essential premise for a correct development of an information system. Reusable conceptual schemas have the potential to facilitate this difficult and time consuming activity. In addition, linking conceptual schemas to reusable software components may provide a way for reusing not only concepts but also software applications.

1. Introduction

Information System (IS) development is a complex task involving modeling several aspects of the enterprise: the enterprise goals, the enterprise business rules, actors and resources, concepts and business processes; in addition technical requirements have to be specified [1, 2].

To facilitate this activity, several issues have been widely debated in the literature: IS development methodologies, modeling techniques guiding the development process, and components, reuse and patterns approaches.

As discussed in [3], claims of the "patterns movement" may be modest, as suggestions for the analyst showing how "similar situations have been represented or solved in other cases, and in that way patterns stimulate thinking and give ideas of particular designs. Other, more optimistic proponents believe patterns and the process of using patterns can be formalized to the extent that, starting out with a description of a current situation or a goal to be reached, the knowledge base of patterns can lead you to a solution of the design problem".

In software engineering, the component issue has been debated widely [4], and the use of software components has its roots in proposals in the 60ties, and found concrete applications with object-oriented approaches. A concrete component based approach is getting established, mainly in the realization of basic functionalities, such as graphical user interfaces, numerical functions, text processing and storage management functions. Proposals for reuse in all design phases, including analysis and design, are starting to emerge [5, 6, 7], based on the concepts of patterns and frameworks.

In IS engineering the concept of pattern is still a research issue. While some aspects are common to the issues discussed in the software engineering area at analysis level, more complex aspects need to be considered, to define patterns and to give guidelines for their reuse, and to relate them to all the different modeling aspects.
Reuse of patterns and components may turn out to be essential for the concrete realization of ISs in complex cases, such as in the development of Cooperative Information Systems (CISs). A CIS is defined, in generic terms, as a large number of cooperating systems, that [8]:

- are distributed over large, complex computer and communication networks;
- work together cooperatively, requesting and sharing information, constraints, and goals;
- support both individual and collaborative human work.

This kind of systems is central in the modern enterprise, allowing the integration of loosely coupled organizations to create a *virtual enterprise*. A crucial problem, not yet completely addressed, is the cooperation of legacy ISs, and the challenge of an evolutionary architecture for the integration of very large heterogeneous systems.
The goal of this paper is to focus specifically at the component, reuse and patterns issues, discussing how to relate approaches proposed for reuse at conceptual level and reuse at software level. A proposal towards the integration of the two fields, IS engineering and software engineering, is presented and its application envisioned. We propose an "optimistic" view towards conceptual and software reuse, in specific application domains, proposing a framework for a possible technical solution which may have a wide impact in some domains, such as public administrations and virtual enterprises.
In Section 2, we illustrate the concept of components, both at the conceptual and software design levels, and of reuse and patterns. In Section 3, an architecture for relating conceptual models and software components is proposed, and in Section 4 its application in the public administration domain is discussed. Finally, in Section 5 possible future research in the area is discussed.

2. Components

In general, a *reusable component* can be defined as a unit of design (at any level), for which a structure is defined, a name identifying the component is associated, and for which design guidelines, in the form of design documentation, are provided in order to support the reuse of the component and to illustrate the context where it can be reused, including constraints, for instance, indicating which other components must be used in combination with the one being considered [9].
From this general definition, two more specific ones follow:

- a conceptual component is a model/schema (or a subset of) to be reused;
- a software component is a a coherent package of software implementation that can be independently developed and delivered, has explicit and well-specified interfaces for the services it provides and for the services it expects from the others, can be composed with other components, perhaps customizing some of their properties, without modifying the components themselves [7].

The reuse of components can be defined from two different points of view:

- *design for reuse*: approaches in which components are developed in such a way that they can be reused in different contexts;
- *design by reuse*: approaches in which an IS is designed assembling and personalizing components.

Main issues to be solved in design for reuse are the derivation (or extraction) of components: design for reuse can be performed either generalizing so called "best practices" or building components starting from the comparison and integration of many different systems having similar characteristics.
To achieve design by reuse the main issue to be solved is an appropriate representation of components and their characteristics for their storage and retrieval from a component base.

2.1. Conceptual Components and Patterns

A conceptual component is a schema (or a portion of a schema) to be reused. Following the object-oriented approach, it may be specified with the Unified Modeling Language (UML) [10, 7]; other more specific models may be adopted in other application areas, such as for instance workflow management [9].
Conceptual components are often related to patterns, that is "a solution to a problem in context", indicating how the skeleton component may be reused, that is instatiated and employed in a specific application. Patterns are generic and abstract (conceptual) design proposals, solutions to recurring problems within the sector of interest, that can be easily adapted and reused [11].
Other approaches, not discussed in detail in the present paper, propose reuse also from the design process perspective: reuse of the *design steps*, such as refinement steps applied during the design of a database schema [REF12]; this approach may help to simplify redesign that occurs when the conceptual schema evolves. This issue is discussed in [13], where conceptual design, logical design and implementation are linked through a sequence of related steps that map the application specification objects into design objects and eventually into the object-oriented code of the new application. In the same line, there are patterns for change management, based on best business practices [14].

2.2. Software Components and Distributed Object Computing

A software component is a a coherent package of software implementation that can be independently developed and delivered.
A modeling aspect of software components is the *granularity* [15] at which components are defined. Recently, the trend is towards components which include several related aspects which are modeled together, providing a so called *application framework* [7]. These are application skeletons which can be customized by application designers, assembled and integrated with other application skeletons until the requirements for the IS are completely defined. Other approaches consider Business Objects as the basic components [16, 17], that is the classes represented in the conceptual model correspond to the software components.
The major obstacle to the definition of software components is the need for a common framework, that is the definition of "the world in which the component will

live in" [6]. Until now the fields in which the component based approach was successful were those ones in which the framework is well defined: e.g., the development of the graphical user interfaces, in which the framework is the operating environment or the virtual machine.
The new approach referred as Distributed Object Computing (DOC) [18, 19] is based on the merge of two trends: the distribution and middleware technologies and the Object Orientation. The computation is performed through messages that objects developed in different programming languages and deployed on heterogeneous hardware and software platforms exchange through the network. The technology of the Component Transaction Monitors (CTMs) offers both the bus allowing various objects to communicate in a transparent way and a standard component model; it is the middleware layer enforcing the important separation of concerns between the design of added-value, business-aware services and their actual deployment. A CTM defines the framework of services and interfaces on top of which it is effectively possible to develop and deploy software components. These components are not restricted to the presentation layer, but they represent business logic to be reused. The component model used by CTMs is standard, thus a component is pluggable on CTMs from different vendors; it is not tied to the platform it was originally developed, but it is portable among different platforms. Currently there are three main component models: the OMG CORBA Component Model (CCM), the Enterprise JavaBeans (EJB) architecture and the Microsoft Component Object Model (COM+). The first is defined by a consortium of vendors and users, the Object Management Group (OMG); the second and the third are de-facto standards, promoted by Sun, IBM, Oracle and others and by Microsoft, respectively.

3. From conceptual models to software components

In this section, we propose an architecture to unify the trends of conceptual components and software components (REFFigure 1). The architecture has the goal of promoting reuse from different points of view, as discussed in the following of the section: reusing legacy applications, providing reference schemas for different applications using the same data, and providing a way to reuse similar applications.
The architecture is multi-layered:

- Back-End layer (on the server domain): where the fragmented and vertical legacy data and applications reside.
- Object Schemas layer (basic components on the server application gateway): implementation of the conceptual model. This layer implements the new cooperative logic by integrating the back-end data to offer the OO schemas.
- Client view layer (2nd level components deployed on the server application gateway to satisfy access needs of client organizations): this layer implements the "navigation" logic of a particular client organization, by performing seamless access to information collected from the OO schemas.
- Client application (on the client domain).

Encapsulation of the IS of each organization through the DOC approach offers a solution for the following problems:

- Cooperation is achieved by making each organization responsible for exporting some views of its own IS as object-oriented schemas (OO schemas), and for describing its services using a standard language, in terms of objects and components.

- Autonomy is respected by offering an integrated view of the information available on the organization IS through OO schemas. Externally these schemas provide a standard interface towards the other cooperating organizations, internally they rely on gateways and/or wrappers [20, 21, 22] over the actual systems inside the organization.

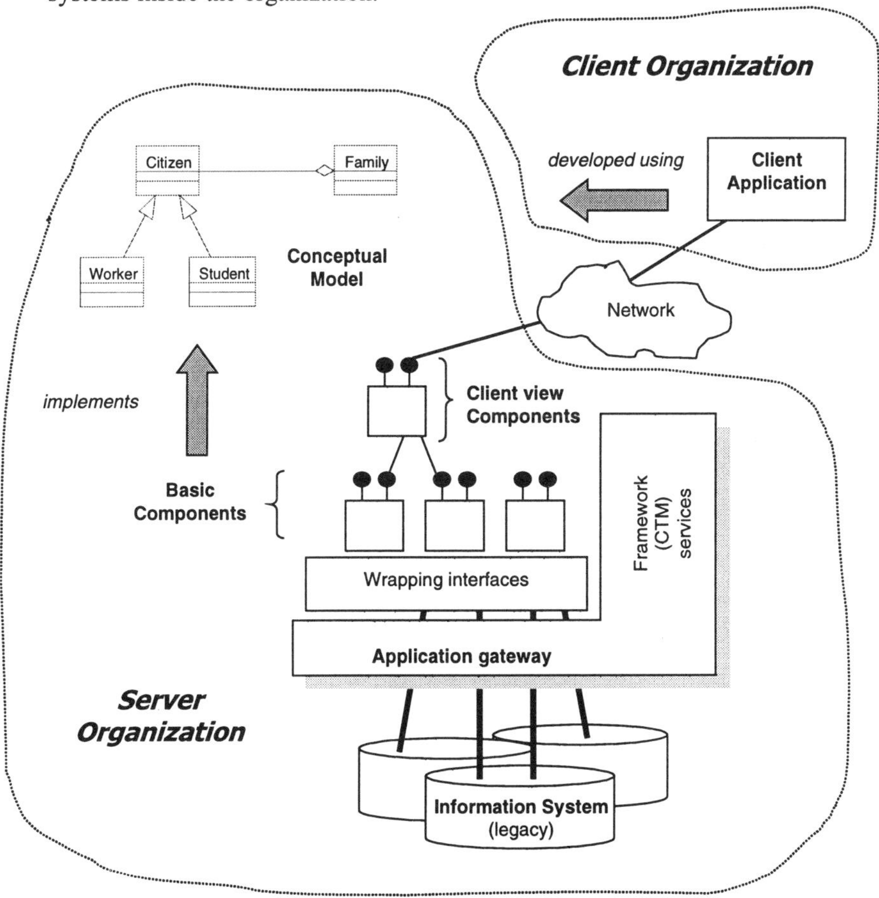

Figure 1. The cooperative architecture based on conceptual models and components

Each cooperating organization is represented as a *domain*; it includes all the computing resources, networks, applications and data belonging to a specific organization, regardless of the technical nature of such IS.

Each domain offers its information asset as a set of OO schemas, deployed and made accessible through *application gateways*. An application gateway is the computing server platform which hosts the components implementing the OO schemas of the domain; it consists of at least a CTM, and eventually all the wrapping logic necessary to access the back-end legacy applications.

The OO schemas are the *conceptual model of the information asset* of the organization, that is the information is modeled as classes and associations, using the typical notation of the UML class diagrams. These schemas offer integrated views over data and services, as if the system were a "virtual" object-oriented database.
The conceptual model is implemented as a set of components, which are hosted on the application gateway. These components are developed according to the standard component model of the CTM used by the application gateway; they offer the objects and links which are instances of the classes and associations from the conceptual model. While in the Business Objects approach the unit of composition is the single business class, in this architecture a component realizes different classes and associations: the unit of composition is a subschema of the conceptual model (pattern).
A client organization, willing to access the information asset, needs to access some objects and to follow the links among different objects. These objects and links are the instances of the classes and associations represented in the OO schemas, and through their properties and operations they export the information of the IS. The act of determining which classes to use and which associations to follow is referred as the "navigation" of the conceptual model. After determining the particular paths, a software application to effectively access the objects running on the application gateway will be developed. The "navigation" of the OO schemas by the client organization is a conceptual step: during the development of a new cooperative application, the client organization checks the OO schemas in order to identify how to access the information exported.
The client view layer offers an interface for a specific application requesting it, built composing basic components in terms of objects and services from the organization conceptual model. In fact, one of the problems associated with the traditional component based approaches is that it is difficult to associate a precise semantics to components, so that applications are able to access them correctly. In addition, usually several components are used together in an application, and it is useful to provide an optimized access to objects based on the "navigation" requirements on the base OO schemas. The client view approach is similar to the concept of view in databases or the concept of role in object-oriented conceptual models [23]. An object, in fact, can provide a different interface in terms of properties and services depending on the role it plays for the application.
According to this architecture, server and client organizations interface on a collection of OO schemas: client organizations can access the schemas that server organizations have exported, both as basic conceptual components and as particular views.
We discuss now the different types of support to reuse in the proposed architecture (Figure 2):

- Reuse of *legacy applications*: in CIS several organizations have to interact, each having its own IS; application gateways provide a way to decouple client and server applications, since it is not possible for the server organizations to develop new applications every time new access paths need to be realized.
- Reuse of the *organization conceptual model.* Each organization can access information exported by others in a not predefined way: client organizations can change their information needs over time. By realizing the components implementing the conceptual model of its information asset, the server

organization creates a decoupling layer from its legacy IS. If new information needs arise in a client organization, the server organization need not to develop new applications accessing the legacy ones, but can use the conceptual model and the components. Another way of reusing the conceptual schema is by adopting it in other organizations dealing with the same kind of information (as discussed later under methodological reuse).

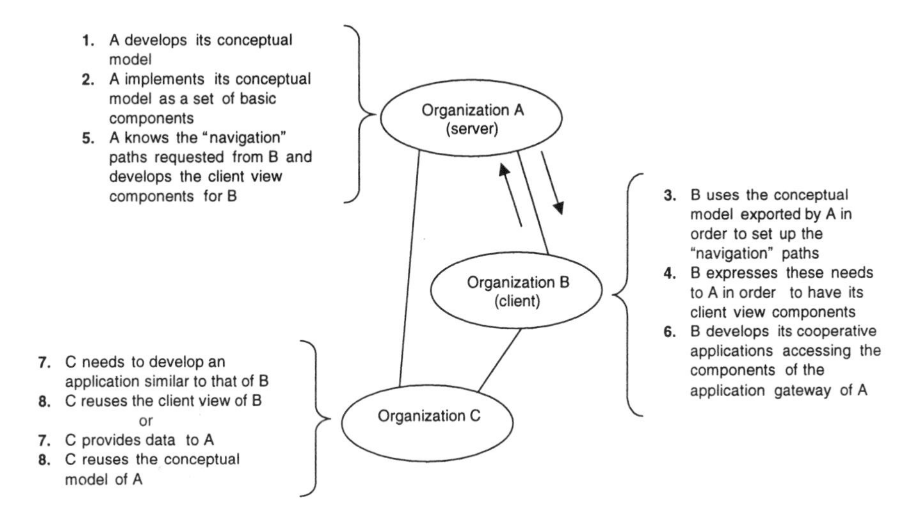

Figure 2. Reuse based on the proposed architecture

- Reuse of *client views.* In many cases, different considerations do not allow the efficient remote access to the server fine-grained objects directly from the client side. These considerations are due to network latency, to resources allocation problems on the application gateway, to low performances, to the need to offer different views of the conceptual model to different client organizations. Therefore the client organization expresses its needs to the server organization, that is the "navigation" it want to perform on the conceptual model. The latter realizes further components on top of the basic ones. These 2nd level components are introduced for addressing those different aspects. The client remote application uses the coarse-grained services implemented by the client view components. The same client view components can be used by several organizations which need to access the same information of an organization. In addition, it is easy to reuse already developed client view components to realize personalized views for different clients requesting them: personalizing a view requires a reduced realization effort, with respect to the effort of creating all navigation paths on the base OO schemas of the organization starting from the basic components, if the requirements of different clients are similar.
- *Methodological reuse.* An interesting application of the previous architecture is when there are many similar organizations: each of them developed its own IS in an independent way, thus today it is very difficult for them to cooperate. Through a common business modeling phase, a common conceptual model for

all the organizations is obtained. This model is designed by using a standard component model. The components, in order to properly implement the OO schemas, need to access the raw data on the legacy systems; low-level interfaces towards the preexisting access mechanisms need therefore to be defined. Such interfaces will be provided by the wrappers of each organizations. The components obtained are deployed on the application gateways of the organizations: because they only rely upon the wrapping interfaces (provided by each organizations) and upon the standard CTM interfaces, they can be effectively reused and plugged on the hosting application gateways. Each organization develops itself the wrapping interfaces and provides the platform for the application gateway hosting the components; the conceptual model and its implementation are developed one time and reused by all the organizations.

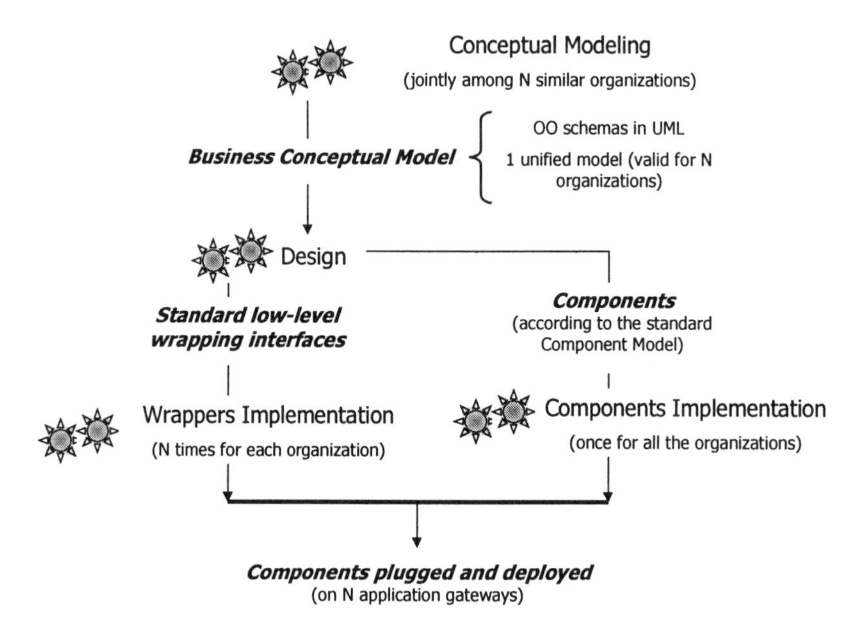

Figure 3. The methodological approach for conceptual and software reuse

As regards wrappers, two cases are possible: for those organizations having their own legacy ISs, it is possible to identify common access mechanisms and to define common wrapping interfaces. This is effective because all these systems were developed to satisfy similar needs. For those organizations without back-end applications, it is very simple to develop new ones according to the wrapping interfaces. These interfaces are at low abstraction level, that is all the integration logic necessary to raise the abstraction level up to the conceptual model is in the components. The wrapping interfaces can be considered as plug-points [7] (see Figure 3).

4. Applications and Examples

Concrete applications of the proposed architecture are in progress in the Italian Public Administration. In Italy each administration can be classified as central or local; among the other differences, it is worthy to note that there is a single instance of each kind of central administration (e.g. one Italian Social Security Service or one Ministry of Justice), while there are a lot of instances of each kind of local administration (e.g. more than 8000 City Councils or 103 Prefectures); each instance of a particular kind of local administration offers similar services, although the internal structures and organizational complexities are different.
Most of the central administrations have their own ISs, developed during the last three decades, thus there is a big amount of legacy applications. Instead many local administrations do not have computer ISs. Only some major City Councils have their own legacy ISs, and only during the year 2000 there will be the first deployment of an integrated and complete application in some Prefectures.
All the administrations are loosely coupled, but they need to exchange information and application services in order to offer services to citizens and to carry out their business processes.
Since 1993, the “Authority for IT in the Public Administration” (AIPA) is promoting the technology innovation and the business process reengineering of the Public Administration.

4.1. The Inventory of the Conceptual Schemas

One of the first activities launched at AIPA has been the project to build an inventory of existing information systems within the Public Administration [24, 25]. Each IS has been reverse engineered and a set of conceptual schemas describing it has been derived. These schemas has been classified and organized in order to query the inventory and to retrieve them.
Some goals that can be achieved with the development of the inventory are data integration, reuse of existing applications and greater availability and accessibility of information.
Due to the lack of coordination in IS developments during the past years, applications pertaining to different administrations has been implemented separately, even if they deal with the same subjects. The creation of a common inventory, containing the reverse engineered conceptual schemas of these applications, allows data and process analysis, to be carried out either with semi-automatic similarity-driven techniques [24], or with human semantic ones.

4.2. The Unitary Network

Tightly coupled with the inventory project, there is the most important and challenging initiative undertaken by AIPA, the Unitary Network of the Italian Public Administration. This project aims at implementing a "secure Intranet" that can interconnect the ISs of different public administrations. The emphasis of the project is on promoting cooperation among the various administrations at the application level. The need for the cooperation was demonstrated by some indices calculated

during an investigation based on the inventory project. For example, 38% of the processes break during their execution because an administration needs information from some others, and currently these needs are satisfied through paper document and ordinary mail.
Each administration is represented as a Domain, and once it is connected on the Unitary Network , each such Domain is modeled as a single entity; externally its architecture is accessible through the Domain Application Gateway, deployed as a component server. The Domain is a powerful tool for addressing the organizational complexities of Italian public administrations at the right abstraction level: consider the difference among a local administration and a peripheral branch office of a central administration. In the first case the architecture consider its own Domain, in the second one the distribution is hidden inside the central administration Domain. Note also that some instances of a peripheral administration have recursively their own branch offices (for example the CCs of Rome and Milan). Of course the same architecture could be recursively used inside a single domain in order to connect all the branch offices.
Currently some pilot projects (e.g. the SICAP project of the Ministry of Justice and the Arconet one of the Italian Social Security Service) are in progress, in order to validate the soundness of the cooperative architecture. Among the projects undertaken by different administrations, in this paper we point out the issues related to the City Councils (CCs). All the CCs have, by law, the duty to offer to other administrations information about the Italian citizens and their social situation (current living address, family status, working rights, etc.). It is not possible to take the census of all the paths according to which other administrations want to access information; therefore it would be useful to export all the information owned by a CC as a *CC model* and leave the client organizations to access and navigate this "running model" according its own paths. The Unitary Network is the bus allowing the communication.
Currently there is a prototypical effort to develop this common conceptual model of the typical CC, through which it is possible to access information and services of a generic CC IS. This model consists of a set of OO schemas depicted in UML.
Each CC can reuse this model to develop its own IS; the use of the common model, besides preventing from redoing the high phases of the software life cycle, guarantees a commonality among CCs.
But the reuse, enabled from the conceptual model, can be further extended from the conceptual level to the design and implementation ones: a collection of EJB components are developed. These components can be deployed on any CTM (independently from the vendors); by invoking them the client organizations can access the information asset according their own paths. The components, in order to properly run, rely on the wrapping interfaces (described in CORBA IDL) that each CC must provide; different technologies allows to develop wrappers for legacy applications, building a thin layer offering CORBA interfaces (thus platform independent) on the "upper side" and interfacing the proprietary legacy environment on the "lower side". These wrappers simply create a decoupling layer from the legacy environment, without any integration logic (which resides in the components implementing the model). In addition, there is a need for developing specific access paths for most common applications, as envisioned with the client view layer.

The use of components allows also a kind of *information systems hosting* for the small CCs not having a computer system and not willing to set up their own Application Gateway. Different organizations, such as consortiums, software houses, or major CCs, could deploy the platform and host the components for other CCs.

5. Future work and concluding remarks

Component based IS development is feasible, provided that the application domain is specified.
An important issue is that future developments in this direction involve both academic and industrial research and public organizations. The involvement of all these organizations would have positive aspects, since it would give the opportunity to work on real applications, of considering both technical and normative aspects, and of developing advanced approaches to link information system conceptual models and software components.
In dealing with components, however, there is a need of considering not only conceptual modeling and architectural issues, but also other aspects which are fundamental in IS development, such as social, organizational and communication aspects. Future research should consider also all these aspects of the problem of IS development. Providing views at the architectural level could provide the basis for taking into account several modeling aspects in addition to pure conceptual modeling.
Another aspect which may be turn out to be fundamental to achieve reuse at the conceptual and architectural level based on the proposed architecture is to associate to such reuse appropriate cost models, which might be based on cost per use, per copy, per unit of time [26], or to establish other means of making such reuse economically practicable, in particular across organizations. To facilitate reuse at all levels, and allow integration, tools are needed to support it: innovative component/pattern repositories, taking advantage of multimedia, internet/intranet repositories [3]. Finally reuse will be really established once techniques to validate actual reusability of components will be available: a possible way according to the proposals presented in this paper is rapid prototyping of applications based on reusable conceptual components and their architectural correspondents, both considering basic components and client views.

References

1. Bubenko jr JA, Rolland C, Loucopoulos P, DeAntonellis V: Facilitating "Fuzzy to Formal" Requirements Modelling. In Proceedings of the IEEE International Conference on Requirements Engineering, Colorado Springs, CO and Taipei, Taiwan, 1994.
2. Bubenko jr JA, Brash D, Stirna J. EKD User Guide. Elektra Project Report, 1998.
3. Bubenko jr JA: Challenges in Information System Engineering. In Proceedings of the Third Baltic Workshop on Data Bases and Information Systems, Riga, Latvia, Institute for Mathematics and Computer Science, University of Latvia, 1998.
4. Persson E: The Quest for the Software Chip. The Roots of Software Components. A Study and Some Speculations. In Proceedings of the First Nordic Workshop on Software Architecture (NOSA'98), Ronneby, Sweden, 1998.

5. Gamma E, Helm R, Johnson R, Vlissides J. Design Patterns. Elements of Reusable Object Oriented Software. Addison Wesley, Reading, MA, 1995.
6. Fowler M. Analysis Patterns. Reusable Object Models. Addison Wesley, Reading, MA, 1997.
7. D'Souza DF, Wills AC. Objects, Components and Frameworks with UML: The Catalysis Approach. Addison Wesley, Reading, MA, 1999.
8. Laufmann S, Spaccapietra S, Yokoi T: Foreword. In Proceedings of the Third International Conference on Cooperative Information Systems (CoopIS'95), Vienna, Austria, 1995.
9. Casati F, Castano S, Fugini MG, Mirbel-Sanchez I, Pernici B: Using patterns to design rules in workflows. To be published on IEEE Transactions on Software Engineering.
10. Booch G, Rumbaugh J, Jacobson I. The Unified Modeling Language User Guide. Addison Wesley, Reading, MA, 1998.
11. Prekas N, Loucopoulos P, Rolland C, Grosz G, Semmak F, Brash D. Developing Patterns as a Mechanism for Assisting the Management of Knowledge in the Context of Conducting Organisational Change. In Proceedings of the 10th International Conference and Workshop on Database and Expert Systems Applications (DEXA '99), Florence, Italy, 1999.
12. Castelli D: "Three levels of reuse for supporting the design of database schemas". In Proceedings of the Entity Relationship Conference, Paris, France, 1999.
13. Bellinzona R, Fugini MG, Pernici B. Reusing Specifications in OO Applications. IEEE Software; vol. 12, no. 2, March 1995.
14. Brash D, Stirna J: Describing best business practices: a pattern-based approach for knowledge sharing. In Proceedings of SIGCPR'99, New Orleans, LA, 1999.
15. Johnson R. Framework = (Components + Patterns). Communications of the ACM; vol. 40, no. 10, October 1997.
16. Persson E: Shibboleth of Many Meanings. An Essay on the Ontology of Business Objects. In Proceedings of the Third International Enterprise Distributed Object Computing Conference (EDOC'99), Mannheim, Germany, 1999.
17. Sims O. The OMG Business Object Facility and the OMG Business Object. OMG Document cf/96-02-03, OMG, Framingham, MA, 1996.
18. Orfali R, Harkey D, Edwards J. The Essential Distributed Object Survival Guide. Wiley & Sons, New York, NY, 1996.
19. Wallnau K, Weiderman N, Northrop L. Distributed Object Technology with CORBA and Java: Key Concepts and Implications. Technical Report CMU/SEI-97-TR-004, Carnegie Mellon University, Software Engineering Institute, Pittsburgh, PA, 1997.
20. Brodie ML, Stonebraker M. Migrating Legacy Systems: Gateways, Interfaces & The Incremental Approach. Morgan Kaufmann, San Francisco, CA, 1995.
21. Umar A. Application Reengineering. Building Web-Based Applications and Dealing With Legacy. Prentice-Hall, Upper Saddle River, NJ, 1997.
22. Weiderman N, Northrop L, Smith D, Tilley S, Wallnau K. Implications of Distributed Object Technology for Reengineering. Technical Report CMU/SEI-97-TR-005, Carnegie Mellon University, Software Engineering Institute, Pittsburgh, PA, 1997.
23. Pernici B: "Objects with roles". In Proceedings of the ACM-IEEE Conference on Office Information Systems, Boston, MA, 1990.
24. Batini C, Castano S, De Antonellis V, Fugini MG, Pernici B: Analysis of an Inventory of Information Systems in the Public Administation. Requirements Eng 1996; 28:47-62.
25. Batini C, Di Battista G, Santucci G. Structuring Primitives for a Dictionary of Entity Relationship Data Schemas. IEEE Transactions on Software Engineering; vol. 19, no. 4, April 1993.
26. Szyperski C. Component Software. Beyond Object-Oriented Programming. Addison Wesley, Reading, MA, 1997.

III.

Concepts for Information Systems

The Plan Recognition/Plan Generation Paradigm

Antonio L. Furtado and Angelo E. M. Ciarlini*
Departamento de Informática, Pontifícia Universidade Católica do R.J.
Rio de Janeiro, Brazil

Abstract

Any computerized application-domain environment, equipped with a Log of the execution of predefined operations, can be regarded as a repository of narratives concerning the activities observed in the mini-world of interest. The analysis of these narratives leads to the construction of a library of typical plans, which can be used by Plan-recognition/Plan-generation algorithms to help prediction, decision-making and the adoption of corrective measures with respect to ongoing activities. The proposed goal-oriented paradigm emerges as a promising research line to help understanding, and hence better modelling and using, so diverse domains as business information systems, literary genres of narratives, and object-oriented software based on frameworks.

1 Introduction

During the early period of growth of database technology, conceptual modelling of information systems was mostly centered on *data analysis*. At a second phase, *function analysis* started to be considered, initially receiving relatively less attention, with an emphasis on workflow methods [1]. In [2], we proposed a different method, based on application-domain operations (such as *hire*, *train*, *assign*), within an abstract data type orientation. The method was especially powerful in the presence of a plan-generation tool having access to a *Log* recording the executions of operations. We showed that such Log, easily implementable as a set of relational tables in correspondence with the pre-defined operations, was a desirable component of temporal database systems. In principle, such tables constituted by themselves a temporal database, since – as we then demonstrated – an inference mechanism was able to answer temporal queries from the records of the operations executed. More recently [3], we developed a more advanced prototype tool, combining plan-recognition and plan-generation, able to match the observed behaviour of agents against a library of typical plans, so as to find out as early as possible what each agent is trying to

* The work of the second author was supported by FAPERJ - Fundação de Amparo à Pesquisa do Estado do Rio de Janeiro - Brazil

achieve, and propose adjustments whenever convenient. Typical plans are those being used in practice, as one can determine by examining the Log.

We then came to realize that any such Log of application-domain operations provides a kind of information of great potential value, since its contents can be regarded as a collection of highly interconnected *plots of narratives* about the trajectories of the various entities involved (e.g. employees' careers). We also saw that the analysis of these narratives to detect meaningful patterns offers a promising, albeit non-trivial to undertake, knowledge-discovery opportunity [4].

On the other hand, narrative is a very general concept, encompassing highly complex literary genres, in addition to the more prosaic real-life ones. In [5], we discussed a method (supported by our tool) to run simulation experiments, thereby allowing to "write future scenario stories" concerning business application domains, according to the directive indicated by Bubenko [6]. Moreover, our successful experiments with one literary genre (folktales) reinforced our belief in the generality of the plan-recognition/plan-generation (**Pr/Pg**) paradigm. Still in the course of these experiments, we also turned our attention to the promising area of computer-software *frameworks* [7].

The present paper demonstrates the generality of the Pr/Pg paradigm, and indicates its relevance to framework design. An example taken from the area of database-oriented information systems is first described in detail, in order to make the presentation more concrete. The basic concept of application-domain operations is reviewed in section 2. Section 3 introduces the key notion of plots of narratives, whose goal-oriented pragmatic aspects are exploited in section 4, in connection with the construction and use of libraries of typical plans, coupled with a plan-recognition/plan-generation software tool. In section 5 we briefly mention our more ambitious experiments with a literary genre. Section 6 outlines a strategy to apply the paradigm to frameworks. Section 7 includes a brief summary and concluding remarks.

2 From Facts to Actions

Database *facts* can be conveniently denoted by *predicates*. Given an application domain whose conceptual schema, expressed in terms of the Entity-Relationship model, involves, among others the entities *employees* and *clients* and a relationship *serves*, the predicates *is-employee(E)*, *is-client(C)* and *serves(E,C)* can be used to express, respectively, that E is an employee, that C is a client, and that employee E is at the service of client C. Specific predicate instances, corresponding to facts of these three types, might be: *is-employee(Mary)*, *is-employee(Peter)*, *is-client(Omega)* and *serves(Peter, Omega)*.

A database *state* is the set of all predicate instances holding at a given instant. A state provides a *description* of the mini-world underlying the database. On the other hand, *actions* performed in this mini-world can be denoted, always at the conceptual level, by *operations*. Assume that, among others, one can identify in our example application domain the operations *hire(E)* and *replace(E1,E2,C)* in order, respectively, to hire E as an employee, and to replace employee E1 by employee E2 in the service of client C. A specific execution of the second operation is *replace(Peter, Mary, Omega)*.

If the set of predicate instances shown before constitutes a state S_i, then executing the above operation achieves a transition from S_i to a new state S_j:

$\underline{S_i}$		$\underline{S_j}$
is-employee(Mary)		is-employee(Mary)
is-employee(Peter)	$\rightarrow$	is-employee(Peter)
is-client(Omega)		is-client(Omega)
serves(Peter, Omega)		serves(Mary, Omega)

An operation can be specified, in a STRIPS-like formalism [8], by declaring its pre-conditions and post-conditions, which characterize the states before and after the operation is executed. Pre-conditions establish requirements, positive or negative, which must hold prior to execution whereas post-conditions express effects, consisting in predicate instances being *affirmed* or *negated*. Pre-conditions and post-conditions can be specified so as, in addition, to ensure that static and transition *integrity constraints* be preserved. Our approach to function analysis is formally described in Part A, chapter 5 of [9].

The definition of *replace* is shown below, in a semi-formal notation:

replace(E1, E2, C):
- pre-conditions: serves(E1, C) $\wedge$ is-employee(E2) $\wedge$ $\sim\exists$ C1 serves(E2, C1)
- post-conditions: $\sim$ serves(E1, C) $\wedge$ serves(E2, C)

The pre-conditions make provision for the constraint that an employee cannot serve more than one client. Notice, however, that some other obvious requirements seem to be missing, e.g. it is not indicated that E1 should be an employee. This kind of simplification is justified if a strict *abstract data type* discipline is enforced. Notice also that the effects indicated via post-conditions refer only to what is changed by execution. It is therefore assumed that anything else that held previously will still hold afterwards (so as to cope with the so-called "frame problem").

3 A Key Concept: Plots of Narratives

At a given instant of time, a factual database provides no more than the description of the current state. Temporal databases [10] allow to keep descriptions of all states reached, without making explicit, however, what actions caused the transitions; this sort of information becomes available if, in addition, a *Log* registering the (time-stamped) executions of operations is maintained [2]. We will then say that, besides static descriptions, temporal databases thus enhanced now contain *narratives*, as will be illustrated in the sequel.

Suppose that, after our example database (owned by a corporation called Alpha) has been running for some time, one extracts from the Log all operations concerning a given client, say Beta, and the employees assigned to its service, whose execution occurred during a given time interval. Let the obtained sequence, kept ordered by time-stamp, be:

Plot 1
a. open(client: Beta)
b. hire(employee: John)
c. assign(employee: John, client: Beta)
d. complain(client: Beta, employee: John)
e. train(employee: John, course: c135)
f. raise-level(employee: John)

The sequence above, named **Plot 1**, summarizes a narrative, which justifies calling it a *plot* [3]. Understood as "the plan of action of a play, novel, poem, short story, etc." (*Webster's New World Dictionary of the American Language*, D. B. Guralnik, ed., New York: The World Publishing Company, 1970, page 1095), a plot should be a way to capture (here, in easy-to-process standard notation) the essential *structure* of some meaningful succession of events. Indeed, **Plot 1** can be read under the expanded form of a natural language text – thus making explicit the underlying full-fledged narrative – such as:

> *Beta became a client of Alpha. John was hired at the initial level, and then assigned to the Beta account. Later, Beta complained of John's service. John participated in training program c135. No further complaints came from Beta. John's level was raised.*

Besides registering past executions of operations, the Log can be further used as an agenda. Future executions, either merely possible or firmly scheduled, can be registered in the Log, of course subject to later confirmation or cancellation.

4 The Pragmatic View of Narratives

Narratives contained in a database Log, far from being fortuitous, exhibit a clear *pragmatic* bent, usually reflecting the *goals* of the several *agents* who promote the execution of operations [11]. In the enterprise model, proposed in [12], goals and agents are contemplated, respectively, in separate Objectives and Actors submodels. It is therefore useful to distinguish, among the possibly many effects of an operation, those that correspond to achieving a recognized goal of an agent; intuitively, they are the reason for executing the operation – which justifies calling them the *primary effects* of the operation – in contrast to other minor, seemingly innocuous, ones.

Sometimes a goal corresponds to the combination of the primary effects of more than a single operation. Even when only one operation would seem to be enough, it may happen that its pre-conditions do not currently hold, but might be achieved by the preliminary execution of another operation. In both cases, a partially ordered set of operations is required, whose execution in some sequence leads from an initial state S_0, through an arbitrary number of intermediate states, to a final state S_f where the goal holds. On the other hand, there are cases in which one finds more than one set of (one or more) operations as alternative ways of reaching a given goal.

Such partially ordered sets (posets) constitute *plans*. The complementary processes of automatic plan generation and plan recognition have been extensively studied, and several algorithms have been proposed for each of them. The distinction between plans and operations becomes blurred as we introduce, besides the operations discussed thus far (to be henceforward called *basic*), others called *complex operations*, which are defined either by the *composition* of two or more operations (part-of hierarchy) or by the *generalization* of different alternatives (is-a hierarchy). Once complex operations have been introduced, one can build with them a hierarchically structured *library of typical plans*. Plan-recognition algorithms detect what plan an agent is trying to perform, by matching a few observations of the agent's behaviour against the repertoire of plans contained in one such library. As part of our ongoing research project, we have implemented a *Plan-recognition / Plan-generation* tool [3], where these two processes alternate. Having recognized that the observed behaviour of an agent indicates that he is trying to execute a library plan, the tool can check whether the plan is executable under the current conditions. If there exist obstacles, the tool can either propose a fix or adaptation (cf. the notion of *reuse* [13], much in vogue in software engineering) or drop the plan altogether and try to generate another plan capable of successfully achieving the agent's goal. If the original plan of the agent is executable but detrimental to the purposes of the corporation, the tool may denounce it as such. More recently, we introduced *goal-inference rules*, in order to model the predicted behaviour of individual agents (or classes of agents). Goals are inferred by identifying problems and opportunities concerning each agent, emerging at real or simulated states.

The plan-recognition part of the tool is based on Kautz's algorithm [14]. Plan-generation uses the techniques developed in the Abtweak project [15]. Both the plan-recognition and the plan-generation algorithms proved to be powerful enough to handle not only plots of simple database narratives but also of fairly intricate folktales [16,17]. The plan-generation algorithm allows the assignment of costs to operations and, although not guaranteeing optimality, it tends to produce the shortest or the least expensive plans first (alternatives are obtained upon backtracking). One version of the tool is written in a Prolog product that offers an interface with SQL for database handling. The libraries of typical plans that we have been using (different libraries for different application domains) were constructed manually from previously defined complex operations. The prototype has been considerably extended [5], but several important capabilities should still be added. We are currently looking at the (semi-) automatic construction of libraries, along the lines indicated in the sequel.

After defining the basic operations and identifying the main goals of the various agents, the plan-generation algorithm itself is an obvious instrument to help finding convenient complex operations, able to achieve the goals. Yet, another attractive strategy is based on the study of plots. Again assuming that the basic operations and main goals were preliminarily characterized, we can let the database be used for some time, allowing the Log to grow to a substantial size. Then a number of plots can be extracted and compared, mainly by a most specific generalization criterion [18], to find how agents have proceeded in practice towards their goals. From this analysis, the complex operations that will form the library would arise.

Delimiting the meaningful plots to be extracted is in itself a problem. In our example, assume that the identified goals of company Alpha are: obtain clients, keep them happy. With respect to the second goal, a situation where a client has complained is undeniably critical. So, operations related to an unsatisfied client, from the moment when his misgivings became patent to the moment when they ceased, seem to form a relevant sequence. We have shown before:

Plot 1

...
d. complain(client: Beta, employee: John)
e. train(employee: John, course: c135)
...

And now consider another plot, everywhere similar to **Plot 1**, except for the solution adopted:

Plot 2

...
d. complain(client: Delta, employee: Robert)
e. hire(employee: Laura)
f. replace(employee: Robert, employee: Laura, client: Delta)
g. fire(employee: Robert)
...

This reveals two different strategies used by Alpha to placate a client. In Plot 1, the operation *train* was used and, in Plot 2, there was a composition of operations *hire*, *replace* and *fire*. This fact suggests the introduction of two complex operations: *reallocate* (composing *hire*, *replace* and *fire*) and *improve-service*: (a generalization of *train* and *reallocate*). Continuing with this sort of analysis, the library of typical plans may finally take the form shown in Figure 1, where double arrows correspond to is-a links and single arrows to part-of links.

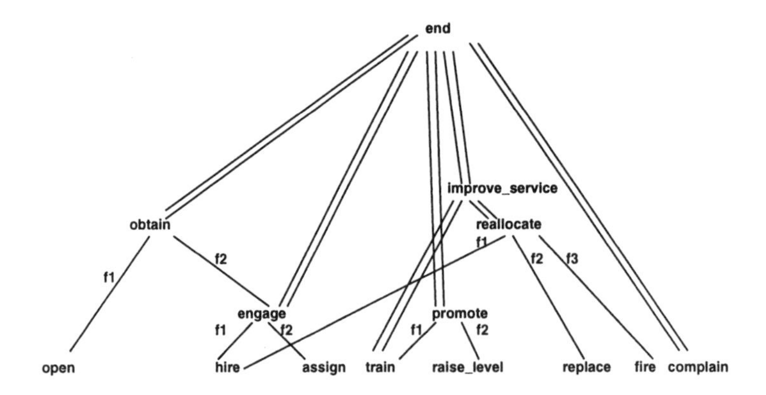

Figure 1: Library of typical plans (complex database operations)

While noting how plots express (or contain) plans of one or more agents, we must bear in mind that such plans are not necessarily optimal, often reflecting policies and practices of the corporation. But, even if for no other reason than becoming aware of these not entirely rational motivations, the detection of typical plans pays off. More generally, the discovery and analysis of narrative patterns appears to be one of the most promising undertakings in temporal database environments. To do it properly, one must look beyond pure database technology, as argued in the next section.

5 The Paradigm in the Context of Genres of Narratives

Our use of operations as linguistic units was largely influenced by the work of the Russian scholar Vladimir Propp [16]. Before him, a common criterion to characterize literary *genres* [19], such as folktales, was to classify as belonging to the genre any text that consisted of a subset of the *motifs* [20] thought to be associated with the genre. Propp objected to that on the grounds of granularity; he argued that motifs (a notion curiously similar to *scripts* [21]) were composites and therefore should not be taken as basic units. For the folktale genre, on the basis of an empirical analysis over a vast collection compiled by Afanas'ev [17], he proposed a repertoire of 31 *functions* (e.g. "villainy", "combat", "return"), attributed to 7 *spheres of action* (personages with different roles, such as "hero" and "villain"). A corollary of Propp's thesis is that by changing the repertoire of functions and the roles of personages one can introduce each genre to be considered, literary or not (e.g. news, project progress reports, etc.).

An early attempt to adapt Propp's ideas to automatic processing resulted in the "story grammars" formalism [22]. A successful parse applying the rules of a story grammar should be enough to accept a given sequence of operations as corresponding to a well-formed story. As a reaction against this allegedly too **syntactical** approach, unduly neutral with respect to the degree of interest that the acceptable stories might arouse, the notion of "points" was proposed [23]. In order to overcome the limitation without discarding Propp's fundamental contribution, our treatment within the Pr/Pg paradigm adds **semantics** through the use of predicates to express states of the world, and to formulate pre-conditions and post-conditions which consistently organize the execution of Proppian functions (formally defined as basic operations). Furthermore, a **pragmatic** guiding aspect is added by identifying meaningful goals as the prime motivation for the behaviour of the various personages.

As a non-trivial benchmark for our prototype, we constructed a library of typical plans for the folktale genre and have been using it to recognize, adapt and generate plots of folktales. The library contains both part-of and is-a links to form several levels of complex operations over Propp's (specialized) functions, thus taking to its full consequences an early trend initiated in [24]. Our experiments with plots of folktales confirm the generality of the Pr/Pg paradigm and provide insights about how to apply it to other contexts.

The key consideration is that, since the notion of *operation* can be introduced in the real world of information systems (as in the previous sections), or in the fictitious world of literary narratives (e.g. folktales), it should in principle be applicable to yet other worlds or contexts, such as that of computer software.

6 Towards the Design of Frameworks

In the world of computer software, operations take the form of *commands* or, at a higher level, of *methods* in an object-oriented environment. Plan generation has indeed been proposed for the synthesis of programs; in particular, generated plans can incorporate conditional schemes as well as iteration or recursion [25].

In the universe of fiction, we saw that operations are deeds in the sphere of action of different categories of personages. Analogously, in the object-oriented computer software world, methods are atached to different *classes*, and, as remarked by M. E. Fayad and D. C. Schmidt [26], object-oriented application frameworks have appeared as a promising technology for reifying proven software designs and implementations in order to reduce the cost and improve the quality of software. A framework is a reusable, "semi-complete" application that can be specialized to produce custom applications [27]. In order to outline our initial proposal for the application of the paradigm to the design of frameworks based on object-oriented methods, we shall discuss in a semi-formal way a simple schematic example. Consider two classes of objects, A and B, and let M1 be a method of class A, and M2 and M3 methods of class B.

The first step in seeing methods from the perspective of the Pr/Pg paradigm is to define each method in terms of its arguments (including the identification of the object executing the method) and appropriate pre-conditions and post-conditions, corresponding to their input/output requirements. For simplicity, let p_0 denote the pre-conditions of M1, and p_1 denote both the post-conditions of M1 and, coincidentally, the pre-conditions of M2 and of M3. Finally, let p'_1 & p_2 denote the post-conditions of M2, and p'_1 & p_3 those of M3. Given these definitions and assuming that p_0 holds at state σ, it is clear that M1 can be immediately executed, thereby enabling the execution of M2 or M3. Thus it should be no surprise that the first or, alternatively, the second trial below can be successfully processed by the prototype:

- Starting from state σ, and given a goal state γ where p_2 must hold, the plan <M1,M2> is generated as in figure 2a.
- Starting from the same state σ, and given a different goal state γ' where p_3 must hold, the plan <M1,M3> is generated as in figure 2b.

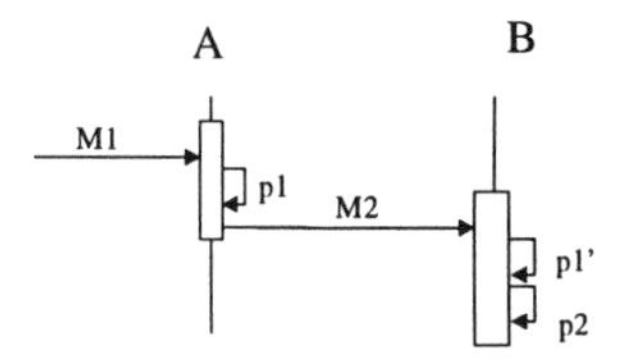

Figure 2a: a plan to generate p_2

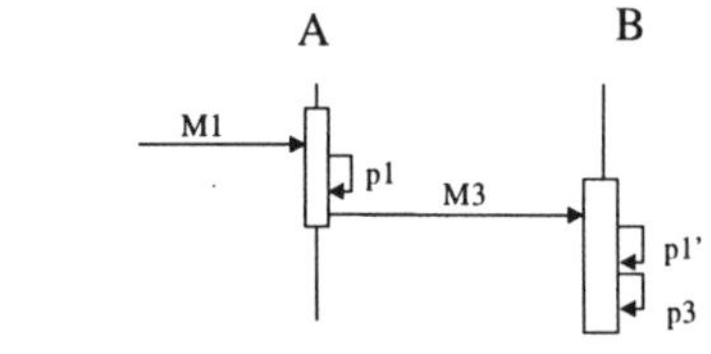

Figure 2b: a plan to generate p_3

The two plans are compatible with (among others) the framework structure represented in figure 3, drawn according to the conventions in [28]. Class A is connected to class B via a link with label "bRef", meaning that bRef is one of the instance variables of class A. The code implementing method M1 finishes with a control statement directing a message to the current instance of B, designated by bRef, to request the execution of either M2 or M3 depending on the value of instance variable aGoal.

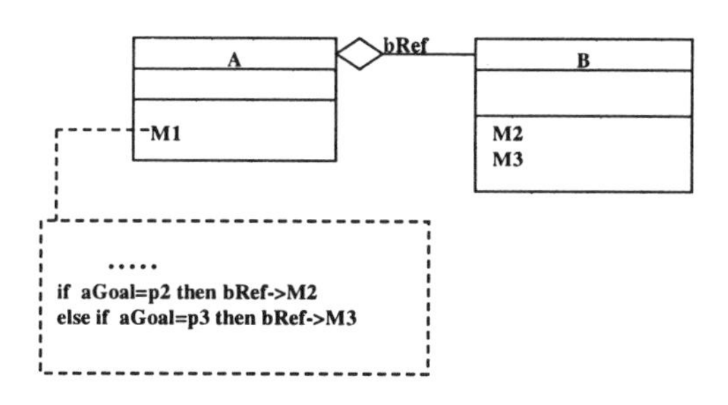

Figure 3: a framework structure

By examining a set of goals and their corresponding generated plans, it might be possible both to check the suitability of particular frameworks and to outline new ones. However, the existence of a library of typical plans, and of an automatic procedure to build it, permits a more direct process.

As prescribed in section 3 for the database context, one can record in a suitable Log sample executions of the methods envisaged. From the Log one can then extract what we called plots — which in the context of methods (and also of simple commands) are usually called *traces* of executions. Again it is useful to identify typical *trace-patterns*, obtained by applying most specific generalization to analogous occurrences, and then to derive *complex methods* by composition and generalization, to be assembled in a library of typical plans similar to that of figure 1. While generalization of methods shows possible hierarchies of classes, composition of methods enables us to identify the chaining of calls for their execution.

Suppose that, when characterizing trace-patterns corresponding to executions of <M1,M2> and of <M1,M3>, it is observed that in most cases their explicitly declared goal was to reach a state γ' where p'_1 alone was required, which suggests that p'_1 is to be isolated as the **primary** intended effect of both M2 and of M3. This would justify the introduction of a complex method M23 as a generalization of M2 and M3; in addition, complex method M0 would be included as the composition of M1 with M23. Figure 4 gives the branch of the library displaying these methods.

The library in turn suggests an alternative framework schema, in which M23 emerges as a *hot spot* method [28]. It calls a method P1', which generates the postcondition p'_1, and a *hook* method to be overriden in two subclasses: B2 and B3. B2's implementation calls a method containing the part of M2's code that generates p_2 and B3's implementation calls a method containing the part of M3's code that generates p_3. Thus, an object of class A does not determine the goal to be achieved by the object of class B pointed by bRef. This alternative, shown in figure 5, makes sense if the choice of the secondary effect depends on the properties of the object pointed by bRef. Yet another possibility is that the choice between M2 and M3 leads to separate framework versions, each one geared to whichever secondary effect (either p_2 or p_3) is preferred. This third design option may prove convenient in certain practical situations where the kind of application determines the appropriate secondary effect.

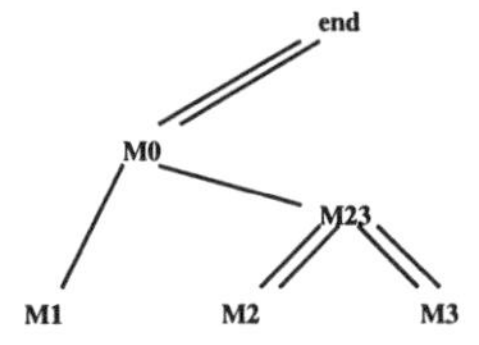

Figure 4: Library of Typical Plans (complex methods)

In general, the proposed study of the behaviour of methods on the basis of the Pr/Pg paradigm may not be sufficient to guide the choice between such alternative framework schemas, which means that other complementary analyses must be undertaken, based on *patterns* [28] and/or on *viewpoints* [29].

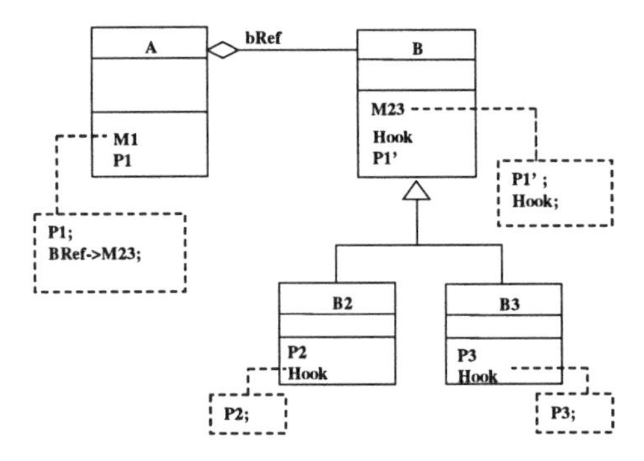

Figure 5: an alternative framework structure

On the other hand, the paradigm remains applicable, after the design of the framework is completed, for verifying its adequacy under different real or simulated conditions. For example, assume that we would like to build an application to solve a problem in the domain of the frameworks derived from our library. We could outline a solution containing the execution of method M1. The fact that M1 is part of the solution can be used by the plan-recognition algorithm, which would infer the intention to execute M0. The Pr/Pg prototype is able in such cases to:

- indicate that there are currently no obstacles to the execution of the various steps of M0.
- or detect that p_1 already holds in the current state, and thus, since there is no need to perform M1, the plan may be adapted to include only M2 or M3;
- or detect that p_0 does not currently hold. Since the intention to execute M0 in turn implies the intention to reach some goal state where p'_1 holds, the plan-generation algorithm would then try to build for this purpose a new plan that does not include M1.

The second case calls for a redesign of the framework, to make provision for situations where this shortened version of the plan should be used. More radically, if in the third case no appropriate plan can be generated, a detailed analysis of the failure will help to determine how to revise (possibly through an expansion) the chosen set of methods.

As the above examples illustrate, cases of plans recognized — with or without obstacles — by the plan-recognition algorithm indicate the possibility of (and contribute useful guidelines for) the *reuse* of existing frameworks. Obstacles may result from various causes, such as changed initial states, unanticipated ways to use the methods, modifications in their definitions, etc.

7 Summary and Concluding Remarks

The Pr/Pg paradigm offers an approach for studying the behaviour of domains where facts are described by predicates and actions by operations, defined in terms of pre-conditions and post-conditions formulated over the predicates. Operations can be combined into plans. The record of the execution of plans related to certain agents and objects is regarded as a plot, summarizing a narrative involving these agents and objects. By a data mining process across plots registered in a Log, meaningful patterns can be discovered and put together to form a library of typical plans.

As explained, plan-recognition matches observed actions against the library, in order to anticipate what an agent is trying to do and the goal he wants to achieve. If obstacles are detected to recognized plans, an effort can be undertaken to either adapt the plans or to generate other plans able to reach the same goal.

The paradigm is quite general, due to the flexibility of the concept of operation. In the real world of information systems, hiring an employee or depositing money in a bank account are examples of operations. In the world of fiction, obtaining a magic sword from a wizard is an operation. In the world of computer software, moving a file from disk to buffer storage is an operation. We have had experience with the paradigm, handling examples taken from the first two worlds through our Pr/Pg prototype, which was developed in a version of Prolog enhanced with constraint programming. Our studies indicate that further investigation in the world of computer software, especially directed to object-oriented frameworks, may lead to useful methods to help in the design and maintenance of highly adaptable software products.

Acknowledgement: We are grateful to Fabio A. M. Porto for very helpful comments and suggestions.

References

1. A. Sheth. From contemporary workflow process automation to adaptive and dynamic work activity coordination and collaboration. In: Proc. of the 8th International Workshop on Database and Expert Systems Applications, September, 1997.
2. Furtado AL and Casanova MA. Plan and schedule generation over temporal databases. In: Proc. of the 9th International Conference on the Entity-Relationship Approach, 1990.
3. Furtado AL and Ciarlini AEM. Plots of narratives over temporal databases. In: Proc. of the 8th International Workshop on Database and Expert Systems Applications. R. R. Wagner (ed.). IEEE Computer Society, 1997.

4. Matheus CJ, Chan PK and Piatesky-Shapiro G. Systems for knowledge discovery in databases. IEEE Transactions on Knowledge and Data Engineering 1993, 5, 6.
5. Ciarlini AEM and Furtado AL. Simulating the interaction of database agents. In: Proc. of 10th International Conference on Database and Expert Systems Applications, (DEXA'99). Florence, 1999.
6. Bubenko jr JA. Challenges in information systems engineering". Invited talk at the 3rd Baltic Workshop on Data Bases and Information Systems, Riga. Technical Report DSV 98-004, Department of Computer and Systems Science, Stockholm Universit, 1998.
7. Wirfs-Brock R, Wilkerson B and Wiener L. Designing Object-Oriented Software. Prentice-Hall, Englewood Cliffs, 1990.
8. Fikes RE and Nilsson NJ. STRIPS: A new approach to the application of theorem proving to problem solving. Artificial Intelligence 1971, 2(3-4).
9. Furtado AL and Neuhold EJ. Fornal Techniques for Data Base Design. Springer-Verlag, Berlin, 1986.
10. Ozsoyoglu G and Snodgrass RT. Temporal and real-time databases: a survey. IEEE Transaction on Knowledge and Data Engineering 1995, 7, 4.
11. Casanova MA and Furtado AL. An information system environment based on plan generation. In: Proc. of the International Working Conference on Cooperative Knowledge Based Systems. Univ. Keele, England, 1990.
12. Berztiss AT and Bubenko jr. JA. A software model for business reengineering. In: Proceedings of Information Systems Development for Decentralized Organizations (ISDO95), an IFIP 8.1 Working Conference. Chapman & Hall, Trondheim, 1995.
13. Gamma E, Helm R, Johnson R and Vlissides J. Design Patterns: Elements of Reusable Object-Oriented Software. Reading: Addison Wesley Publishing Company, 1995.
14. Kautz HA. A formal theory of plan recognition and its implementation. In: Reasoning about Plans. J. F. Allen et al (eds.). Morgan Kaufmann, San Mateo, 1991.
15. Yang Q, Tenenberg J and Woods S. On the Implementation and Evaluation of Abtweak". Computational Intelligence Journal 1996, 12, 2: 295-318.
16. V. Propp. Morphology of the Folktale. Laurence Scott (trans.). University of Texas Press, Austin, 1968.
17. Afanas'ev A. Russian Fairy Tales. N. Guterman (trans.). Pantheon Books, New York, 1945.
18. Furtado AL. Analogy by generalization and the quest of the grail. ACM/SIGPLAN Notices 1992, 27, 1.
19. A. Jolles. *Formes Simples..* Paris: Éditions du Seuil, 1972.
20. Aarne A. The Types of the Folktale: A Classification and Bibliography. Translated and enlarged by Stith Thompson. FF Communications, 184. Suomalainen Tiedeakatemia, Helsinki, 1964.
21. Schank RC and Abelson RP. Scripts, plans and knowledge. In: Proc. of the Fourth International Joint Conference on Artificial Intelligence, 1975.
22. Rumelhart DE. Notes on a schema for stories. In: Representation and Understanding - Studies in Cognitive Science. D. G. Bobrow and A. Collins (eds.). Academic Press, New York, 1975.
23. R. Wilensky. Points: a theory of the structure of stories in memory. In: Readings in Natural Language Processing. B. J. Grosz, K. S. Jones and B. L. Webber (eds.), Morgan Kaufmann, San Mateo, 1986.
24. Greimas AJ. Sémantique Structurale. Librairie Larousse, Paris, 1966.
25. Warren DHD. Generating conditional plans and programs. In: Proc. of the AISB Summer Conference, Edinburgh, 1976, pp. 344-354.
26. Fayad ME and Schmidt DC. Object-Oriented application frameworks. Communications of the ACM 1997, 40, 10.

27. Johnson RE and Foote B. Designing reusable classes. Journal of Object-Oriented Programming 1988, 1, 5.
28. Pree W. Design Patterns for Object-Oriented Software Development. Addison-Wesley, Wokingham, 1995.
29. Fontoura MFMC, Haeusler EH and Lucena CJP. A framework design and instantiation method based on viewpoints. Technical report MCC35/98, Dept. Informatica, PUC-Rio, 1998.

Frameworks of Information Modelling: Construction of Concepts and Knowledge by Using the Intensional Approach

Hannu Kangassalo
Department of Computer and Information Sciences, University of Tampere
Tampere, Finland
E-mail: hk@cs.uta.fi

Abstract: Information modelling can be performed by using different approaches. Most methods developed emphasise set theoretical, extensional approach. The intensional approach emphasises the content and structure of concepts. In this work we study conceptual modelling and construction of concepts, conceptual models, and factual knowledge, by using the intensional modelling approach.

1 Introduction

There are several kinds of information modelling approaches applied, and conceptual modelling seems to be one of the most important of them. We regard conceptual modelling as a process of:

1. creating, recognition, or finding the relevant concepts and conceptual models which describe the Universe of Discourse (UoD) of the information system (IS), and
2. constructing or selecting the relevant conceptual content of information to be contained in the IS, based on this UoD.

This characterisation emphasises information, concepts, and knowledge, i.e. the content of the information system, instead of the physical implementation of it. Often the conceptual model, or the conceptual schema, can be regarded itself as a desired system, too. In principle, the conceptual content of information contained in the IS includes the derived and deductive information, too, although it may be in an implicit form.

The first attempts to develop a high level specification of an information system were made forty years ago. The infological approach to data bases and information systems was developing in Scandinavian countries early, on 1960's and 1970's. Janis Bubenko was actively developing the conceptual aspect to modelling and information systems (e.g. [1, 2]). In 1978 the first conceptual level graphical user interface was implemented [3]. The development is still continuing (e.g. [4]).

Most methods for conceptual modelling emphasise a set theoretical, extensional approach. It is based on the set of instances. Concept formation and conceptual modelling are based on intensional aspects which start to form much earlier, before we have any instances or even hypothetical instances, yet. The intensional approach emphasises the content and structure of concepts used to model phenomenon in the actual or hypothetical world. In this work we study some aspects of construction of concepts, conceptual models, and knowledge by using the intensional modelling approach.

We will first study the foundations of conceptual modelling. Then concept formation and conceptual modelling are considered from the point of view of members of the user society and the professional modeller, and from the point of view of the researcher studying the modelling process. After that, some aspects of concepts, definitions, and concept structures in the intensional approach to conceptual modelling are described. Finally the structure of an intensional concept system is briefly described.

2. Foundations of Conceptual Modelling

There are many theories about the nature and origin of concepts, e.g. [5, 6, 7]. They are basic notions for conceptual modelling and scientific research, which are closely related processes. According to some theories, concepts are regarded as cognitive 'tools' necessary to human thinking. They result from human cognitive processes and indicate what kind of conceptual content a person has received or formed for himself, concerning the subject matter, which may be the UoD or some abstract epistemological constructs. According to some others, e.g. realistic theories, concepts are abstract entities existing independently of any human being observing them. In this work we apply both approaches.

People use concepts generally without giving any closer thought to where concepts originally come from and how they behave. Usually they do not think about what kind information is included in concepts, how that information changes and develops, or how the users' knowledge of the entire application area develops. In conceptual modelling these questions are important.

The fundamental notion, *concept*, is defined to be a possibly named, independently identifiable structured construct composed of knowledge primitives and/or other concepts [4]. It can also be regarded to be an intensional knowledge structure which encodes the implicit rules constraining the structure of a piece of reality (cf. [8, 9]). It is a basic epistemological unit of knowledge. It has two roles at the same time:

1. it composes and organises information regarded as necessary or useful for structuring and understanding some piece of knowledge, and
2. it characterises some features of objects which fall under it.

A *basic concept* is a concept which cannot be analyzed using other concepts of the same conceptual system. Concepts are not classified e.g. into objects, entities, attributes, relationships, events, or processes. This kind of classification is not an intrinsic feature of knowledge - it is a superimposed abstract scheme into which systems of concepts and knowledge are forced, and which depends on our ontological commitments.

There are various ontological schemes which are commonly used, and each of them is a basis for the different ontology, i.e. the set of model concepts, rules and principles of how knowledge is composed. Sometimes this scheme has to be changed and the concepts should be re-classified. Therefore, we will not emphasise this kind of classifications of concepts, although we have to apply them.

A *derived concept* is a concept (definiendum) the characteristics of which have been derived from the characteristics of other concepts in the way described in the definition of that concept. The defining concepts are in turn defined by other concepts until the level of undefined, basic concepts is reached. [4]

The result of conceptual modelling, i.e. conceptual content of concepts and conceptual models, depends on:

- information available about the UoD,
- information about the UoD, regarded as not relevant for the concept or conceptual model and therefore abandoned or renounced,
- additional knowledge included by the modeller, e.g. some knowledge primitives, some conceptual 'components', mathematical structures, etc.,
- ontology used as a basis of the conceptualisation process,
- epistemological theory which determines how the ontology is or should be applied in the process of recognising adequate conceptual models or theories,
- the purpose of the conceptual modelling work,
- the process of the practical concept formation and modelling work,
- knowledge and skill of the person making modelling, as well as those of the people giving information for the modelling work.

A concept, and the conceptual model as well, is a result of a concept formation process, in which the structure of it is constructed from other concepts and of more primitive pieces of knowledge. In some cases concept formation is a quite 'natural' process, and sometimes a concept is constructed with a specific purpose in mind. In some approaches to conceptual modelling the conceptual model is itself a single structured concept defining the whole UoD (see [4]).

Conceptualisation, i.e. human concept formation, can be seen as a process in which individual's knowledge about something is formed and gets organised (see e.g. [10]). This view is applied much in education, but it is not complete and precise. On the basis of the characterisation of the notion of a concept, the result of the conceptualisation process is regarded as an intensional knowledge structure which encodes the implicit rules constraining the structure of a piece of reality (cf. [8, 9]). These results are concepts or more weakly organised conceptions.

The content of a concept is its *intension*. It consists of other concepts, knowledge primitives, and the structure they form in the concept. If a conceptual model is a concept defining the whole UoD, then the whole conceptual content of the model is the intension of the same concept. The concepts, knowledge primitives, and the structure they form in the intension of a concept are called its *characteristics*. A basic concept cannot have other concepts as its characteristics.

For example, the concept of an authorised doctor contains the concept of an adult person, the concept of an education passed by that person and required to become a doctor, and the concept of a license required to run the profession. In fact, the concept of the authorised doctor is an epistemologically abstract concept which can be expressed as an intensional sum of these concepts (see more details in [11, 12]).

Observe, that all these three concepts are (intensionally) contained into the concept of the authorised doctor, but only the concept of the adult person is related to the concept of the authorized doctor with the is-a relation. What is the relationship between the concept of an education passed by that person and the authorised doctor, or the relationship between the concept of the license to run the profession and the authorised doctor?

The set of objects (as well as data representing these objects) to which the concept applies is called its *extension.* A concept has always the intension, but its extension can temporarily or always be empty. The elements of the extension are called *instances* of that concept.

The *human conceptual model* (including its intension) is important for determining what we are able to recognise in the UoD, because it contains the results of all conceptualisation processes made by the person. However, we have to take into account, too, that a person may have several conceptual models which are not necessarily coherent with each other.

When a new information system is being developed, then first a human conceptual schema should be developed or re-organised. On the basis of it, an external concept descriptions and relationships between them should be constructed and located into the external conceptual schema. The external conceptual schema is a construct based on the human conceptual schema, on the representation language, and on the semantics of the representation language. By means of the external conceptual schema the conceptualisations of relevant concepts and the rules between concepts are made 'visible'. With it the recognition, structure, behaviour and functions of the object system, and the information content based on this object system are externally described and shared.

In general philosophy, ontology is a theory of existence. In formal ontology and conceptual modelling *ontology* can be regarded as a logical theory which gives an explicit, partial account of the conceptualisation. Guarino and Giaretta say also that ontology can be regarded as a synonym of conceptualisation, (see above, and cf. [8, 9]). The difference between a concept (or a conceptual schema) and an ontology is very small, if there is a difference at all. In a concept the set of those implicit rules constraining the structure of a piece of reality is unique for each concept. An ontology is an intensional knowledge structure encoding the implicit rules, possibly a number of times, like a theory applied in different cases. An example of this kind of ontology is the ER model. A theory consists of semi-abstract model concepts and rules between them. They are called model concepts because they form a (semi-abstract) schema to which all actual knowledge about the UoD must conform, when the mapping to the real world is done. The rules determine how the model concepts of the ontology are related.

There are two alternatives of how the notion of ontology can be interpreted. Usually the model concepts in an ontology contain only relatively few characteristics and none of them contain a direct reference to the actual and concrete UoD, i.e. the concepts are epistemologically semi-abstract. Only when actual knowledge about the UoD is added, i.e. when the ontology is applied to the actual UoD, these characteristics are concretised, like in the ER model. The other interpretation is that an ontology is a conceptual model describing the conceptual content of the UoD completely.

Epistemology is a branch of research that studies knowledge. It is the study of nature of knowledge and justification: specifically, the study of (a) the defining features, (b) the substantive conditions, and (c) the limits of knowledge and justification [13]. It attempts to answer the basic question as what distinguishes true or adequate knowledge from false or inadequate knowledge? Practically, this question translates into issues of scientific methodology; how one can develop theories or models that are better than competing theories or models. That question is exactly the same as the basic question for selecting the best conceptual content of the information system.

Epistemology of conceptual modelling is a branch of theory that studies concepts and knowledge in conceptual models and in the conceptual modelling process, especially questions as what distinguishes adequate concepts from inadequate concepts in conceptual models, and how one can develop theories or conceptual models that are better than competing theories or conceptual models. It is thus interested in principles of which kind of knowledge is needed and used in creating an adequate conceptual model of the UoD, and how that knowledge should be formed or synthesised.

3. Modelling Process and Modelling Situations

3.1 Modelling: the Beginning

On the initial stage conceptual modelling and concept formation are cognitive actions of a human being. Usually the work starts from observations concerning some details of the UoD. The set of users may recognise an epistemic, conceptual difficulty, which they try to handle by using concepts and knowledge at their possession so far. If they can not solve that difficulty, then a conceptual change may be needed. The system of concepts, using which the persons are modelling and structuring the UoD, has to be changed and structured in a better way.

Often a professional modeller is invited to handle the case. The modeller is compelled to recognise, analyse, and design, with the users, the concepts of an application area, the content and structure of concepts, and how these concepts and their structures are applied in users' thinking.

The users are utilizing the concepts developed to be most suitable for their own views. Often the modeller's and the users' views on the UoD, or even on the same concept, may differ from each other quite a lot. It may sometimes be difficult to say whether the users, or the modeller and the users, are using the same concepts at all.

The only way to find it out, is to analyse the intension of concepts in details. In this analysis the concepts regarded as necessary or useful to construct the concept in question are studied until the level of undefined, basic concepts is reached.

The modeller is often compelled to develop new concepts, the content of which must fit into, and be consistent with, the earlier developed concepts. Since conceptual modelling process often initiates the knowledge development process both in user's and modeller's mind, the modelling work is iterative and lateral.

3.2 Different Models in Different Situations

Modelling concerns always at least three parties: the object of modelling, the modeller, and the model. Two cases can be separated: a user is making a model of the object, or a professional modeller is making the model for the user. Concepts are created by the human mind, partly on the basis of information structures grounded on observations, partly on the basis of abstract reasoning, and sometimes, partly, on the basis of communications with other people within the society.

We describe some cases which concern co-operation in the concept formation and modelling work, e.g. that a user is the modeller, that there are many users, that the user and the modeller are different persons, and finally that there exists an extensive environment on which modelling is performed in many ways. Analysis of differences between these cases reveal that concept formation is quite different in each case and the role of intension of concepts is important. The intensional approach to modelling is applied in each case.

A diagram in which a person is paying attention to a phenomenon in the UoD, possibly as a result of an observation, and constructing concepts, is shown in Figure 1. Observation is first conceived as a mental image in the human mind. Later on, a concept may be constructed, in which information in the mental image may be included, possibly when more information has been accumulated. There are several theories about concept formation both in cognitive science and psychology (see e.g. [7, 14]).

From the point of view of this work, two (extreme) approaches are interesting. The exemplar view [15] is assuming that a summary description of instances is not created to represent concepts, but instance information is accepted as an instance on the concept if it is similar to other examples of the concept. That clearly extensional view is suitable for representing instance information, e.g. information of instances of cars in a town, but it does not deal with concepts.

The other extreme is that concepts are based on knowledge and world theories that constrain the construction and organisation of concepts [7], i.e. representations whose structure consists in their relations to other concepts (see e.g. [6, 16]). This approach seems to be a more appropriate candidate as a starting point for modelling human concepts in information modelling.

According to the constructivistic personal construct psychology [17], a concept structure in a person develops on the basis of his experiences, including his own observations, communication with other persons, his own reasoning, reading, organised learning, etc.

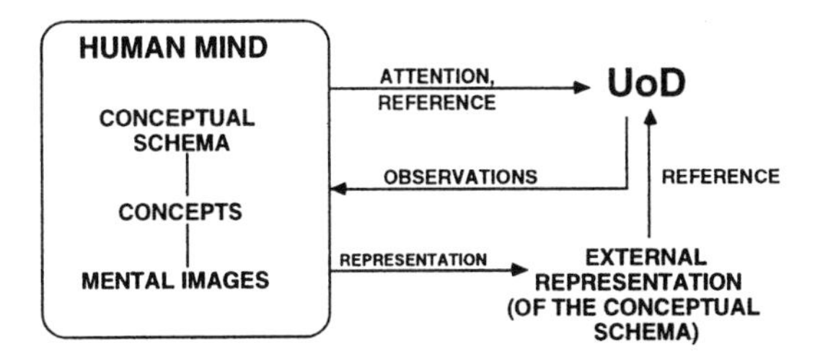

Figure 1. A phenomenon in the UoD, concepts, a conceptual schema, and an external representation

The external representation of a conceptual schema is represented by using more or less formal notation. The external representation of a concept is used e.g. for communication. The user of it must master the concept or the concept system itself before he can meaningfully use the symbol. On the other hand, the 'reader' of the symbol must have enough information to grasp the concept, or the meaning of the symbol.

A human concept is always a subjective thing. Similarly, a conceptual schema of the UoD of a person is a subjective construct, which reflects the personal perspective of that person who has created it. However, people live in societies and therefore need common concepts and conceptual schemata about their environments. That need (among other things) triggers the processes of learning and sharing of concepts and conceptual schemata with other people. Otherwise communication between them would not be possible at all.

Figure 2 illustrates the situation in which people are sharing at least some of their concepts and make some common external representations (only one is shown). All the community members have some concepts which are 'their own' and some concepts which are shared with other people. The community members each have their own views of the phenomenon in the UoD, based on their own observations, and they organise their knowledge on the basis of their own history, their own skills of thinking and forming concepts and knowledge, and on the concepts they were using in constructing their views.

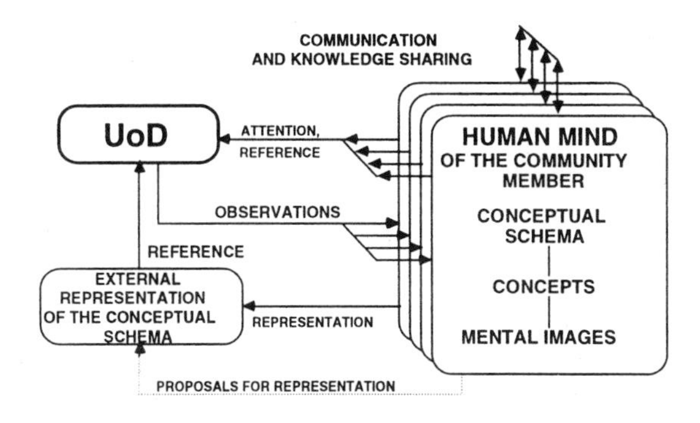

Figure 2. A person shares some of his concepts within the community.

Sharing a concept means that two persons posses the token of a concept the intension of which is (at least nearly) the same as the intension of the concept of the other person. Both persons have the token of the shared concept in their conceptual schemata, i.e. they posses (at least nearly) same knowledge. However, both tokens belong to a different conceptual schema, and therefore the result of reasoning in which these tokens are applied, may mean different things to both people. Therefore analysis of tokens of shared concepts and systems of concepts should be performed carefully, down to the finest details of the intensions, and taking the neighbouring concepts into account as well.

Differences in concepts and in the way how the communicating parties are using words in the communication process may have the effect that the receiver understands only some fraction of the message in the way the sender thought it should be understood. There are several methods how the communicating parties may try to improve understanding of the message but complete understanding can hardly be achieved. However, learning and the extensive education systems tend to standardise some of the concepts, and the use of language, but not all of them. [18]

When people are representing their private or common concepts an additional collection of concepts and conceptual schemata is accumulating. That collection is represented by EXTERNAL CONCEPTS in Figure 3. They may effect in several ways to the creation and evolution of concepts developed later, e.g. prevent the evolution of newer concepts, or extend the 'life-time' of useful concepts. Stored external concepts can be studied and applied, modified and refined in making new conceptual schemata.

The modeller and each member of the community have a more or less different concepts and different conceptual schemata. Their concepts may be based on different epistemological theories, too. Therefore the analysis and sharing of different schemata may require much work and a strong investment e.g. in a company. We will not study these problems in this paper.

The external conceptual schema may be represented in a computer. Figure 3 reveals that on the basis of one UoD several (at least 5) collections of private or common concepts or systems of concepts may evolve. The human conceptual schemata are evolving (all the time) and therefore the whole complex of schemata must be maintained.

Both the modeller and users should pay attention to the applicability of the external/computerised concepts to modelling the current UoD. If the external concepts are represented by using the surface structures of the language, only, then the (descriptions of) intensions of concepts may not be understood properly. If the concepts are epistemologically abstract, i.e. they are on the high level in the concept construction hierarchy, the user of the concepts must, in principle, know the other concepts in the hierarchy, down to the basic level. It is important that in collections of external concepts the descriptions of intensions of concepts should be stored and maintained, i.e. that complete definition are stored, whenever possible. The external conceptual schema contains the *description* of concepts and rules, and the data base (not shown in the figure) contains the extension corresponding to the intension of the conceptual schema.

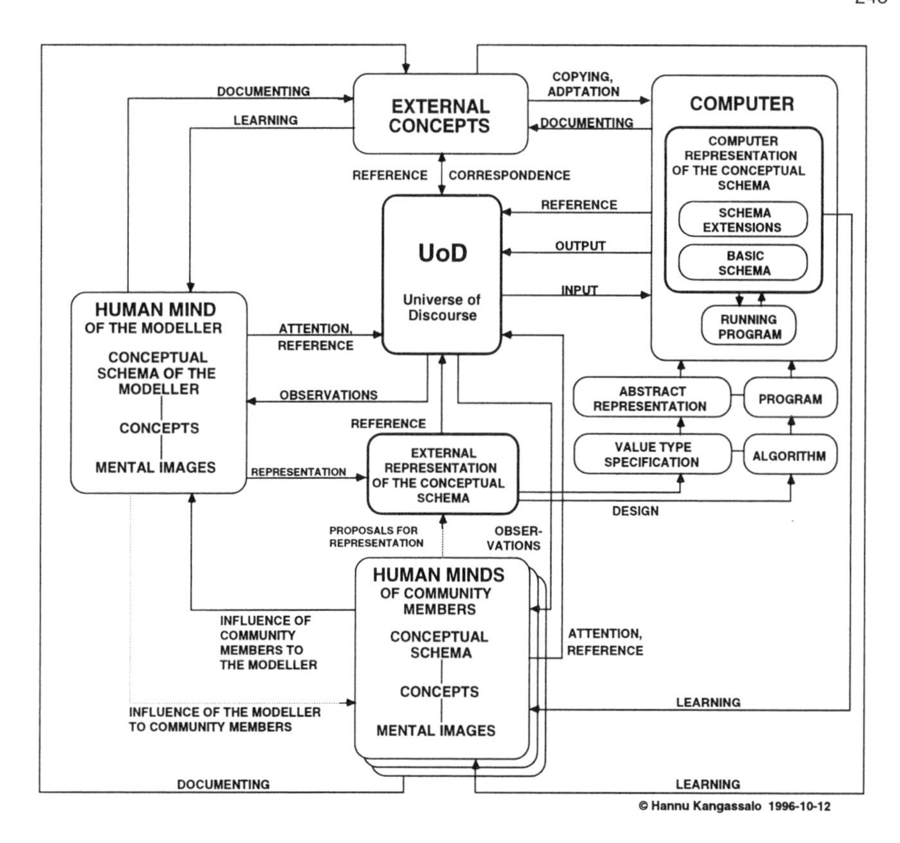

Figure 3. Human mind, linguistic constructs and a computer as a 'media' of concepts

4 Concept Structures

Structurally the derived concept is always a directed a-cyclic graph based on the relation of intensional containment (IC) (for IC, see [4, 11, 12]). The graph contains the definiendum and all information needed to define it. Intensionally contained concepts are defining concepts, i.e. concepts used in the definition, but all the defining concepts shall not necessarily be contained concepts. In addition, the graph contains intensional containment relations which relate the definiendum to defining concepts, and possibly knowledge primitives which add some characteristics to the concept. A concrete example can be seen in Figure 9 of [4].

A concept can be described by using the characteristics of it. This description is called a *definition* which is a set of rules or linguistic instructions which specify how the knowledge forming the defined concept is to be constructed from the knowledge given in the defining concepts and in the definition itself [4]. Because a definition may contain conceptual operations, then in order to find out the intension of the defined concept the definition must be evaluated, i.e. the definiendum must be actually constructed.

A definition and the defined concept are different things. Because of that several different definitions may all evaluate to the same concept. [4] The definitions can even be of different types, i.e. the type of one definition may be an aggregate, and the type of another definition may be a generalisation, but they may evaluate to the same concept. This is important from the users' point of view. If two (or more) users give different definitions to the same concept, users can be allowed to use their own definitions, which may be very helpful for their work.

A *concept structure* is an external diagram which represents a definition of a concept. It consists of a defined concept and of its definition hierarchy, in which the characteristics of the definiendum derive from the characteristics of basic concepts [4]. The graphic layout is meaningful in a concept structure diagram. The definiendum is on top of the hierarchy and concepts defining it are on the next or lower levels of the hierarchy. Concept structure diagrams can be used to define concepts of the working environment of users and the modeller. They may describe (concepts of) customers, products, other objects, events, processes or relationships, data requirements, reports, data structures. They can be as large and complex as needed to explain the essential content of concepts.

5 Structure of an Intensional Concept System

When abstraction type of generalisation was taken into use the representation method based on the work of Quillian [19] was adopted. It usually shows general constructs at the top of the diagram and specialised constructs at the bottom. However, that extensional representation scheme seems to be inappropriate in many cases. For example, in the description of a company, the highest concepts are not necessarily the most general concepts, rather the contrary.

A schematic example of the structure of the intensional concept system is described in Figure 4. In the diagram, concepts are represented by circles and intensional containment relations between concepts by arrows (see more details in [4]). If a concept is defined by a value transformation (function), then the intensional containment relations are combined and a symbol (Fi) for function specification in included.

The highest concept is the concept which intensionally contains the whole conceptual system. It is also the most specific concept. Usually it is an intensional sum of the concepts on the next or some lower level (see [11, 12]).

Each concept in the system is constructed in the same way: the concept contains intensionally the defining concepts, which contain its defining concepts, etc., until the level of basic concepts is reached. A concept may contain intensionally two contained concepts (intensional sum), or two concepts may have a common concept (intensional product). Kauppi has defined the notions of intensional negation, intensional difference, and intensional quotient, too [11, pp 53-71].

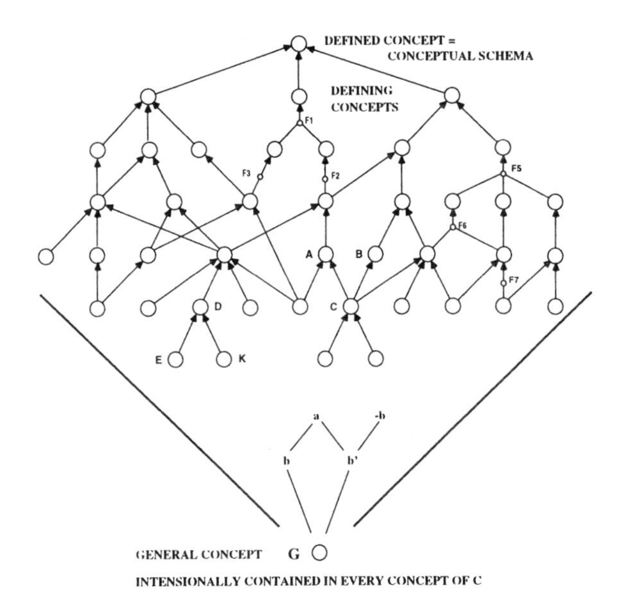

Figure 4. Schematic structure of a conceptual system

The lowest concept is the General Concept, which is intensionally contained to all other concepts. Any other concept may contain generalised concepts which are based on information contained into the defined concept, e.g. the concept of a doctor contains intensionally the concept of a person.

The diagramming technique requires that all the information contained into the concept is represented under the concept (and its name). Therefore the system of concepts is very well organised and it is easy to read.

Theoretical justification for the intensional structure has been presented in [12, pp. 94-97]. Palomäki has concluded that the formal structure is a complete semi-lattice (when intensional sum, intensional product, and intensional negation are taken into account). The development of the theory and experiments are continued.

References:

1. Bubenko J.A. Jr. The Temporal Dimension in Information Processing. In Nijssen GM (ed) Architecture and Models in Database Management. North-Holland, 1977, pp. 93-118
2. Bubenko J.A. Jr. IAM: An Inferential Abstract Modeling Approach to Design of Conceptual Schema. ACM-SIGMOD-77, pp. 62-74.
3. Senko M.E. Foral LP: Design and Implementation. In Yao S.B. (ed) Proc. Intl Conf on Very Large Data Bases. West-Berlin, 1978, pp. 255-267
4. Kangassalo H. COMIC - A System and Methodology for Conceptual Modelling and Information Construction. Data & Knowledge Engineering 9 (1992/1993) pp. 287-319

5. Bunge M. Treatise on Basic Philosophy. Vol. 1. Semantics I: Sense and Reference. D Reidel, Dordrecht, Holland 1974
6. Laurence S, Margolis E. Concepts and Cognitive Science. In Margolis E, Laurence S (eds.) Concepts - Core Readings. MIT Press, London, 1999
7. Medin D.L, Goldstone RL. Concepts. In Eysenck MW. (ed.) The Blackwell Dictionary of Cognitive Psychology. Basil Blackwell, Oxford, UK, 1990, pp. 77-83
8. Guarino N, Giaretta P. Ontologies and Knowledge Bases: Towards a Terminological Clarification. In Mars, NJI, (Ed), Towards Very Large Knowledge Bases, IOS Press, 1995, pp. 25-32
9. Guarino N. Formal Ontology and Information Systems. In Guarino N. (ed) Formal Ontology in Information Systems. IOS Press, 1998
10. Klausmaier HJ, Conceptualizing. In Jones BF, Idol L. (eds) Dimensions of Thinking and Cognitive Instruction. Lawrence Erlbaum, Hillsdale, N.J., 1990, pp. 93-138
11. Kauppi R. Einführung in die Theorie der Begriffssysteme. Acta Universitatis Tamperensis, Ser.A, Vol.15, Tampereen Yliopisto, Tampere, 1967
12. Palomäki J. From Concepts to Concept Theory. Discoveries, Connections, and Results. PhD thesis. Acta Universitatis Tamperensis. Ser. A, Vol. 416, University of Tampere, Tampere, 1994
13. Moser PK. Epistemology. In Audi R. (General editor) The Cambridge Dictionary of Philosophy. Cambridge University Press, 1996
14. Margolis E, Laurence S. (eds) Concepts - Core Readings. MIT Press, London, 1999
15. Smith EE, Medin DL. The Exemplar View. In Smith EE, Medin DL. Categories and Concepts. Harvard University Press, Cambridge, Mass. 1981
16. Murphy GL, Medin DL. The Role of Theories in Conceptual Coherence. Psychological Review. 92, 1985, pp. 289-316
17. Kelly GH. The Psychology of Personal Constructs. W.W. Norton & Company, New York, 1955, Two volumes.
18. Kangassalo H. Are Global Understanding, Communication, and Information Management in Information Systems Possible? - A Conceptual Modeling View - Problems and Proposals for Solutions. In Chen PP, Akoka J, Kangassalo H, Thalheim B. (eds) Conceptual Modeling: Current Issues and Future Directions. Springer-Verlag, 1999, pp. 105-122
19. Quillan MR. Semantic Memory. In Minsky M. (ed) Semantic Information Processing. MIT Press, Cambridge, MA., 1968

Towards a Useful Information Concept

Paul Lindgreen
Professor emeritus, Copenhagen Business School
Copenhagen, Danmark

Abstract

It is sad that there is no concensus among professionals about what the word 'information' means - or should mean. Looking at it from six perspectives the paper describes *one particular concept* to serve as the *fundamental common sense* of the word, and it argues that it would be common sense to adhere to this common sense.

Where is the knowledge,
we have lost in information?
T.S.Eliot

1. To Escape a Chaotic Situation, . . .

In Informatics*) the word 'information' is probably the one homonyme that causes most problems in professional communication. Alone and in combination with other words it is used as a term for a vast number of individual concepts most of which are rather irrelevant and some even senseless. It is a severe problem that everybody disagree about what 'information' means - or rather what it should mean to serve as a proper term for a *useful* concept. On the contrary: There is mere chaos.

To illustrate the Babylonian confusion the following is a minor part of a long list of definitions informally collected from dictionaries, textbooks, standards and by asking: what is information?

- Information = $\log_2(M)$.
- Information is evaluated data.
- Information is the level above data.
- Information is exchanged knowledge.
- Information is data put into a context.
- Information emerges when data is processed.
- Information is facts which are communicated.
- Information is embodied or mediated knowledge.
- Information is data that has been automatically processed.
- Information is a subset of data present in the form of knowledge.

---

*) This paper is based on a chapter from a dissertation-draft describing *a framework of fundamental concepts* to serve as a foundation of a proper professional language for the areas of communication, datalogy, computer science, and so-called information-systems - here all in one called *Informatics*.

- Information is knowledge recorded by and/or communicated between agents.
- Information is the knowledge-increment brought about by a receiving action in a message transfer.
- Information is the meaning currently assigned to data by means of the conventions applied to that data.
- Information is any kind of knowledge about things, facts, concepts, etc. of a universe of discourse that is exchangeable among users.
- Information is any distinct signal element forming part of a message or communication, especially one assembled and made available for use by automatic machines.

The list is representative of the many, sometimes curious senses of the word, and it clearly demonstrates the tendency to focus rather on very special senses instead of aiming first at clarity about a general concept. This "going for the general" as regard a useful information concept, is exactly what is attempted in the following. And fortunately, *there is already one single concept* that can - and seriously should serve as the basis for whatever specialization one may need. The different roles this concept plays in practice are all well-known, but few people seem to realize that they are played by *one and the same concept.* By looking at it from different perspectives the paper will describe and justify it as *the most appropriate fundamental concept to deserve being called 'information'.*

The Common Language Perspective

Since most professional languages have a base in common language (CL), we shall start considering the CL-sense of 'information'. At first sight this sense seems to be rather close to that of the word 'knowledge'. For example, saying: 'we must have knowledge of the girl's address' means the same as: 'we must have information about her address'. Common for the two is also that when used in practice, they usually refer to certain *portions* of knowledge/information. Information about the girl's address is one such portion, and although knowing what characterizes an equilateral triangle is quite different from knowing to play chess correctly, both are portions of knowledge.

Sometimes, however, 'knowledge' and 'information' are not used in quite the same way indicating an important difference. Aspects of this is revealed in two ways: by some of the *connotations* usually associated with 'information', but *not* generally with 'knowledge', and by *the two verbal forms* - 'to know' compared with 'to inform'.

The connotations in CL specific for 'information' comprise «fact», «well-determined», «exact», «rational», «formal», «explicit», «precise», «discrete», etc. We shall discuss some of them further, but so far they just indicate that it may be useful to think of information as a special kind of knowledge.

The verb *'to know'* indicates *a mental state* - the portion known at a certain time. In case of a single person it is a subjective state, and when it is knowledge shared by a group of persons it is an inter-subjective state. Fair to the CL-sense we shall define:

The knowledge of a person is that state of the person's mind which over time has been established - by *experience* from individual or social behaviour, by *transfer* from other persons, or by creative or deductive *thinking* - and which potentially may *influence* any *future* behaviour of the person.

In contrast, the other verb *'to inform'* indicates *an action*. It may be one aimed at informing oneself - acquire a portion of knowledge - for example by reading the address of the girl somewhere, or by asking a specific question at an information counter. The aim could also be to inform someone else, e.g. by telling a taxi driver to go to the girl's address. But although most people daily experience succesful communication, there seems to be some kinds of knowledge which it is impossible in practice to transfer properly. When we succeed it is usually with knowledge for which the connotations «well-determined» or «precise» are appropriate characteristics.

Thus, the two verbal forms and the connotations indicate the difference: Although a person in principle can acquire any *kind* of knowledge, it is different when a person is being informed of something. In order to succeed in a communication, the partners must be sure that the portion of knowledge the sender intends to transfer is *precisely the same* as that which actually is received. But then the portion must be well-determind. And the connotations indicate that information is exactly the kind that can be transferred with succes.

Therefore, the CL-perspective reflects that it may be useful to distinguish this way:

Information is a kind of knowledge that in practice can be *correctly* transferred between persons *in well-determind portions.*

The Intellectual Knowledge Perspective

Given knowledge as defined above, we shall now be more explicit about the *kind of knowledge* that is claimed to be relevant as information: It should be obvious that a person cannot transfer a certain portion of knowledge unless he or she knows it. But that is not enough. The person must also *be conscious* of it. Therefore, we shall first define:

Intellectual knowledge is knowledge which a person *potentially* can become conscious of.

However, a person can very well be conscious of some intellectual knowledge, but it can still be of a kind that another person cannot become precisely informed of, even if both persons have all intentions to achieve this, and even if they both understand and can use the same language. Whatever they try, this kind of knowledge cannot be conveyed *in any definite or exact way.*

Consider, for example, a person who immediately after a concert is asked by a friend: «How did the pianist play?» The person thinks a while and answers: «Oh, she played wonderfully in a soft floating manner and with a warm sense of the instrument. I believe she really understood the feeling the composer wanted to express in the composition.»

From the answer it seems that the person is fairly conscious of the performance and also is able to express it in some manner. But is the portion of knowledge the friend acquired by interpreting the answer really the same as the one the person intended to convey? Is there, in fact, any realistic way to be certain about it? The answer to both questions is: «No!».

The reason is that all references to specific performance-aspects applied in the answer are highly subjective and wide open for interpretation. It could also be because there are no appropriate words for most aspects of the experience. Many feelings that consciously can be recalled in mind cannot be properly described by whatever means.

While there are many situations where the transfer of intellectual knowledge can be accomplished with success, there are many cases where it is not so. Consider for example a scientist or an inventor who try to describe how they "by intuition" or "by a hunch" got the right idea that mentally "guided" them to a certain important scientific realization, respectively to an epoch-making invention. The persons are fully conscious of the situations, but they cannot convey what caused the idea to emerge. Also when an experienced doctor shall explain the result after examining a patient - how a complex and far from unique pattern of symptoms together with memory of similar cases resulted in a correct diagnosis. The doctor is fully aware of the diagnostic situation, but cannot express the crucial aspects such that *other doctors* can apply the knowledge *with certainty* in their own future diagnoses.

It doesn't mean that one cannot acquire and apply this kind of knowledge. *Neither can we conclude* that it is impossible to transfer at least some portions or some derived "essence" of it to other persons, but the receiver can never be sure which things and which types of aspects the sender really implied in the transferred knowledge.

We shall use the word *'empation'* for this kind of intellectual knowledge.

Empation is intellectual knowledge that can *not* be reproduced in a *certain way.*

The term 'empation' is coined from Greek 'empatheia' derived from 'em' . «in» and 'pathos' . «feeling», cf. the English word 'empathy', the German 'Einfühlung' and the Danish 'indfølelse'. There are several connotations to the word and many examples of empation in practice - feelings, intuition, "chemistry" between people, sense of humor, pain, joy, the delicious wine yesterday, the smells in a Middle-east souk, the extreme beauty of Grand Canyon, etc.

We shall therefore end this perspective by stating:

Information is the complement to empation in respect to intellectual knowledge.

However, although this distinction is very important for understanding "the nature" of information, it does not say that information generally should be more important for our life and behaviour. The examples surely indicate the importance also of empation. Empation may even be the key to answering T.S.Eliot's question at the beginning.

The Perspective of Well-Determined Enquiring

After this indirect approach we shall now focus on *the most crucial aspect of Information* – one that is closely connected with the connotations «well-determined» and «fact». But to avoid philosophical discussions of how fact is related with the possible senses of the word 'truth', we shall just regard a fact as something that *in a certain social domain* can be *relied on* as "being the case".

A traveller asking an arbitrary person at Hamburg central station from where the next train to Cologne will depart may get the answer: «I'm not sure. I believe they used to leave from track 13». Here the traveller has acquired a certain portion of intellectual knowledge (now the traveller knows part of what the other person believes), but it may cause problems to rely on the answer as a *fact* about train departures. The question is well-determined (in the given station-domain), but the answer is quite uncertain. The traveller receives no useful information pertaining to the question.

But if the traveller instead directs the question at a station counter signed 'INFORMATION', the answer could be: «It departs from track 14». In this case the traveller gets a portion of *factual knowledge* and the following behaviour governed by it can be expected to be succesful. The traveler has recieved information, because both the question *and* the answer are *well-determined.*

A fact is a portion of intellectual knowledge that is quite different from empation, because something cannot be a fact, unless it is well-determined. But what is it that determines the well-determinedness? Again one of the connotations - here «explicitness» - is a key to understanding: Consider the following examples of questions in CL, which all are well-determined (at least in some implied domain): Shall we sell or buy? Will 10.45 be too late? How are the rules for chess? Where is the book department? How long is that pencil? Who was the first president in Israel? Are the first digits of her telephone number '24' or '25'?

In all the questions the phrasing determines *explicitly* what the information is about. For example, one question concerns a certain instance of the type of public profession: *«president of a nation»*. It is not about somebody who is carpenter, judge, school teacher, or secretary of foreign affairs. Moreover, the question is about the president of *Israel* - not of Egypt or Argentina or any other nation. Furthermore, it concerns the *first* president - not the second or the current one. It is about the *name* of the president - not about the year of birth or the possible spouse of this person.

In another question the required information is about the particular type of game known as *«Chess»* - not of «Monopoly» or «Trivial pursuit», and it is about the *rules* for playing the game - not about when or where the game was invented or first played.

Well-determined answers to well-determined questions can typically be expressed as:

- a choice between a set of *unambigous possibilities*, eg.: {«yes», «no»}, {«up», «down»}, {«Denmark», «Norway», «Sweden»}, {«Monday», «Tuesday», ...}, etc.
- by means of *an identifier* or other kind of referential expression that is *unique* in the given domain, e.g. «Jupiter» or «The greatest planet in the solar system»
- as one or more *numbers* indicating a time, distance, weight, frequency, etc. or corresponding to results of counting, deciding, calculating, etc.

Very often answers are well-determined, when they apply or refer to *commonly recognized conventions* or *identification schemes*, typically as those known from the use of: Calendars, time-schemes, land-registers, postal addressing schemes, air traffic lanes, country or area-codes, ISBN-numbers, longitude/lattitude coordinates, etc. Generally, answers conveying information are often available from *common "information sources"*, for example, libraries, databases, dictionaries, thesauri, manuals, time tables, blueprints, directories, maps, etc.

As a conclusion from this perspective we can state:

Information is always about *well-determined aspects of well-determined things.*

The Entity Perspective

The word 'information' is derived from Latin 'informare' . «to give form to» coming from 'in' . «in» and 'forma' . «form». This etymology together with another of the connotations - discreteness - indicate that there always is a certain *formalism* involved if a portion of knowledge shall be regarded as information.

From the examples above it may seem that the information provided in an answer to *an elementary question* - one that cannot be divided into separate simpler questions - always is about: *a certain aspect* of *a certain instance* of *a certain type of thing*. And a special interpretation of such a discrete "unit of information" - such an *information element* is, in fact, a fundamental constituent of the formalism. The formalism is exactly what makes it possible to regard a certain portion of information as *a set of information elements*. Information is inevitably bound in the structure of the *things that are considered relevant in a certain domain*, because they play a role in the activities of that domain. In the station example such things as «destination», «train departure», «track», and «departure time» are relevant, and behind the rules-for-chess question are «player», «chess-piece», «position on chess-board», «move», etc.

But nearly always in practice matters turn out to be more complicated than first anticipated. Many relevant things are mental, i.e. cannot be physiologically sensed, and often things are compositions of physical and mental things, e.g. an organisation or a chess tournament. Furthermore, things are not always what they at first sight appear to be: Although a person who buys a product from some supplier may be a customer, it could very well be that the buying person just is someone who is employed by an organisation, which then is the real customer.

Before anything can be said with certainty about what the information concerns, it is necessary to carry out *a careful analysis of the domain*, where it is decided *exactly* - not only which aspects of the involved things the information concerns, but also what it is that gives *identity* to the things.

And this brings us to the important connection with the concept *entity*. Or rather to a certain useful sense of 'entity', because in Informatics this word is also a severe homonyme associated with a number of dubious or vaguely defined concepts. For an entity concept to be useful it is necessary to take two pragmatic circumstances into consideration: First *entities are not absolute*. They are always *relative* to a certain (cultural / social / organisational / etc.) domain where things are *interpreted* in a certain way that is relevant for those who are active there. Secondly, *not all things have identity a-priory*. Mental things definitely not.

The notion of something having identity ensures that the existence of the thing, as interpreted, is *well-determined in the domain*, such that the thing *pragmatically* can be distinguished *with certainty* from all other things in the domain - in particular from all other things of the same type. But since most relevant things in practice do not have identity a-priory, it is necessary in the analytic process *socially to decide* exactly what it is that gives identity to the recognized entity types. Many people, in particular those influenced by the relational DB-model, believe that things get identity from a possible identification, and - the other way round - that relevant things must have a "primary key" as an explicit identification.

But this is a tool-influenced, unsatisfactory way of thinking. Generally it is much more adequate to base the identity on the set of aspects chosen as relevant, and then let *an appropriate subset of these aspects* determine the identity. This has the further pragmatic advantage that it forces the analyst to become conscious of what characterize the relevant entity types.

Therefore we shall define:

An entity is a *socially established interpretation* of *a thing in a certain domain* as being relevant for some purpose(s) and accordingly characterized by *a certain set of aspects* of which a subset uniquely determines *the identity* of the thing in the domain.

Then we can express the essense of the formalism behind information as follows:

An information element is a certain aspect of a certain instance of a certain type of entity.

And we can state:

A portion of information is always a non-empty set of information elements.

2. ... to Consider a Useful Concept, ...

After having so far encircled and described the information concept announced at the beginning, it remains to justify that it really is so useful as claimed. We shall do that by studying the concept from *two other perspectives* - each one pointing out the importance of information in situations that are very common and relevant throughout in practice.

The Action Perspective

The first "proof" of usefulness concerns a situation type encountered by most people many times every day: *the execution of actions* in order to achieve something considered important. It will be shown, partly by a simple example, that a certain portion of information always is *necessary to execute actions in a definite and purposeful manner.*

But before we can do that properly, we must once more take into account the lack of concensus as regard the sense of commonly used professional terms. Experience shows that people in Informatics (as well as in other professional areas) hardly ever associate the same concept with each of the following words: 'job', 'task', 'work', 'act', 'operation', 'procedure', 'behaviour', 'activty', and - 'action'. They may partly consider the same concepts, but will name them differently, and only very few will care to relate them in a sensible and useful way to each other. Therefore, we shall first by means of two specialization steps present a useful meaning to the concept called 'action'.

On the most generel level the word 'behaviour' is used simply as the term for *the cause of any change*. Then, as the first step we specialize in order to restrict us to *discrete* behaviour that is *not* going on by itself:

An activity is a behaviour that has *a well-determined start and termination*, that involves one or more *explicit operands*, results in a *discrete change*, and is carried out *by an explicit agent.*

The agent is typically a person or an artifact (e.g. a computer or robot) or any systemic composition of such elements. The agent of a certain activity must generally have the nessessary skills and ressources to be able to cause the change. From that we specialize once more to incorporate the pragmatic aspect *intention*:

An action is an activity for which there is *an explicit goal* which the agent is loyal to during the execution.

The goal is a well-determined (possibly combined) *state* of one (or more) of the involved operands - a state that the agent intents to achieve by executing the action. The goal is specific for each action type and most often in practice also for each instance of it. Therefore, whenever an action is to be carried out the agent must not only know what is necessary for executing that particular *type* of action, which typically is about the structure of possible sub-activities and about which *types of operands* are involved. The agent must also have or acquire exactly the *portion of information* that is specific for the particular *instance* of the action-type. It concernes which *particular* operands are involved and which *state* each of them should be brought into. Unless such situation-specific information is available for the agent, it is impossible *with certainty* to carry out the action according to the goal. One part of that information will determine the instance(s) of the involved *operand type(s)*, and another part the resulting state of the operands.

Consider for example, the action-type «drill hole in plate» where a skilled mechanic with a fully equipped drilling machine is the agent and where the goal is *to drill a specified hole in a certain plate.* Every time an instance of this action-type shall be executed, the agent must have *three specific information sets* available:

- one determining in which plate the hole shall be drilled
- one determining the location of the hole on the plate
- one determining the diameter and possible quality aspects of the hole.

Each of these information sets may comprise one or more information elements depending on the actual conditions in the machine shop. For example, is it just an instance of a standard type of plate or is an individually manufactured one? Also the location of the hole may require several information elements depending on the geometric shape of the plate and on the conventions for origo and orientation applied to determine it.

This is just a very simple example, but everywhere in private organisations and in public administration there are lots of complicated business procedures involving actions where laws, regulations, company policies etc. require that they must be carried out *strictly in well-determined manners*, such that the result in each case is *independent* of whom (or what) is chosen as the agent. But if the actions in any way depend on empation, this is not possible, because it cannot be verified after the action if the goal really has been achieved. The situation-specific knowledge must always be a certain portion of information, i.e. one or more sets of information elements concerning instances of previously recognized entity types.

The Data Perspective

In this final perspective we shall discuss some fundamental issues concerning the relationship between information and the equally important concept *data*. Well, again here, only if we associate a sensible meaning with the word.

Behind the wish for a useful data-concept are *two unavoidable circumstances* that influence our individual and social life:

- Most people tend to *forget* some of the already acquired knowledge, or at least they are often incapable of becoming conscious of it when needed
- For many good reasons people are often *doing things together*, but purposeful *co-actions* always require *communication* between the involved agents. But *it is a commonly recognized fact* that *we cannot rely on telepathy* as the means for the transfer of knowledge.

The essence of these conditions is that generally in practice there is a need to be able to *reproduce information.* (To talk about reproducing empation is senseless, because it is in contradiction with the definition).

To reproduce a portion of information is *to provide* the information for *a receiving person* at *a later time* and/or at *another place* such that the receiver in practice can acquire it with certainty.

If the portion of information - *the message* - is provided *for the same person* from whom it originates, we are talking about the *preservation* of information (outside mind) until a later time. If the message is provided for other persons, we talk about *communication.*

And that is where there is a need for a useful data-concept:

It is based on the fundamental semiotic notion of *representing* - intentionally to let *something stand for something else.* In our case: to let something of physical nature stand for something that is mental. For example, in the main part of this paper the written manifestation - the word 'information' represents *the particular concept* introduced and encircled here. Similarly, the fragmentary description presented in the paper represents a part of the *meaning* of that concept.

Now, with the purpose of being more explicit we shall define:

A message is a certain information set that with the purpose of having it reproduced *is represented* in an appropriate *portion of data* - in a **data-package** where:

Data is *a structure of physical phenomena* that according to a convention in a certain domain represents information.

Then, by knowing the *message type* and the *representation rules* applied by the sender, the receiver can reproduce the information in his or her own mind. Although this semiotic approach in principle is simple, reliable message reproduction may be more difficult in practice and will usually require some further representation principles known from datalogy to be applied.

An intended communication from one person to another is one of the most common social situations. It occurs at home or in business whenever people are co-acting or when otherwise they are dependent on each other for some reason. The general situation is that one agent needs a certain portion of information (but does not have it at the moment) while another one who already has it (e.g. from previous activities) is willing to provide it for the partner.

To succeed in this they must apply a certain, mutually agreed set of rules - *a communication protocol* - organized in three parts:

1: *A message-part* determining the *types of information sets* and *information elements* that are to be transferred (possibly with reference to a properly established *information model* specifying the structure of relevant entity types).

2: *A data-part* determining the physical and structural means of representation used. (This is where a considerable part of the data-types well-known from datalogy are involved, e.g. file, record, field, pixel, array, list, character-set, etc.).

3: *A transfer-part* determining how the data-package is moved to or otherwise provided for the receiver. (This is where references could be to everything from traditional messenger or postal services to advanced data-transmission protocols and systems).

The other kind, message preservation, is based on a similar set of rules, *a preservation protocol*, where the two first parts are the same as above, while the last, *the storage part*, determines where and how the involved data-package physically is kept, such that it can be retrieved when someone needs to have the message reproduced.

In both cases the two last protocol parts are rather well-established in practice, mostly as a consequence of the extensive use of modern data-technology. (NOT the nonsense phenomenon implied by the unfortunate mis-term 'information technology')! As regard the message part there are still considerable problems, because of the delinquencies as regard satisfactory analysis caused by the lack of insight and by confusion about proper concepts, as illustrated with the examples at the beginning. The information concept described here is intended to help taking a first step in improving message protocols.

3. ... and to Focus on the Usefulness of Concepts.

After having described and fairly well justified the information concept from six different points of view, we can now integrate the crucial aspects into the following definition:

Information is intellectual knowledge that in a certain domain can be acquired, preserved, transferred and applied as non-empty *sets of information elements*, such that each element determines *a certain aspect of a certain entity.*

In a shorter, but less constructive version the essence can be expressed:

Information is the kind of intellectual knowledge that in a certain domain can be relied on as being well-determined, but *independent of the persons knowing it.*

Trying to communicate a fair part of the meaning of this information concept, it has been a severe handicap - as also occationally indicated - that *there is no adequate professional language in Informatics* to rely on. If there is, it is certainly not commonly recognized. And I am afraid that the conceptual and terminological confusion will not come to an end before scientists and other professionals take another more *pragmatic* view of what I prefer to call *concept engineering* - the explicit process of constructing an adequate structure of well-determined and useful concepts.

Therefore, as a by-product of the study of information I want to finish by presenting one more perspective this time *on a meta-level* - a perspective of *the concept «concept»*. It can be expressed in this way:

A concept is a conception of a thing *characterized in such a way* that it thereby becomes *useful for understanding other things* - in particular for understanding other concepts.

Information as described in the paper is one such concept.

Tropos: A Framework for Requirements-Driven Software Development

John Mylopoulos[1] and Jaelson Castro[2]
[1]Department of Computer Science, University of Toronto,
Toronto, Canada
jm@cs.toronto.edu
[2]Universidade Federal de Pernambuco, Centro de Informática
Recife, Brazil*
jbc@di.ufpe.br

Abstract

Traditionally, software development techniques have been implementation-driven in the sense that the programming paradigm of the day dictated the design and requirements analysis techniques used. For example, structured programming led to structured analysis and design techniques in the '70s. More recently, object-oriented programming gave rise to object-oriented analysis and design. In this chapter we explore a software development methodology which is *requirements-driven* in the sense that the concepts used to define requirements for a software system are also used later on during design and implementation. Our proposal adopts Eric Yu's *i** framework [1], a modeling framework for early requirements, based on the notions of `actor` and `goal`. We use these notions as a foundation to model late requirements, as well as architectural and detailed design. The proposed framework, named Tropos, seems to complement nicely current proposals for agent-oriented programming platforms.

1 Introduction

Software development techniques have traditionally been inspired and driven by the programming paradigm of the day. This means that the concepts, methods and tools used during all phases of development were based on those offered by the pre-eminent programming paradigm. So, during the era of structured programming, structured analysis and design techniques were proposed [2, 3], while object-

* This work was carried out during a visit to the Department of Computer Science, University of Toronto. Partially supported by the CNPq – Brazil under grant 203262/86-7.

oriented programming has given rise more recently to object-oriented design and analysis [4, 5]. For structured development techniques this meant that throughout software development, the developer can conceptualize her software system in terms of functions and processes, inputs and outputs. For object-oriented development, on the other hand, the developer thinks throughout in terms of objects, classes, methods, inheritance and the like.

Using the same concepts to align requirements analysis with software design and implementation makes perfect sense. For one thing, such an alignment reduces impedance mismatches between different development phases. Think what it would be like to take the output of a structured analysis task, consisting of data flow and entity-relationship diagrams, and try to produce out of it an object-oriented design! Moreover, such alignment can lead to coherent toolsets and techniques for developing software (and it has!). As well, it can streamline the development process itself.

But, why base such an alignment on implementation concepts? Requirements analysis is arguably the most important stage of software development. This is the phase where technical considerations have to be balanced against social and personal ones. Not surprisingly, this is also the phase where the most and costliest errors are introduced to a software system. Even if (or rather, when) the importance of design and implementation phases wanes sometime in the future -- thanks to COTS, software reuse and the like -- requirements analysis will remain a critical phase for the development of any software system, answering the most fundamental of all design questions: "what is the system intended for?"

This paper speculates on the nature of a software development framework, named Tropos[1], which is requirements-driven in the sense that it is based on concepts used during early requirements analysis. To this end, we adopt the concepts offered by *i** [1], a modeling framework offering concepts such as `actor`, `agent`, `position` and `role`, as well as social `dependencies` among actors, including `goal`, `softgoal`, `task` and `resource` dependencies. These concepts are used in a small example to model not just early requirements for an insurance claim management system, but also late requirements, architectural design and detailed design.

The proposed methodology spans four phases of software development:

- Early requirements, concerned with the understanding of a problem by studying an existing organizational setting; the output of this phase is an organizational model which includes relevant actors and their respective goals;
- Late requirements, where the system-to-be is described within its operational environment, along with relevant functions and qualities;
- Architectural design, where the system's global architecture is defined in terms of subsystems, interconnected through data and control flows;
- Detailed design, where each architectural component is defined in further detail in terms of inputs, outputs, control, and other relevant information.

Section 2 introduces the primitive concepts offered by *i** and illustrates their use with an example. Sections 3, 4, and 5 sketch how the technique might work for late requirements, architectural design and detailed design respectively. Throughout, we assume that the task at hand is to build generic software to support back office

---

[1] The name "Tropos" is derived from the Greek "tropé", which means "easily changeable", also "easily adaptable."

claims processing within an insurance company. Finally, section 6 summarizes the contributions of the paper, offers an initial self assessment of the proposed development technique, and outlines directions for further research.

2 Early Requirements Analysis with *i**

During early requirements analysis, the requirements engineer is supposed to capture and analyze the intentions of stakeholders. These are modelled as goals which, through some form of a goal-oriented analysis, eventually lead to the functional and non-functional requirements of the system-to-be [6]. In i* (which stands for ``distributed intentionality''), early requirements are assumed to involve social actors who depend on each other for goals to be achieved, tasks to be performed, and resources to be furnished. The i* framework includes the *strategic dependency model* for describing the network of relationships among actors, as well as the *strategic rationale model* for describing and supporting the reasoning that each actor has about its relationships with other actors. These models have been formalized using intentional concepts such as goal, belief, ability, and commitment (e.g., [7]). The framework has been presented in detail in [1] and has been related to different application areas, including requirements engineering [8], business process reengineering [9], and software processes [10].

A strategic dependency model is a graph, where each node represents an *actor*, and each link between two actors indicates that one actor depends on the other for something in order that the former may attain some goal. We call the depending actor the *depender* and the actor who is depended upon the *dependee*. The object around which the dependency centers is called the *dependum*. By depending on another actor for a dependum, an actor is able to achieve goals that it is otherwise unable to achieve, or not as easily, or not as well. At the same time, the depender becomes vulnerable. If the dependee fails to deliver the dependum, the depender would be adversely affected in its ability to achieve its goals.

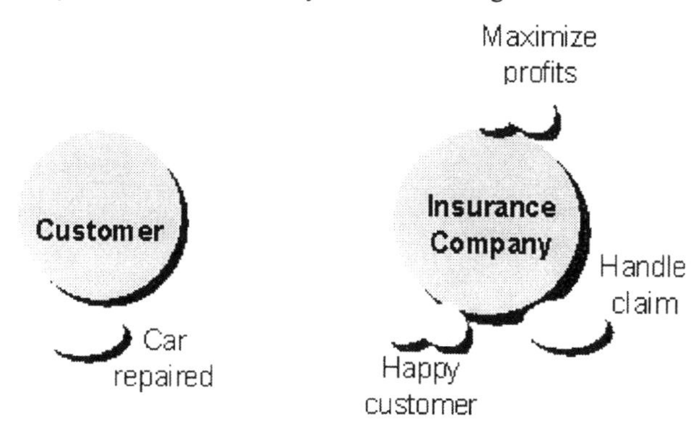

Fig. 1. "Customers want their cars repaired, while the insurance company wants to maximize profits, settle claims and keep customers happy"

Figure 1 shows the beginning of an *i** model consisting of two relevant actors for an automobile insurance example. The two actors are named respectively

Customer and Insurance Company. The customer has one relevant goal CarRepaired, while the insurance company has goals Settle claim, Maximize profits, and keep Happy customer. Since the last two goals are not well-defined, they are represented in terms of softgoals (shown as cloudy shapes.)

Once the relevant stakeholders and their goals have been identified, a means-ends analysis determines how these goals (including softgoals) can actually be fulfilled through the contributions of other actors. Let's focus on one such goal, namely Handle claim.

As shown in figure 2, the analysis is carried out from the perspective of the insurance company, who had the goal in the first place. It begins with the goal Handle claim and postulates a task Handle claim (represented in terms of a hexagonal icon) through which the goal might be fulfilled. Tasks are partially ordered sequences of steps intended to fulfill some goal. The task we have selected is decomposed into sub-tasks Verify policy, Prepare offer, Finalize deal which *together* can complete the handling of a claim. It should be noted that the same goal (Handle claim) might have several alternative tasks that can fulfill it. Likewise, there may be several alternative decompositions of a task into sub-tasks. Figure 2 only shows one set of decompositions which collectively can fulfill the root goal.

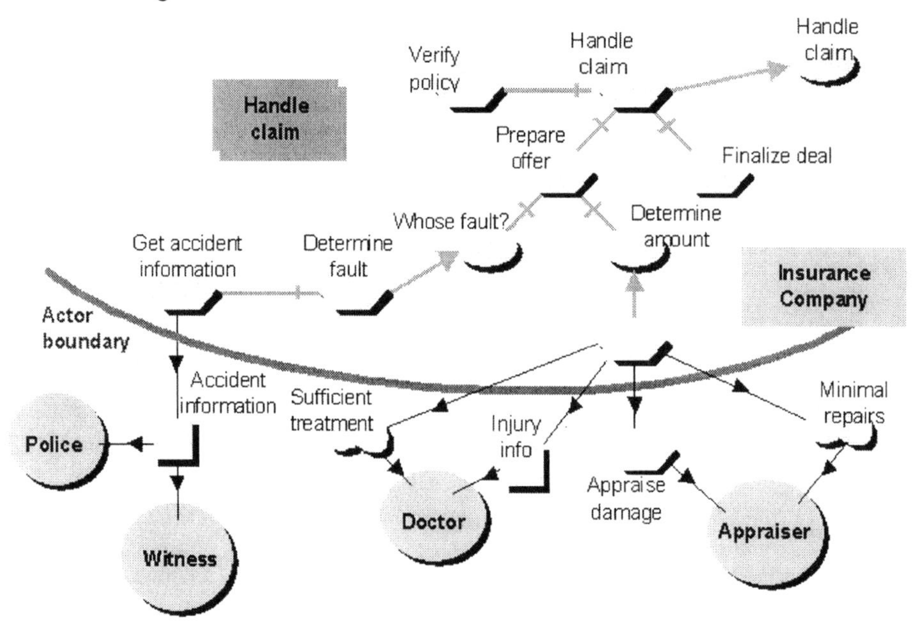

Fig. 2. Means-ends analysis for the goal Handle claim.

Tasks can also be decomposed into goals, whose fulfillment accomplishes the task. For example, the task Prepare offer is decomposed into goals Whose fault? and Determine amount. Representing a task component as a goal means that there might be several possible ways of accomplishing that component.

Decompositions continue until the analysis can identify an actor who can fulfill a goal, carry out a task, or deliver on some needed resource. Such dependencies for the `Handle claim` goal include:

- Resource dependencies on actors `Police` and `Witness`, who are expected to deliver accident information;
- Another resource dependency on `Doctor` for injury information;
- Task dependency on `Appraiser`, who is expected to carry out the standard appraisal task;
- Softgoal dependencies on `Doctor`, who must make sure that the patient receives adequate treatment, and the `Appraiser`, who is expected to minimize the amount of the appraisal.

The result of such means-ends analyses for the initial goals leads to the strategic dependency model mentioned earlier. Fragments of such a model for the insurance claim example are shown in figure 3.

According to this model, the customer depends on the appraiser for a fair appraisal. However, the appraiser can be expected to act in the interests of the insurance company because of his dependence on the latter for continued employment. The customer, in turn, depends on the body shop to give a maximal estimate, while the body shop depends on the customer for continuing business.

Although a strategic dependency model provides hints about why processes are structured in a certain way, it does not sufficiently support the process of suggesting, exploring, and evaluating alternative solutions. That is the role of the Strategic Rationale model. A strategic rationale model is a graph with four main types of nodes -- goal, task, resource, and softgoal -- and two main types of links -- means-ends links and process decomposition links. A strategic rationale graph describes the criteria in terms of which each actor's selects among alternative dependency configurations.

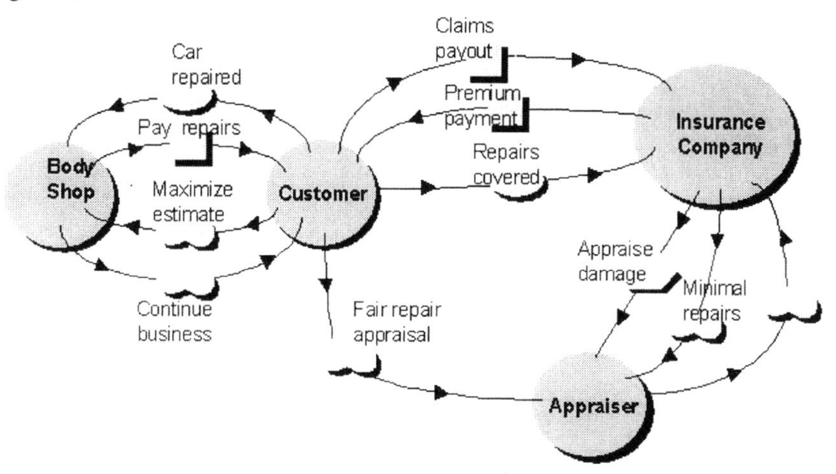

Fig. 3. Partial strategic dependency model for the handling of insurance claims.

3 Late Requirements Analysis

Late requirements analysis results in a requirements specification document which describes all functional and non-functional requirements for the system-to-be. In Tropos, the system is represented as one or more actors which participate in a strategic dependency model, along with other actors in the system's operational environment. In other words, the system comes into the picture as one or more actors which contribute to the fulfillment of stakeholder goals. For example, the system may be introduced in the strategic dependency model in order to support the goal `Process claim`, as well as the softgoal `Fast processing` of insurance claims, which contributes positively to both the `Maximize profits` and `Happy customer` softgoals (figure 4). Of course, as late requirements analysis proceeds, the system is given additional responsibilities, and ends up as the depender of several dependencies. Moreover, the system is decomposed into several sub-actors which take on some of these responsibilities. To obtain this decomposition, `Process claim` is first reduced into subgoals, such as `Select process` (i.e., what sequence of steps will be used to process the claim), `Process claim` and `Report status`, using the kind of means-ends analysis illustrated in figure 2, along with a strategic rationale analysis. The result of this analysis is a set of (system and human) actors who are dependees for some of the dependencies that have been generated.

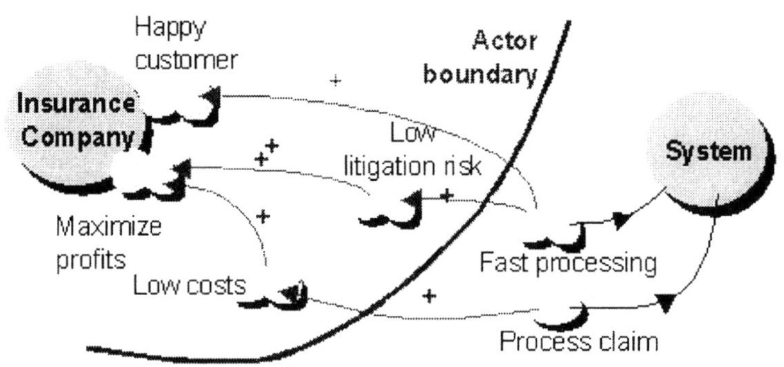

Fig. 4. The insurance company depends on the system for fast processing of insurance claims

Figure 5 suggests one possible assignment of responsibilities. In particular, `Process Selector` decides what kind of processing will be done for a given claim, and relies on a clerk to carry out this process. We assume that different insurance companies using the software may be processing various types of claims (e.g., large vs small) differently. `Tracker` keeps track of the status of claim and needs information from the processing clerk in order to do so. `Reporter` reports to the claims manager, or the customer on the status of a claim following a given script (hence the task dependency), while `Trouble shooter` is looking for signs of problems ahead. `Tracker` and the `Trouble shooter` are introduced in order to contribute to the fulfillment of the `Fast processing` softgoal.

Resource, task and softgoal dependencies correspond naturally to functional and non-functional requirements. Leaving (some) goal dependencies between system actors and other actors is a novelty. Traditionally, functional goals are "operationalized" during late requirements [6], while quality softgoals are either operationalized or "metricized" [11]. For example, `Fast processing` may be operationalized during late requirements analysis into particular business processes for processing claims. Likewise, a security softgoal might be operationalized by defining interfaces which minimize input/output between the system and its environment, or by limiting access to sensitive information. Alternatively, the security requirement may be metricized into something like "No more than X unauthorized operations in the system-to-be per year".

Leaving goal dependencies with system actors as dependees makes sense whenever there is a foreseeable need for flexibility in the performance of a task on the part of the system. For example, consider a communication goal "communicate X to Y". According to conventional software development techniques, such a goal needs to be operationalized before the end of late requirements analysis, perhaps into some sort of a user interface through which user Y will receive message X from the system. The problem with this approach is that the steps through which this goal is to be fulfilled (along with a host of background assumptions) are frozen into the requirements of the system-to-be. This early translation of goals into concrete plans for their fulfillment makes software systems fragile and less reusable.

For our example, we have left two goals in the late requirements model. The first goal is `Trouble shooting`, because we propose to implement `Trouble shooter` as an intelligent agent who "learns on the job" (...so to speak...) by using machine learning techniques. Also, `Select process`, because we want to include in the system's architecture a number of components which reflect different types of claims processing done in the insurance industry. So, instead of operationalizing this goal during requirements analysis, we propose to do so during architectural design.

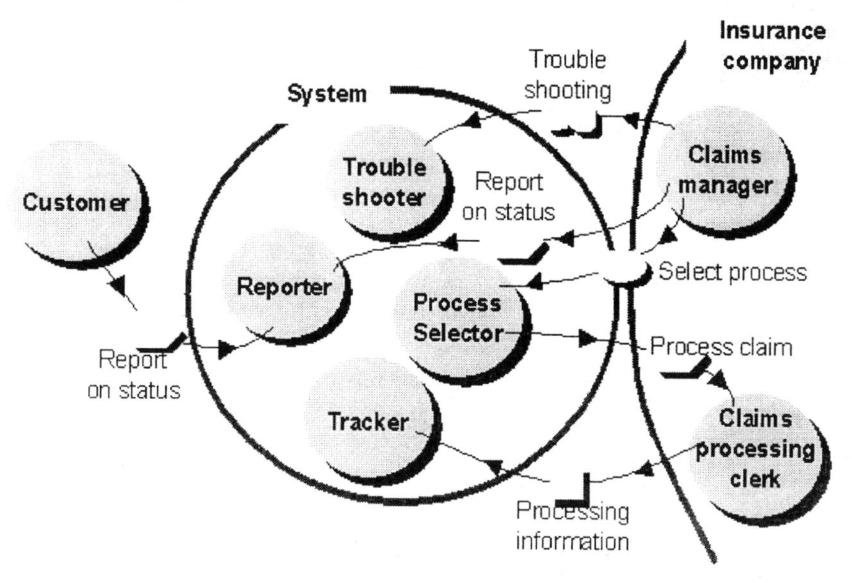

Fig. 5. The system consists of four actors, each with external dependencies.

4 Architectural Design

Architectural design has emerged as a crucial phase of the design process. Initially a software architecture can be considered to be the realization of early design decisions made regarding the decomposition of the system into components. A software architecture constitutes a relatively small, intellectually graspable model of system structure, and how system components work together. Software architects have developed comprehensive catalogues of software architectural styles (see, for example, [12]). Such styles range from *Independent components* (such as *Events-driven architectures* and *communicating processes*), *Call-and-return* (e.g., *object-oriented systems*, *layered*, *main program* and *subroutine* architectures), *Data flow* (for example, *batch sequential*, *pipes-and-filters*), *Data centered* (e.g., *repository* and *blackboard architectures*), as well as *Virtual machine architectures* (*rule-based systems*, and *interpreters.*) Most of these apply to software systems, even when the basic software components are intentional actors rather than subsystems and modules. For instance, a pipe and filter architectural style corresponds to an agent assembly line, while the blackboard style has been used extensively in the agent programming literature.

Architectures are influenced by system designers as well as technical and organizational factors. During architectural design we concentrate on the key system actors, defined during late requirements analysis, and their responsibilities. There would be a set of desired functionality as well as a number of quality requirements related to performance, availability, usability, modifiability, portability, reusability, testability, etc. The functional requirements can be handled by many standard technologies, such as structured analysis and design, or object-oriented design methods. However, quality requirements are generally not addressed by such techniques [13].

Suppose that in addition to the requirements of figure 5, we have an "easily modifiable" requirement for the `Process selector` actor imposed by the `Claims manager actor,` to make sure that it can accommodate ever changing variations on how an insurance claim is processed. Likewise, in order to fulfill the requirement of good response time, imposed by the claims processing clerk on process selector, a "good performance" softgoal is introduced. To cope with these goals, the software architect, who is another (external) actor, goes through a means ends analysis comparable to what was discussed earlier. In this case, the analysis involves refining the softgoals to sub-goals that are more specific (and more precise!) and then evaluating alternative architectural styles against them, as shown in figure 6 [14].

In the figure, the two softgoals take `Process selector` as argument, meaning that the quality requirements they represent apply specifically to this system component (rather than the whole system). The first of the two softgoals has been AND-decomposed into subgoals `Modifiable[Process]`, `Modifiable[Data representation]`, `Modifiable[Function]`. This analysis is intended to make explicit the space of alternatives for fulfilling the top-level quality softgoals. Moreover, the analysis allows the evaluation of several alternative architectural styles. The styles are represented as goals (saying, roughly, "make the architecture of the new system repository-based/object-oriented/...") and are evaluated with respect to the alternative quality softgoals as shown in figure 6.

The evaluation results in contribution relationships from the architectural goals to the quality softgoals, labelled "+', "-", "++", etc.

As with late requirements, the interesting feature of the proposed analysis method is that it is goal oriented. Goals are introduced and analyzed during architectural design, and guide the design process.

Apart from goal analysis, this phase involves the introduction of other system actors which will take on some of the responsibilities of the key system actors introduced earlier. For example, to accommodate the responsibilities of the `Reporter` actor of figure 5, we may want to introduce a `Data selector` actor, who selects the data to be presented, a `Transformer` actor, who performs computations that transform the input data to useful information for the `Customer` and the `Claims manager`, and a `Presenter` actor who presents these dàta in a suitable format. Of course, this analysis is nothing but good old functional decomposition and will not be discussed in any detail here.

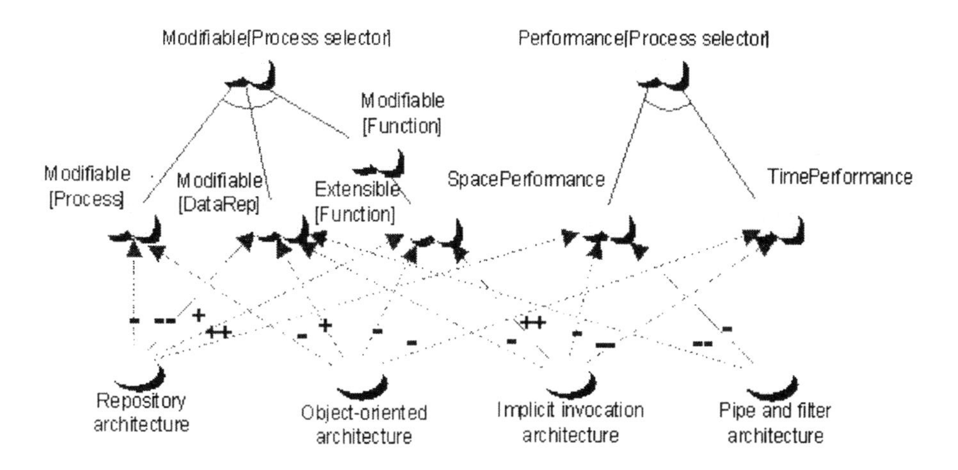

Fig. 6. A strategic rationale model, from the perspective of the process selector actor.

An interesting decision that comes up during architectural design is whether fulfillment of an actor's obligations will be accomplished through assistance from other actors, through delegation ("outsourcing"), or through decomposition of the actor into component actors. Going back to the `Reporter` example, the introduction of other actors described in the previous paragraph amounts to a form of delegation. `Reporter` retains its obligations, but delegates subtasks, subgoals etc. to other actors. An alternative architectural design would have `Reporter` outsourcing some of its responsibilities to some other actors, so that `Reporter` removes itself from the critical path of obligation fulfillment. Lastly, `Reporter` may be refined into an aggregate of actors which, by design, work together to fulfill `Reporter's` obligations. This is analogous to a committee being refined into a collection of members who collectively fulfill the committee's mandate. It is not clear, at this point, how the three alternatives compare, nor what are their respective strengths and weaknesses.

5 Detailed Design

The detailed design phase is intended to introduce additional detail for each architectural component of a software system. In our case, this includes actor communication and actor behaviour. To support this phase, we may be adopting agent communication languages, message transportation mechanisms, ontology communication, agent interaction protocols, etc. from the agent programming community. One possibility, among admittedly many, is adopt one of the extensions to UML [5] proposed by the FIPA (Foundation for Intelligent Agents) and the OMG Agent Work group.

For our example, let's concentrate on the `Fast processing` goal dependency, which might involve a detailed design on *agent interaction protocols* (AIP). Such a protocol describes a communication pattern among actors as an allowed sequence of messages, as well as constraints on the contents of those messages. To define such a protocol, we use AUML - the Agent Unified Modeling Language [15], which supports templates and packages to represent the protocol as an object, but also in terms of sequence and collaborations diagrams. In AUML inter- and intra-agent dynamics are also described in terms of activity diagrams and state charts.

Figure 7 depicts a protocol expressed as a UML sequence diagram for `Select process`. When invoked, a `Claim manager` actor sends a `Call-for-Proposal-Process-claim` to a `Process Selector` actor who is willing to participate in processing the claim.

The `Process Selector` actor can then choose to respond to the `Claim manager` by a given deadline by submitting a proposal for a suitable `Processing clerk` actor to deal with the processing (for example an expert on small claims). Alternatively, `Process selector` may decide to refuse to process the claim or indicate that it does not understand. If a proposal is offered, the `Claim manager` actor has a choice of either rejecting or accepting the proposal. When `Process selector` receives a proposal acceptance, it will contact the appropriate `Claims process clerk` actor and place a request regarding (small) process claims. Based on the returned information, `Process selector` can inform `Claims manager` about the proposal's execution. Additionally, the `Claim manager` actor can cancel the execution of the proposal at any time.

Of course the sequence diagram in Figure 7 only provides a basic specification for an agent claim processing protocol. More processing details are required. For example, a `Claims manager` actor requests a call for (process claim) proposals (CFP) from a `Process selector` actor. However, the diagram stipulates neither the procedure used by the `Claims manager` to produce the CFP request, nor the procedure employed by `Process Selector` to respond the CFP. Yet, these are clearly important details at this stage of the software development process.

Such details can be provided by using *leveling*, i.e., by introducing additional interaction and other diagrams which describe some of the primitive action of the one shown on figure 7. Each additional level can express *intra-actor* or *inter-actor* activity. At the lowest level, specification of an actor protocol requires spelling out the detailed processing that takes place within an actor in order to implement the protocol. Statecharts and activity diagrams can also specify the internal processing of actors who are not aggregates.

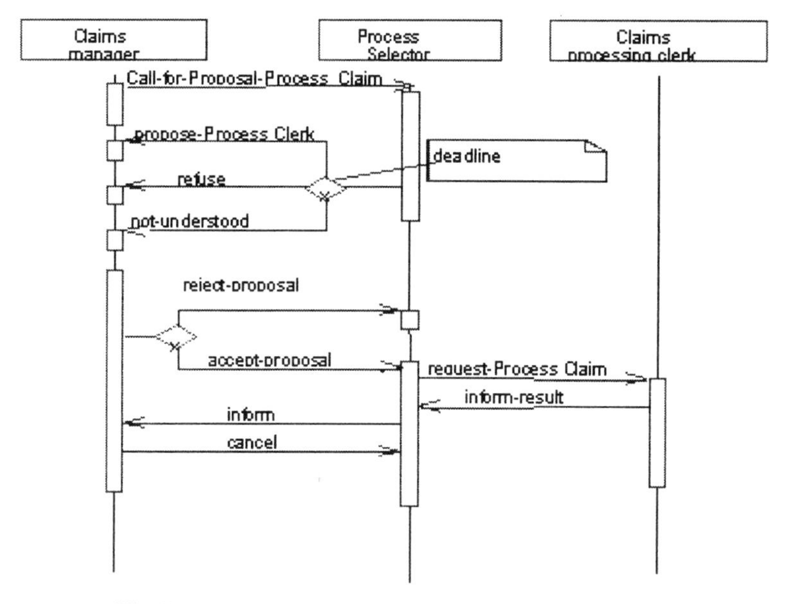

Fig. 7. An actor interaction protocol for processing claims

6 Conclusions and Discussion

We have argued in favour of a software development methodology which is founded on intentional concepts, such as those of `actor`, `goal`, (`goal`, `task`, `resource`, `softgoal`) `dependency`, etc. Our argument rests on the claim that the elimination of goals during late requirements, freezes into the design of a software system a variety of assumptions which may or may not be true in its operational environment. Given the ever-growing demand for generic, component-ized software that can be downloaded and used in a variety of computing platforms around the world, we believe that the use of intentional concepts during late software development phases will become prevalent and should be further researched.

The Tropos project is only beginning and much remains to be done. We will be working towards a modelling framework which views software from four complementary perspectives:

- **Social** -- who are the relevant actors, what do they want? What are their obligations? What are their capabilities?...
- **Intentional** -- what are the relevant goals and how do they interrelate? How are they being met, and by whom?...
- **Process-oriented** -- what are the relevant business/computer processes? Who is responsible for what?...

- **Object-oriented** – what are the relevant objects and classes, along with their inter-relationships?

In this paper, we have focused the discussion on the social and intentional perspectives because they are novel. As hinted earlier, we propose to use UML-type modelling techniques for the others.

Of course, diagrams are not complete, nor formal as software specifications. To address this deficiency, we propose to offer three levels of software specification. The first is strictly diagrammatic, as discussed in this paper. The second involves formal annotations which complement diagrams. For example, annotations may specify that some obligation takes precedence over another. These could be used as a basis for simple forms of analysis. Finally, we propose to include within Tropos a formal specification language for all built-in constructs, to support deeper forms of analysis. Turning to the organization of Tropos models, the concepts of *i** will be embedded in a modeling framework which supports generalization, aggregation, classification, materialization and contextualization. Some elements of UML will be adopted as well for modeling the object and process perspectives.

Like other requirements modelling frameworks proposed in the literature, we recognize that diagrams are important for human communication, but are imprecise and offer little support for analysis. Partially formal annotations can help in defining some forms of analysis, and they serve as bridges between informal diagrams and formal specifications. Finally, formal specifications serve as foundation for a formal semantics, as well as a range of analysis techniques, including proofs of correctness, process simulation, goal analysis etc.

Tropos constitutes the last leg of a trilogy on modelling languages. The first language in the trilogy, Taxis [16], was intended as a design language for information systems. Its main novelty was the adoption of semantic network representation techniques to offer a modelling framework which was object-oriented and emphasized taxonomic organization for data, transaction and exception classes. Telos [17] focused on the use of classification to offer meta-modelling facilities where concepts such as `goal`, `activity`, etc. could first be defined at the metaclass level before being used at the class level. Telos was intended for software modelling, where one could represent requirements, design, implementation and other information about a software system within a single modelling framework. Tropos is probably the most ambitious undertaking in the trilogy in that it aspires to influence not just the modelling of different types of information about a software system, but also the software development process itself.

Most appropriately, this preview of Tropos has been written on the occasion of Janis Bubenko's sixty-fifth birthday. Janis has made significant research contributions to conceptual modelling, databases, and model-based software development. His early work [18] was an inspiration for our own work on RML [19], and we have benefited by following his research ever since. Just as importantly, Janis has served as role model for younger generations of researchers and academics around the world.

Acknowledgements

Many colleagues contributed to the ideas that led to this paper. Special thanks to Eric Yu, whose insights helped us focus our research on intentional and social concepts.

The Tropos project includes as co-investigators Eric Yu (University of Toronto) and Yves Lesperance (York University); also Alex Borgida (Rutgers University), Matthias Jarke and Gerhard Lakemeyer (Technical University of Aachen.) The Canadian component of the project is supported in part by the Natural Sciences and Engineering Research Council (NSERC) of Canada, and the CITO Centre of Excellence, funded by the Province of Ontario.

References

[1] Yu, E., *Modelling Strategic Relationships for Process Reengineering*, Ph.D. thesis, Department of Computer Science, University of Toronto, 1995.

[2] DeMarco, T., *Structured Analysis and System Specification*, Yourdon Press, 1978.

[3] Yourdon, E. and Constantine, L., Structured Design: Fundamentals of a Discipline of Computer Program and Systems Design, Prentice-Hall, 1979.

[4] Wirfs-Brock, R., Wilkerson, B., Wiener, l., *Designing Object-Oriented Software.* Englewood Cliffs, NJ; Prentice-Hall.

[5] Booch, G., Rumbaugh, J., Jacobson, I., *The Unified Modeling Language User Guide*, The Addison-Wesley Object Technology Series, Addison-Wesley, 1999.

[6] Dardenne, A., van Lamsweerde, A., and Fickas, S., "Goal–directed Requirements Acquisition," *Science of Computer Programming, 20*, 3-50, 1993.

[7] Cohen, P. and Levesque, H. Intention is Choice with Commitment. *Artificial Intelligence, 32(3).*

[8] Yu, E., "Modeling Organizations for Information Systems Requirements Engineering," *Proceedings First IEEE International Symposium on Requirements Engineering,* San Jose, January 1993, pp. 34-41.

[9] Yu, E., and Mylopoulos, J., "Using Goals, Rules, and Methods to Support Reasoning in Business Process Reengineering", *International Journal of Intelligent Systems in Accounting, Finance and Management 5(1),* January 1996.

[10] Yu, E. and Mylopoulos, J., "Understanding 'Why' in Software Process Modeling, Analysis and Design," *Proceedings Sixteenth International Conference on Software Engineering*, Sorrento, Italy, May 1994.

[11] Davis, A., Software Requirements: Objects, Functions and States, Prentice Hall, 1993.

[12] Bass, L., Clements. P., Kazman, R., *Software Architecture in Pratice,* SEI Series in Software Engineering, Addison-Wesley, 1998.

[13] Chung, L. K., Nixon, B. A., Yu, E., Mylopoulos, J., *Non-Functional Requirements in Software Engineering*, Kluwer Publishing, 2000.

[14] L. Chung, D. Gross, E. Yu , *Architectural Design to Meet Stakeholder Requirements,* in Software Architecture, Patrick Donohue, ed., Kluwer Academic Publishers. 1999. pp. 545-564. (TC2 First Working IFIP Conference on Software Architecture (WICSA1), 22-24 February 1999, San Antonio, Texas, USA.)

[15] Parunak, H. Van Dyke, Suater, J., Odell, J., *Engineering Artifacts for Multi-Agents Systems*, ERIM CEC, 1999.

[16] Mylopoulos, J., Bernstein, P., and Wong. H. K. T., "A Language Facility for Designing Data-intensive Applications," *ACM Transactions on Database Systems 5(2)*, 1980.

[17] Mylopoulos, J., Borgida, A., Jarke, M., and M. Koubarakis, M., "Telos: Representing Knowledge About Information Systems," *ACM Transactions on Information Systems*, 1990.

[18] Bubenko, J., "Information Modeling in the Context of System Development," *Proceedings IFIP Congress '80*, 395-411, 1980.

[19]Greenspan, S., Mylopoulos, J., and Borgida, A., "Capturing More World Knowledge in the Requirements Specification," *Proceedings Sixth International Conference on Software Engineering*, Tokyo, 1982.

On Levels of Business Modelling, Communication and Model Architectures

Björn E. Nilsson
Adera +
Stockholm, Sweden

Abstract

The emergent new business logic requires, to some extent, new approaches to modelling. Within what kind of framework should we work and develop new methods? Does the new logic even expose weaknesses in the definition of business as such? We will tie our reasoning to a layered, multi-perspective model architecture used in analysis and design of businesses and their information systems. Questions will mainly be handled from a practitioner's viewpoint and in most cases be tied to a business communication perspective.

1 A Line of Reasoning

The core of business is the addition and transfer of value.

With this in mind, we will let a small set of questions pilot us in reasoning and also give structure to the text. We will also take a look at the essence of the new business logic.

1.1 Questions on Modelling

We will follow a simple line of reasoning based upon four simple questions:

- Can we, in practice, benefit from using the same generic modelling perspectives in the analysis and design of different realms as market, customer, business operation, information support system and infrastructure for data processing?
- How do we secure that models of different realms interface properly and how do we follow requirements propagation between realms of analysis?
- How do we, during modelling, make sure that efficient value transfer between processes of different realms becomes possible?
- How do we adapt the contemporary process concept to more flexible interaction patterns?

Before we probe these questions, perhaps it is prudent to ask what the essence of the, so-called, new business logic might be.

1.2 To e or not to be – A new Business Logic?

The concept corresponding to the term *new business logic* is rapidly evolving. Frequently, definitions mix appearing new conditions with assumed effects thereof. Information technology and, especially, the internet are seen as either the core of the concept or as a driving enablers. From a practical viewpoint, we perceive the logic of business as such, to be fairly stable, even when new patterns of interaction emerge. A summary of the mechanism is presented in figure 1.

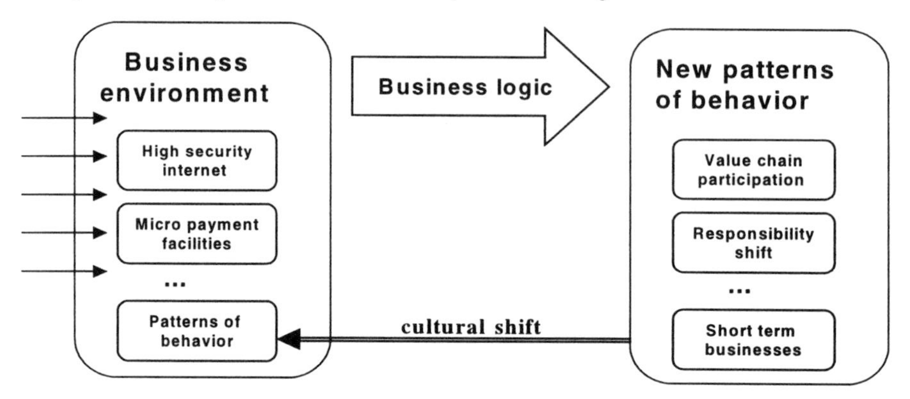

Figure 1: Common business logic leading to the emergence of new business interaction patterns

Business logic is grounded on creating value and on transfer and exchange thereof. In the gestalt of what is called the new business logic, we may over the last decade detect some important evolutions of enterprise behaviour:

- New patterns of interaction emerge between actors involved in value chains [4]. Examples: The customer controls production by configuring desired products. Suppliers take responsibility for design and production as well as distribution. Transparent, broken value chains are created.
- Virtual enterprise components are employed. Example: Virtual factory approaches radically transforms the demands an enterprise can meet and minimise the capital bound in production facilities.
- High-speed business development. Examples: In some option markets, a four-hour launch advantage is enough to establish a market lead and block competition. By introducing changes in the rule base for a broker application, within hours, a new business may be exposed to the market. One shot businesses emerge with very short term alliances and commitments.
- One to one interaction. Example: Based on behavioural patterns, adaptive applications control customer exposure to products.
- Market transparency. Example: Customers may directly or via brokers overview product pricing and cover markets globally.

One of the few examples of a new logic is the payment of low value products with low-cost distribution. Payment for shareware or music becomes a question of honour or satisfaction rather than obligation. However, even if micro payments lead to a change in the basic logic of business on an actor to actor basis – this is not the case on a market basis, where statistical properties reign.

In parallel to a shift to short lived businesses, often based on virtual components, we also detect the opposite trend in changing from operating in the conventional value net to operations in more integrated structures. Over time, transaction considerations of a conventional value net transforms into process integration considerations between actors. The consequences for modelling the enterprise relate mainly to interaction aspects [6]. We will apply the theory of semiotics [11] to enhance analysis patterns within this realm. Before we look at interaction, let us study the context in which interaction occurs.

2 On Generic Perspectives

As the presented results have practical implications, we will use a model architecture, which has been developed and thoroughly tested in practical analysis and design over a couple of years, in a variety of large projects [8]. In this model architecture, the same generic perspectives are applied on all levels of analysis and design. This means that the analyses of the customer, the business and the IT based systems follow the same patterns. Moreover, the same generic modelling perspectives are also used to control the development in all phases of the life cycle i.e. in analysis, design, implementation, operation and change. Which are these generic perspectives?

2.1 Generic Analysis and Design Perspectives

As stated, business focuses on the addition and transfer of value. Normally, in business, value is created in co-ordinated action. The basic generic architecture in figure 2 illustrates the conditions for and results of action.

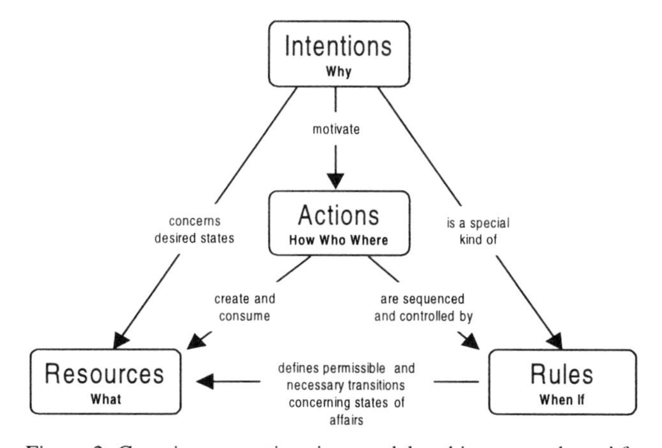

Figure 2: Generic perspectives in a model architecture, adapted from [8]

Intentions define expected values and motivate actions. To create value means to create or maintain resources or to change their state. In theory, this is best seen as state changes of business objects. In the process of creating value other resources are consumed i.e. have their state changed or are blocked from alternate usage. Rules determine permissible and necessary transitions of state of affairs in general, including business objects [9] as well as business processes.

To expose some of the motives behind the generic structure, we will apply it on an individual and managerial level.

2.3 Some Motives behind the Generic Perspectives

A sound architecture for models as well as basic concepts for modelling has to be based on reasons adherent to:

- Individual action. On what premises do we act as individuals – what are the basic motivational forces?
- Management principles. How do we direct and control business affairs – how is motivation and thereby action obtained and controlled on a larger scale?
- Philosophical and formal principles. Is there a good theory to guarantee a reasonably controlled usage?

Let us briefly, in figure 3, take a look at the structural basis for human action and compare it to management models. For a more thorough presentation see [8].

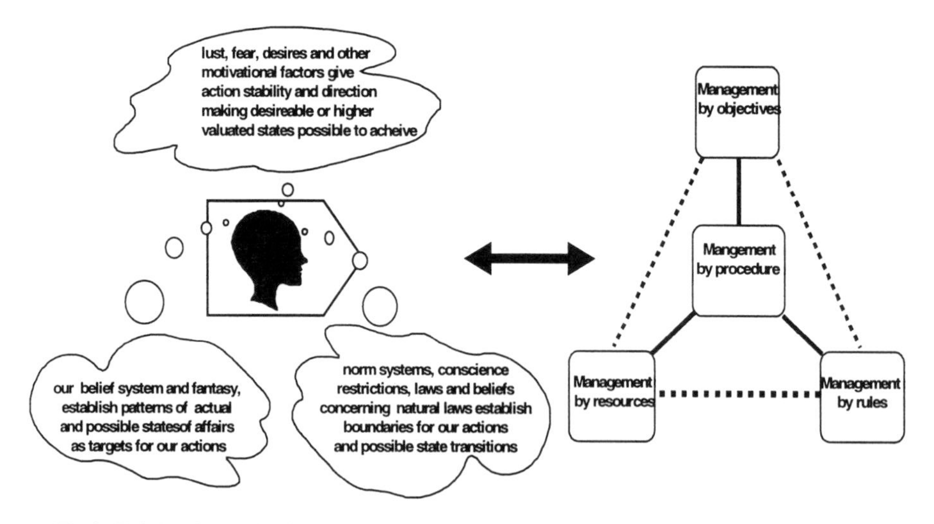

Fig 3: Driving forces for human action and the corresponding management perspectives.

A person's motivation for value creating action, both in terms of desires and fears, are linked, on one hand, to beliefs concerning actual state of affairs and their evolution over time and, on the other, to conceptions of desirable state of affairs over time. Action is performed on the basis of such beliefs and aims to change, or conserve certain states. Action is also restrained by what we perceive as norms, natural laws etc.

Management trends, such as management by goals or management by resources, have focussed on one of the aspects mentioned at a time. In practice, a mixture of these aspects is used in most cases.

On a more formal level, applying AI terminology, the perspectives correspond to goal-state, facts, rules and inference mechanism or logic. What happens if we apply the generic perspectives in modelling the business?

3 On Applied Perspectives in a Layered Architecture

To facilitate reasoning about the perspectives, we will simplify each perspective to a single model type. From a practitioner's point of view, this is a rather brutal oversimplification. For instance, to handle the perspective of action we use a multitude of models like process models, functional models, actor models, organisation models, geographic models, competence models etc [3].

Modelling the business is defining a social contract on how to view relevant phenomena and how to communicate about them[1]. Which are our instruments?

3.1 The Business Operations in Context

Modelling business operations, the perspectives translate to the following model types:

- What is seen as value is determined by business intentions. These kinds of phenomena are handled in goal models.
- Value addition is seen as state changes of business objects. Business resources are handled in object class models.
- Intentional changes of states for business objects, i.e. the transformation mechanisms, are handled in process models.
- The conditions for change, the business rules, are handled in rule models.

Concerning the information system, we find the following correspondences:

- A process model of the business has similar transformational characteristics as an application or functional model of the information support system.
- The major resource for value addition of an application or a system function would be its data i.e. objects describing the business objects.
- In the same way as state transitions in the business is regulated, state transitions in data is also regulated by rules. In some paradigms, these rules would be handled in methods bound to objects. This makes no difference in reasoning.

In business, value is created in co-ordinated action. Co-ordination is enabled by communication, the exchange of information, between process.

- The system intention, i.e. the goals to be met by an information system, is best seen as communication or information requirements of the business processes. These demands are to be met by information system services.

This means that the goal model for the information system could be seen in interface terms as in the lower part of figure 4.

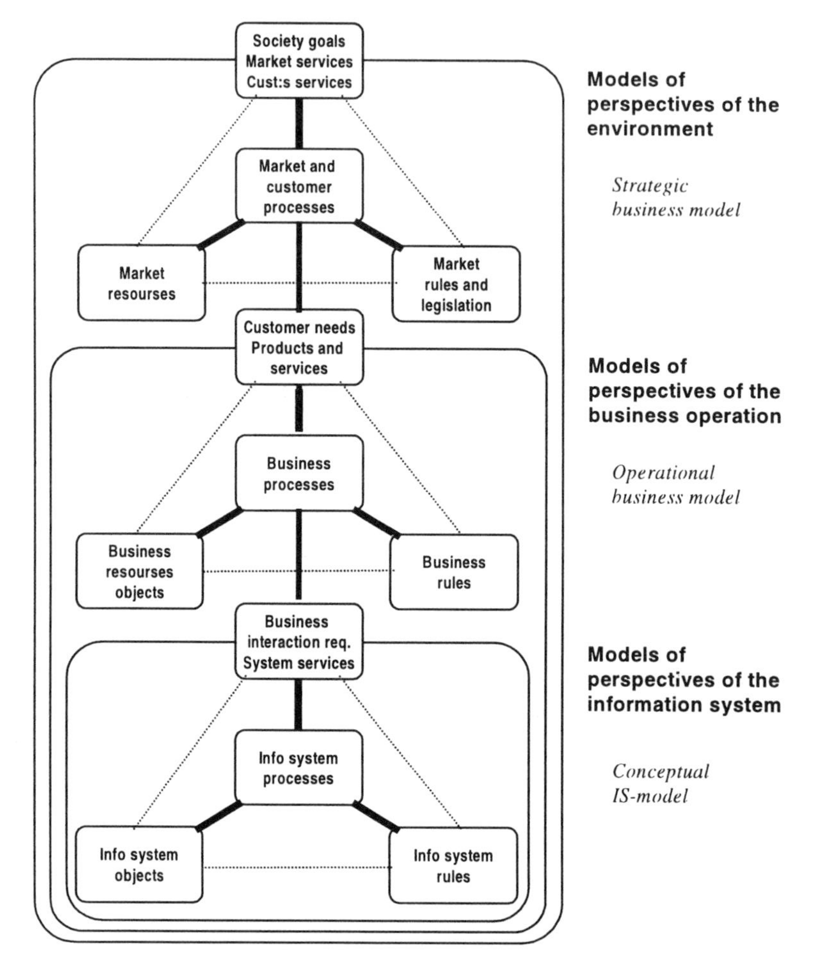

Figure 4. The business operations (normal business processes) model in context, from [8]

The business interaction and processing requirements should match the services provided by the information system. The value of services provided for by the information system are related to the value of (the extension of) different classes of messages used to control the business processes at hand. If we turn to the external environment of the business operations, we could apply the same perspectives. For example, when producing strategic business models:

- The goal-perspective of the business operation requires analyses of market and customer processes, with their demands, and how these are satisfied by services.

- The resource perspective involves analysis of competing and supplementing products on the market – as well as their producers and production facilities - searching possible partnerships, evaluating competition etc.
- The rule perspective leads to analysis of international market and product legislation, ethics etc.

In the model architecture, as illustrated in figure 4, the model realms encapsulate one another. Between all realms we also exercise the same interface thinking. As an example, a potential or actual customer only sees the products and services provided, including advertising and such, but not the internals of the business. Likewise, the actor in a business process only sees the services provided by the information system, not the internals thereof.

- In modern patterns of interaction, different external actors have access directly into the information system of an enterprise. Formally, this might be seen as these actors being used as resources in the local processes.

Aspects of security, permitted operations etc are still under control of the enterprise itself. In direct system communication, similar assumptions have to be made. Naturally, information handling might well be the value creating core process.

3.1 The Standard five Layer Architecture

As the intentional model is of an interface nature, the architecture is easily adaptable. In the non-configured development model, we work with a five-layer architecture as in figure 5 – in practical usage, levels are configured to the problem at hand.

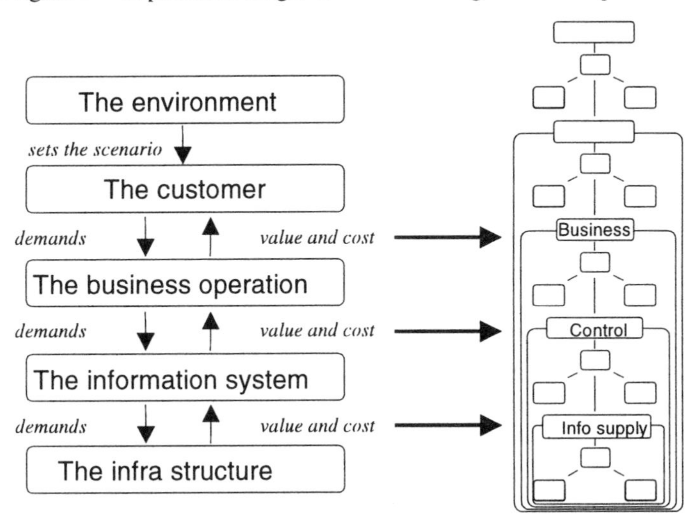

Figure 5: Simplified standard pattern of analysis, using the model architecture.

The benefits of using the same perspectives on all levels are threefold. We may use the same kinds of concepts in communication and reasoning, we have the basis for a reasonably simple tracking mechanism for requirements and it is very easy to configure the analysis to different problems.

3.2 On the Mapping of Levels

Apart from sharing the interface model, two adjecent realms are bound together by other relationships. Some important ones, between the business operation and its information system, are illustrated in figure 6. In practice, all mappings have to be handled, but the principles are the same.

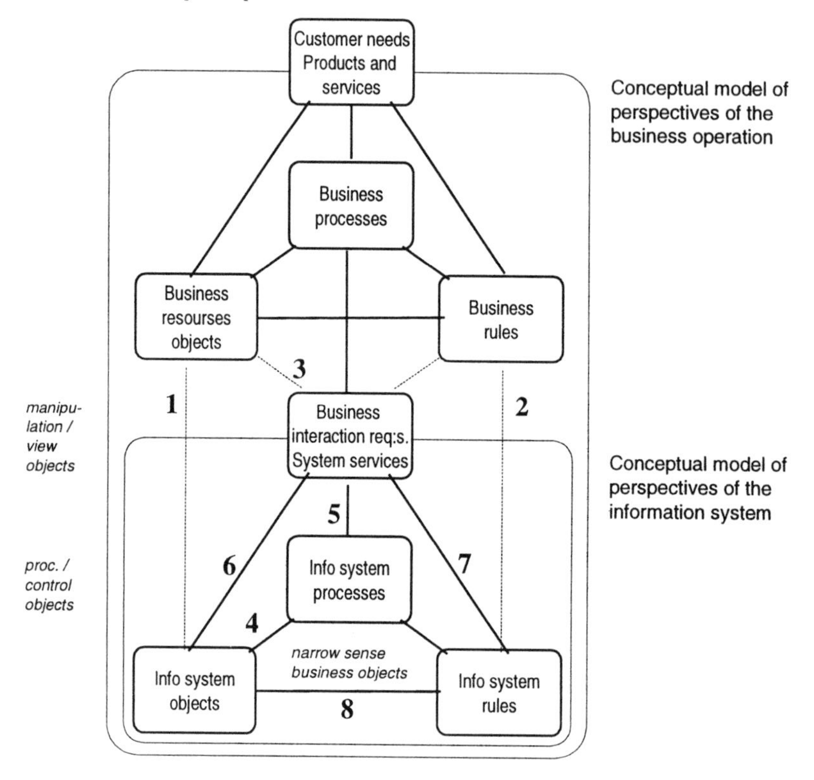

Figure 6: Examples of classes of mappings between models

- 1and 3: The correspondence of concepts and their representations
- 2: The correspondence of business rules to application rules
- 4 plus 5: Required functional data transformations to accomplish a service
- 6: Abstract data transformations
- 7: Interface sequencing and data independent behavioural properties

If we, as is normal in practice, use a purely conceptual view of the information system, the mappings shift a little. Mapping 6, for example, would be used in checking if the information required is possible to derive. The other mappings follow the same line of argumentation.

4 On the Definition of Business

As new interaction patterns emerge, the common concept of business becomes less and less useful. As the business operations themselves may change considerably over time, while providing for the same broad categories of services, the separation of these two foci seems advantageous. We have found the following definition to be useful as a guiding device in analysing and designing the business:

- The business is defined as the total exchange or interplay between the enterprise at hand and its customers / market.

The phenomenon of business might be studied on a variety of abstraction levels. To compare with the information system interface, table 1 includes some examples for both levels.

	Value abstraction	**Information abstraction (type level)**	**Physical abstraction (type level)**	**Mightiness (abstract extension)**
Business (Market interface)	The value transfer for all transactions over the business interface	Flow in terms of possible classes of information to and from the market / customers	Flow of types of products and services, also in the form of data, to and from market / customers	Set of possible deliverable physical manifestations and set of acceptable demands
Control (IS-interface)	The value transfer implied for all transactions over the IS-interface	Flow of possible classes of information to control and monitor processes	Flow of input and produced message types, and the transaction (sequence) patterns.	Set of possible generatable sentences together with all acceptable requests and input sentences

Table 1: Examples of abstraction levels in business and control analysis

The major benefit from applying the new definition lies in the clear separation of the operational and transactional perspectives. Moreover, the value aspect is enforced in analysis and design.

Note that the mightiness of the information system interface actually corresponds to the definition of an information base – all derivable abstract messages, which is a consequence of the information content and the access mechanisms at hand. The mightiness of business follow the same pattern.

5 Interface Modelling

We will make the theory of semiotics [11] easy to grasp in the form of a pattern for analysis and design [3]. The thinking is generic to any level of interface, but we will illustrate the analysis pattern, using the information system interface as used in controlling action.

When a message, with the intension of getting something done, reaches a receiver, i.e. an actor in a business process, many things can go wrong. Every step of the semiotic ladder [3] will act as a filter, as illustrated in table 2. To take an analogy, one faulty link will break the chain. We will use the six steps of the conventional ladder, add two more steps from contemporary business to business practice and express ourselves in terms of economies. We will use a trivial subject perspective.

	Sender	**Receiver**
Satisfaction economy (external)		(What will the end customer feel for the deliverable and the deliverer in the long perspective?)
Business economy Aspects of value addition	Do I have a clear-cut understanding of what I want?	Will the receiver see the value of the action proposed – and act as intended?
Culture economy Social aspects:	Do my intentions match the value system of the receiving structures?	Will the action and the consequences thereof fall within the norm system of the receiver?
Communication economy: Pragmatical aspects	Do I have the authority / credibility to demand the action?	Will the sender of the message be regarded as having enough authority or credibility to issue the message as interpreted?
Information economy Semantical aspects	Is the meaning of my intention clear?	Will the interpretation make sense?
Representation economy Syntactical aspects	Have I formulated my intention efficiently?	Will the message be formulated as to be interpretable?
Storage and transmission economy Empirical aspects	Will I be heard above the noise?	Will the contrast be high enough; will the distortion be low enough?
Encoding and detection economy: Physical aspects	Will what I express be seen as a message demanding action?	Will the receiver at all understand that there is a message arriving?

Table 2: Questions illustrating steps on the semiotic ladder

These basic steps of the so-called semiotic ladder might be applied to the business proper, i.e. the market interaction as well as the business process interaction with the information system. In the first case, a question might be would this advertising of a product be distinguishable from other companies advertising or may be a question about profiling the product itself. In fact, with modifications, we have found the generic pattern of analysis to be applicable for all layers of service interfaces and process inter-communication. Only the interface is illustrated in figure 7.

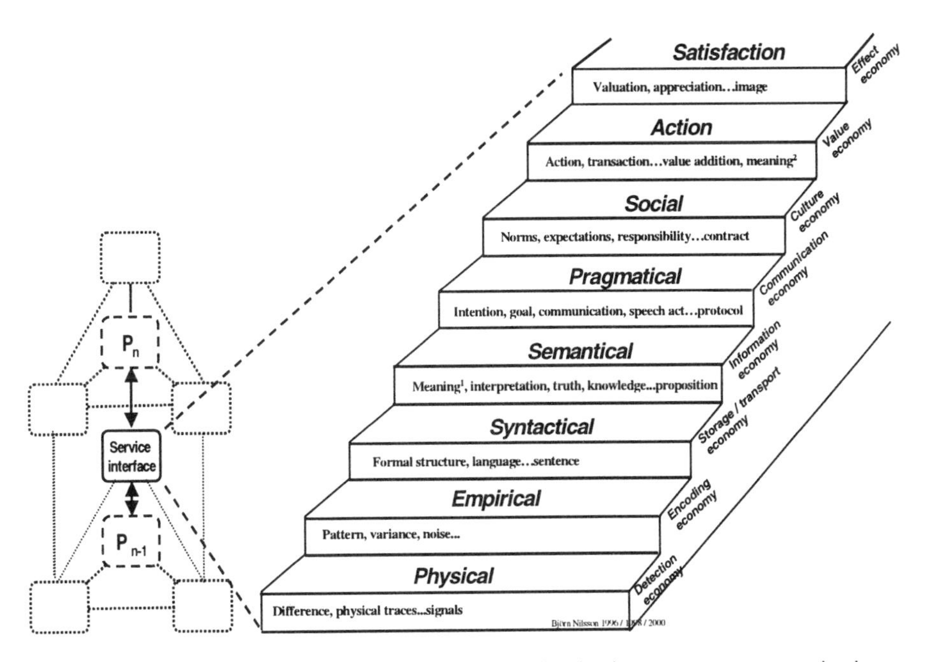

Figure 7: The semiotic ladder extended and applied to business process communication

The impact of one model linking different realms of analysis and design is hard to overestimate. It effects the mappings between realms as well as the development process itself. Applied between the business operations and information system levels, the interface model is not only bridging the gap between the IT- and the business-expert but is actually forcing the parties to perform a shared task. In practice, development processes where requirements from the business realm are developed by business analysts and delivered to be implemented by IT specialists does not work, has never worked and will never work.

The interface model, which may have an abstract or physical orientation, is a shared responsibility between the business and IT arenas. In modern business and IT development, the competence from both parties is needed through the whole development process. It is sometimes also very hard to draw the lines between IT as enabling and driving force. In the architecture, the interface is linking processes both between layers and within layers. What determines the structure of processes in an enterprise – and what is a process actually?

6 Business Operation Process Modelling

The value in a definition lies in its power to assist thinking. On the market, we find an abundance of definitions for the concept of process. Early definitions gave a new focus for analysis –customer value. [5] What, then, the pattern of value added?

- The pattern of a process corresponds to a desired result structure.

A trivial observation, but a revolution in practical analysis. As results are products of resource transformations, a process model defines an architecture, where the transformational elements are coupled by interfaces.

The value added, as viewed in a result structure, is dependent on goals defining value. Applying trivial systems theory, as a process is recursively decomposable, a multiple level process structure corresponds to a goal hierarchy. In simple terms:

- On every level of decomposition, goals are broken down into contributing goals – each of which can be satisfied by a structured sequence of transformations.

As decomposition is misused and misleading, it might be easier to regard the task as a recursive construction or design of necessary functions to satisfy higher goals. The sequence is normally defined in terms of pre and post conditions for the production of results. Naturally , parallelism and alternate paths must be allowed.

In common process modelling, sequencing is exaggerated and often hampers flexible solutions. To reduce this effect, a certain degree of desynchronisation has been accomplished by introducing a partitur or resource plane[8]. Process modelling formalisms are still far too static. Unfortunately, we have not found the speech act based formalisms [4] very workable either. Still, with the help of interaction diagrams [2], often using the gross pattern of Flores[7], we employ the basic thinking when applicable.

To base the architecture on a humanistic foundation – and focussing on value as created in processes - is probably what has lead to its survival over the years as well as to its robustness in evolution.

The architecture has greatly helped us bridging the gap between the business and IT communities in practical work by providing a basis for high quality modelling. It has also helped in breaking barriers between different levels in organisations.

Acknowledgements

I would like to thank Dr. Eva Lindencrona for her conception of the research project TRIAD on information administration. I was given the opportunity to develop the initial two-layer model architecture. I would, in this context, also like to express my gratitude to Prof. Janis Bubenko – at that time our common manager at the SISU research institute – for his invaluable support of the project.

Professor Ron Stamper, University of Twente, most kindly piloted me over the depths of modern semiotics. This has lead to a systematic method for the analysis of service interfaces. It has also connected nicely with speech act theory [10].

My former colleague Stefan Sandh inspired the method of integrating use cases in the model architecture. I am also greatly in debt for the reusability control permitted by complex mappings between system services and underlying system functions.

The introduction of abstraction levels in the definition of business would never have come about without the intervention of my dear colleague, Christer Nellborn during our writing of [8]. His dislike for, and opposition against, the layered architecture has promoted much of its evolution and enabled its successful usage in practice.

Recent thinking has been influenced by my work with the command and control system for the Swedish army. I am greatly in debt to my masters Jan Röjerdal and Ove Gardelius for their generous attitude to testing new theory in an ongoing project.

Finally, I would like to thank Magnus Penker for his adoption of some of the ideas of the architecture as a basis for the released UML business extension stereotypes [2], probably to become an OMG standard. Without these extensions, UML would be a much less suitable tool for practical business analysis.

References

1. Berger P, Luckmann T, The Social Construction of Reality, Doubleday, New York, 1966
2. Eriksson H, Penker M, Business Modelling with UML: business patterns at work, OMG Press, Wiley Computer Publishing, New York 2000
3. Falkenberg E D, Hesse W, Lindgreen P, Nilsson B E, Oei J L H, Rolland C, Stamper R.K, Van Assche F J M, Verrijn-Stuart A A, Voss K. A Framework of Information System Concepts, The IFIP WG 8.1 Task Group FRISCO, International Federation for Information Processing, Geneva, 1996
4. Flores F, Graves M, Hartfield B, Winograd T. Computer Systems and the Design of Organizational Interaction, ACM Transactions on Office Information Systems, 1988 Vol. 6, No. 2.
5. Hammer M, Champy J, Reengineering the Corporation: A Manifesto for Business Revolution, Harper Business Books, New York 1994
6. Leavitt H, The Volatile Organisation: Everything Triggers Everything Else. Managerial Psychology. The University of Chicago press, Chicago 1972
7. Medina-Mora R, Winograd T, Flores R, Flores F. The Action Workflow Approach to Workflow Management Technology In: Turner, J. & Kraut, R. (eds.) Proceedings of the 4th Conference on Computer Supported Cooperative Work, ACM, New York, 1993
8. Nilsson A G, Tollis C, Nellborn C, Perspectives on Business Modelling, Springer Verlag, New York 1999
9. OMG. Object Constraint Language Specification, version 1.3. OMG, Framingham Massachusetts 1999,
10. Searle J.R. A Taxonomy of Illocutionary Act, Expression and Meaning: Studies in the Theory of Speech Acts, Cambridge University Press, Cambridge, Massachusetts, 1979
11. Stamper R. K Information in Business and Administrative Systems, Wiley, New York, 1973.

Time and Change in Conceptual Modeling of Information Systems

Antoni Olivé
Departament de Llenguatges i Sistemes Informàtics
Universitat Politècnica de Catalunya
Barcelona (Catalonia)

Abstract

We survey the main approaches to change definition (or behavioral modeling) that have been developed and used in conceptual modeling. We characterize the approaches, show their relationship with temporal aspects, and review a few research efforts that have contributed to those approaches. Based on the characterization and previous work, we then point out a few research issues.

1 Introduction

A conceptual schema defines the general knowledge about a domain that an information system (IS) needs to know in order to perform its required functions. A large and significant part of a conceptual schema is related to the definition of how a domain changes. This definition includes the possible changes and their effects on the Information Base (IB).

Time and change are the two faces of the same coin. If the universe did not change, there would be no time. Any sort of event that can be described or thought of has a corresponding time [1]. Therefore, any approach to change definition in conceptual modeling has a temporal aspect.

This paper surveys the main approaches to change definition that have been developed and used in conceptual modeling and related fields. We characterize the approaches, show their relationship with temporal aspects, and review a few research efforts that have contributed to those approaches. Based on the characterization and the previous work, we then point out a few research issues.

There are in the literature many good reviews of time in conceptual modeling. We mention only four of them in the following. Bubenko [2] gives a detailed analysis of how the temporal dimension is dealt with in relational approaches, and also in four information modeling approaches. Bolour, Anderson, Dekeyser and Wong [3] is a widely referenced bibliography of some seventy early references, with many detailed annotations. Theodoulidis and Loucopoulos [4] is a detailed review of the treatment of time in nine conceptual models. The models are compared with respect to the semantics of time they use and the temporal functionalities they offer. Gregersen and Jensen [5] survey eleven temporal ER data models. Each model is characterized with respect to properties including time semantics, model semantics, temporal functionality and user-friendliness.[1]

This paper complements the studies reported above. Our focus is more on the specification of change than on the way facts may be associated with time. On the other hand, we deal mainly with approaches, and less with particular conceptual models.

2 Time and Facts

We review here the main temporal concepts that will be used in the paper, and the relationships between facts and time. Some of the terms used below are adopted from the excellent glossaries developed in the temporal databases field [7,8].

2.1 Time

Several models of time have been studied in science and philosophy. There are some important differences among them and several variations. The most important difference is the nature of the primitive unit of time: instant, interval or both. Most conceptual models of information systems assume instants as the primitive time unit, and then an interval is defined as the time between two instants. In the following, we will consider that the primitive time unit is an instant.

A time domain is a pair $(T;<)$ where T is a non-empty set of instants and $<$ is a precedence relation on T having the properties of transitivity and irreflexivity. Additional properties give rise to a number of variations. Time may be linear or branching; bounded or unbounded; and discrete, dense or continuous [9].

Normally, users structure the time domain by means of calendars. Conceptual schemas must include a model of the calendars. A calendar is a set of granularities over a single time domain, that includes a bottom granularity. A granularity is a mapping G from the set of integers to subsets of the time domain. Each non-empty subset $G(i)$ is called a granule of the granularity. The granules of the bottom granularity are called chronons. Usually, each granule has a textual representation, called its label. A label mapping defines the mapping between labels and integers.

Temporal references are expressions that denote instants or intervals in a given granularity G. Temporal references may be determinate or indeterminate. An instant reference is determinate if it denotes a granule of G. An interval reference is determinate if its two endpoints are determinate. An instant reference is indeterminate if it denotes a determinate interval, with the meaning that the instant occurs sometime during that interval. A determinate instant reference in granularity G is indeterminate with respect to all finer granularities. An interval reference is indeterminate if some of its two endpoints is indeterminate. Similar definitions can be given for temporal references to a continuous time domain.

Temporal references may be also absolute or relative. A temporal reference is relative if it includes a relationship with some other time.

---

[1] See also [6] for a review of time in Artificial Intelligence.

2.2 Valid Time of Facts

In the information systems field, we make the assumption that a domain consists of entities and relationships between them, which are classified into the concepts that we describe in a schema. The state of a particular domain, at a given time, consists then of the set of entities that exist, and the set of relationships that hold, at that time. Concepts may be represented in the schema by predicates, and then the state of the domain is represented, in the IB, as instances of those predicates.

We call fact the classification of an entity into some entity type or the classification of a relationship into some relationship type. A fact type is an entity or a relationship type. The valid time of a fact is the time when the fact holds in a domain. Usually, the same fact has several valid times. The state of the IB at time t is the set of facts with valid time t. We will call temporal fact a fact with a valid time. Valid times are represented by temporal references in some granularity. In information systems, valid times of all facts of the same fact type are usually in the same granularity. However, different fact types may have different granularities.

When valid times are determinate, the IB may be viewed in several different but equivalent ways [10]. The most usual are the fact (or model-theoretic) view and the timestamp view. In the fact view, each predicate has an additional argument (usually the last), which is a determinate instant reference to the valid time. In this view, when references are to a continuous time domain, the IB may be infinite. In the timestamp view, predicates have also an additional argument, but now it is either a determinate interval reference or a set of them. The meaning is that the fact holds at all instants included in the intervals.

When valid times are indeterminate, the preferred view of the IB is the timestamp. Each predicate has also an additional argument, which is again either a determinate interval reference or a set of them, but now the meaning is that the fact holds sometime during the intervals. Indeterminate valid time has not been studied very much in conceptual modeling, and even less relative valid time.

Related to determinate valid time there is the concept of structural event, which plays a key role in behavioral modeling. For every fact type there are two main structural event types: insertion and deletion. An insertion event occurs when a fact holds at a (valid) time and it did not hold immediately before. A deletion event occurs when a fact holds at a time, and it does not hold immediately after.

2.3 Temporal Properties of Fact Types

Many temporal properties of fact types have been studied in the literature. Perhaps the most basic one is the durability, which has two values: durable and instantaneous. Other names often used are state and event. Informally, a fact type is durable if its facts are assumed to persist in the IB from the time they are inserted until they are deleted by structural events. A persistence axiom for durable fact types defines the facts that hold at any time in terms of the structural events that have occurred. A fact type is instantaneous if its facts only hold at the time when they are inserted.

2.4 Belief Time of Temporal Facts

The belief time of a temporal fact is the time when the system believes (or knows) it. In temporal databases, it is called transaction time, defined as the time when the fact is current in the database and can be retrieved. Normally, belief times are considered to be determinate instants at some granularity. It seems reasonable to assume that the belief time granularity is the same for all facts in an IB. Belief times are represented in the IB in a way similar to valid times.

In the most frequent case, if a system believes some temporal fact at time *t*, it will believe it during the rest of the system's lifespan. However, it is a matter of fact that corrections and revisions of the past happen sometimes. On the other hand, belief time is usually equal or greater than valid time, but sometimes it is considered plausible that the system may believe future facts.

Most work in conceptual modeling assumes that valid and belief times of a fact are the same and that, if belief time is needed in some application, it can be introduced in a schema as an afterthought.

3 Approaches to Behavioral Modeling

The behavioral (sub)schema is the part of the schema that defines which facts hold in the IB at any time of the system lifespan, and the allowed events. In this section, we characterize the main approaches that have been studied and used in conceptual modeling, and we review some of the research works that have contributed to them.

3.1 Events

The causes of the changes in the IB are the external events and the passing of time. In general, one can distinguish among at least three kinds of events: (1) External, which are those communicated to the IS by the environment, through the I/O interface; (2) Internal, which are induced by external (or other internal) events, or detected by the system when some condition is satisfied; and (3) Structural events, as indicated above. Structural events are induced by external and internal events. External and internal events are defined in the behavioral schema. Structural events are defined implicitly.

An external event occurs in the domain at a given time, called its valid (or occurrence) time. The valid time may be a determinate or an indeterminate instant. The system believes an external event at a determinate instant (when it is notified of the event). An event is allowed to occur if a given condition is satisfied. Normally, this condition is defined with the event type, but sometimes it is an implicit composition of several conditions defined in different parts of a schema.

If an internal event is induced by external events, then they give its valid and belief times. If the internal event must occur when some condition is satisfied, then the valid time must be given by the condition, and the belief time is the time when the system detects it (normally, the earliest possible time). The valid and belief times of a structural event are the same as those of the event that induces it.

3.2 Basic Approaches

We review in this section the basic approaches to behavioral modeling. We focus on base fact types. We will assume that valid times are determinate, that the granularity of all fact types is the same and that valid time coincides with belief time. Later on, we comment on some variations that relax these assumptions.

In information systems, a behavioral schema is usually very large. In order to be able to develop and maintain that schema, it must be structured in some way. This is achieved by means of a structuring concept. The idea is to structure the behavioral schema in a number of behavioral units (or, for short, units), one for each instance of the structuring concept, and in such a way that the overall behavioral schema is given by the set of its units.

The most widely used structuring concept is the event type. We call event-centered the approach to behavior modeling that uses this structuring concept. In this approach, there is a unit (usually called operation or transaction) for each event type. The approach "centers" on an event type and defines all the effects that can happen when an event of that type occurs. However, the event type is not the only structuring concept that has been used in conceptual modeling. Other main approaches used or investigated are the object-centered, the temporal fact-centered and the cause-effect rule-centered. We try to characterize these approaches in the following.

Some approaches use several structuring concepts in a behavioral schema. Often, one of them is the primary one, used to structure the main part of the schema, and other concepts are used to structure the remaining part. On the other hand, there are approaches that, besides the primary structuring concept, use one or more secondary ones to give an additional substructure to the units.

3.2.1 Event-Centered Approach

In this approach there is a unit for each external or internal event type. A unit defines the name E of the event type, the event parameters $\mathbf{X}$, the condition the events must satisfy (precondition), the effects of an event occurrence on the IB, and the internal events to be induced (if any). The intuitive idea of the approach can be summarized by the expression:

"If event E($\mathbf{X}$) occurs at time T and precondition α is satisfied
then the new IB state is defined by β."

Many possibilities exist with respect to how α and β can be defined and with the relationship between T and the valid time of the facts of the new IB state. We can distinguish two main subapproaches, called structural event and postcondition, depending on how the new IB state is defined (β). In the structural event subapproach, β is a more or less declarative expression that gives all the structural events induced by E. The same structural events may be induced in different units. In the postcondition subapproach, β is a condition (called postcondition) that must be true in the new IB state. The frame problem, which is specific to this subapproach, is the problem of ensuring that β characterizes precisely the new IB state.

3.2.2 Temporal Fact-Centered Approach

In the temporal fact-centered approach there is a unit for each temporal fact type. In this approach, a unit corresponding to fact type *F* is a rule that defines the necessary and sufficient conditions for a fact *F(**X**)* to hold at a time. The intuitive expression of the approach is:

"Fact F(**X**) holds at time T iff condition α is satisfied"

Condition α involves one or more events that have occurred at time *T* or before. It may involve also facts of the IB with valid time *T* or before. The arguments ***X*** and *T* are free variables in α. The same event may appear in several units.

Additionally, there is a unit for each event constraint and for each internal event type (defining its occurrence condition).

3.2.3 Cause-Effect Rule-Centered

The two previous approaches provide complete behavioral units centered on an event type or a temporal fact type. They are complete in the sense that their units define all that can be said from the point of view of their respective center. In contrast, the cause-effect rule-centered approach gives units which are, in principle, partial. Now a unit consists of a rule stating that if a given condition is satisfied (the cause) then some effect must be produced. The cause is the occurrence of an external or internal event in a situation satisfying some condition. The effect may be an internal or structural event, or a post condition. Event preconditions are defined separately. The intuitive expression of the approach may be:

"If event E(**X**) occurs at time T and condition α is satisfied
then the effect β is produced."

Condition α may involve facts of the IB and other events with valid time *T* or before. The same event may appear in several rules, and the same effect may be caused by several rules.

In principle, we can find here again two subapproaches, depending on how β is defined. In the event subapproach, β is an expression that induces one (or more) internal or structural event. In the postcondition subapproach, β is a condition that must be satisfied in the new state of the IB.

3.2.4 Object-Centered Approach

Most object-oriented conceptual models center the behavioral schema on object types. In this approach, there is a unit for each object type. This unit defines the possible states of the objects, the event types that are relevant to objects of that type, with their preconditions, and the effects of these events on the population of the object type, and on the state, attributes and relationships of its objects. A unit cannot define effects for objects of other types. The same event type may be relevant to several units.

The object type is the primary structuring concept in this approach, but there is a secondary one that provides an additional structure inside each object type. In principle, any of the three previous structuring concepts could be used as secondary. If it is the event type, then a subunit defines the effect of each relevant event. If it is the temporal fact type, then a subunit defines the facts of that type in terms of relevant events. Finally, if it is the cause-effect rule, then a subunit defines the effects when a relevant event occurs and some condition is satisfied.

Independently from the secondary structuring concept, one can distinguish two subapproaches in the object-centered approach, depending on how objects interact. In the first, object interaction is implicit. All event types may be in principle relevant to all objects, and each object defines the specific event occurrences that are relevant to it. The same event occurrence may be defined as relevant to many objects of the same or different types. On the other hand, objects may induce internal events, which become potentially available to all objects. In the second subapproach, which is the most popular, object interaction is explicit. Each event has a sender and a receiver object. The object that induces an event must know the identity of the receivers. The same event may be sent to several objects.

3.3 Variations

The above discussion has focused on base fact types. For derived fact types, the usual approach is to define a derivation rule with the general form:

"Fact F(**X**) holds at time T iff condition α is satisfied"

which is the same as in the temporal fact-centered approach, but now α involves only IB facts with valid time T or before. The temporal fact-centered approach can be seen as one that assumes that event types are (instantaneous) fact types of the IB, and then it only needs to define ordinary derivation rules.

Several problems appear when event valid times may be different from their belief times. Assume that it is allowed to receive an event at time *t2* reporting some domain event at time *t1*. For example, an allowed event at time *15* may be "A deposit of amount *a* is made into account *acc* at time *10*". Now, the effects of events are not limited to changing what the IB believes at time *t2* about the state of the domain at time *t1*, but it also needs to change what it believes (at *t2*) about the states between *t1* and *t2*, and act accordingly. In the example, the balance of account *acc* at times *10..15* must change, but it may be necessary also to change past interests already paid, automatic transfers between accounts and so on. Moreover, some events believed prior to *t2* might be rejected now on the basis of the new beliefs.

Some subtle problems may arise when the granularity of event valid time is finer than that of the valid time of its affected facts. For example, assume the events *enroll* and *drop* in granularity seconds, and the fact type *EnrolledIn (student,course)* in days. In such cases, care must be taken to avoid that a fact hold and do not hold at the same time. For example, when a student enrolls and drops a course on the same day.

In general, indeterminate event valid times give rise to indeterminate valid times of the affected facts. The same happens when the granularity of event valid time is coarser than that of the valid time of the affected facts. For example, when valid time of payments is given in days, and customers' balances have granularity seconds. Very little is known about how to deal with indeterminacy in behavioral modeling.

3.4 Review of contributions

It is not possible to give in a short space a fair account of the impressive amount of contributions that have been made to behavioral modeling. We describe in the following (in approximate chronological order) a few of the most significant or particularly relevant to our characterization. Other important contributions can be found in the references.

3.4.1 Conceptual Information Model

The explicit consideration of the role of time in conceptual modeling started around the end of the 70's. Bubenko was among the pioneers [2,11,12]. He developed an approach to conceptual modeling, called Conceptual Information Model (CIM), in which entities, attributes and relationships are time-varying. Entities exist at some (valid) time. Similarly, attributes and relationships hold at some (valid) time. CIM uses a continuous time model based on instants. CIM provides also constructs for the definition of calendars. Time references may be a determinate instant of the underlying time domain or a determinate instant at some granularity. Fact types with different granularities are allowed.

Changes to the IB are due to the occurrence of events. Events may be external or internal. The definition of an internal event includes an occurrence rule, which indicates when the event is induced.

CIM was the first to use a temporal fact-centered approach to change definition. The population of an entity type *E* at time *t* is defined by means of a derivation rule, with the general form:

E(e,t) if condition

where *condition* defines the events that must have occurred, and the facts of the IB that must hold, at time *t* or before, for *e* to be an instance of *E* at *t*. For example, the rule stating that *x* is an author on day *d* if he has submitted a letter of intent (event *LetterOfIntent*) or a paper (event *PaperArrival*) or if he is an invited speaker could be:

$$\begin{aligned}\text{Author}(x,d) \leftarrow\ & (\text{LetterOfIntent}(y) \wedge \text{mainAuthor}(y) = x \wedge \text{day}(y) \leq d) \vee \\ & (\text{PaperArrival}(q) \wedge \text{mainAuthor}(q) = x \wedge \text{day}(q) \leq d) \vee \\ & \text{InvitedSpeaker}(x,d)\end{aligned}$$

Time-varying attributes and relationships are defined also by means of similar derivation rules. There is a derivation rule for each entity type, attribute and relationship type.

3.4.2 *Remora*

Rolland and Richard [13] describe the Remora methodology. We focus here on its contribution to behavioral modeling. There are three kinds of events: external, internal and temporal. An internal event is associated with an object type and a condition. An internal event occurs when an instance of that object type suffers a state change satisfying that condition. Events trigger operations. An operation type can modify at most one object type, and an operation instance can modify at most one object instance. The triggering of operations may be conditioned or iterative. In the first case, the condition must be specified; in the latter, what is specified is the set of objects to be modified.

An event may trigger several operations, and an operation may be triggered by several events. If the pairs (event, operation) are specified independently, then the approach used is that of cause-effect rule-centered. Given that operations are elementary actions, the subapproach used is very close to that of structural event. On the other hand, Remora includes also the concept of dynamic transition, which includes an event, all the operations it triggers and the objects modified by these operations. If the behavioral schema is specified as a set of dynamic transitions, then the approach used is that of event-centered/structural event.

Remora was used as a basis for the Rubis system [14], which -among other things- extended the concept of temporal event. A temporal event occurs when a predefined temporal assertion becomes true. The assertion, which uses the gregorian calendar, may involve absolute and relative references. For example, a valid assertion could be the determinate relative reference "3 days after the occurrence time of event Ev5".

3.4.3 *Formalization of the Event-Centered/Structural Event Approach*

Veloso and Furtado [15] give a simple and precise formalization of the event-centered/structural event approach. The specification of an operation (event type) includes a set of conditions, a set of facts to be added and a set of facts to be deleted. These sets of facts can be seen as sets of structural events. The application of an operation to a state σ of the IB is valid if the conditions are satisfied in σ, and it is productive if the facts to be added do not hold (and the facts to be deleted do hold) in σ.

Application of operations is subjected to two general rules: the inertness and the frame rules. The inertness rule states that if the application of an operation to a state is not valid or not productive, then the state remains unchanged. The frame rule states that upon application of an operation to state σ yielding τ, the facts that did hold in σ and that have not been deleted by the operation remain in state τ. Note that these rules assume that the IB fact types are durable.

3.4.4 *CANDID*

The CANDID project was concerned with the logical modeling of databases. We focus here on its temporal aspects, as presented in Lee, Coelho and Cotta [16]. Facts are represented by predicates with a (valid) time argument. CANDID assumes a continuous time dimension, but temporal references may be only end points of determinate common calendar dates (minutes, months, etc.).

The CANDID approach to change definition is cause-effect rule-centered/structural event. For each fact type there is an insertion and a deletion structural event type. Changes to the IB are due to the occurrence of external events. The valid time of events may be indeterminate. Occurrence of an external event induces one or more structural events, in the way defined by logical rules. For example, the rule:

RD(d): (Own(x,z)!Own(y,z) ← RD(d): OwnershipChange(x,y,z)

means that the occurrence sometime during the time span *d* of external event *OwnershipChange(x,y,z)* induces two structural events (occurring also sometime during *d*): deletion of *Own(x,z)* and insertion of *Own(y,z)*. An external event may induce also internal events, which eventually induce structural events. A set of general persistence rules define the facts that hold at a time in terms of the structural events that have occurred.

3.4.5 Situation calculus

The situation calculus is a language designed for representing changing worlds (domains). It has been widely studied and used in Artificial Intelligence. The basic conceptual and formal ingredients of the situation calculus are due to John McCarthy in 1963. Reiter has adapted the situation calculus to the formalization of "database evolution" under "update transactions". The following comments are based on Reiter's work [17, 18].

In the situation calculus, the world is at any instant in a given situation (state). A situation is characterized by a set of fluents (time-varying facts), denoted by predicates taking a situation term as one of their arguments. All changes to the world are the result of actions (external events). Occurrence of an action *a* in a situation *s* leads the world to a successor situation, denoted by *do(a,s)*. Actions have preconditions. An action can only occur in a situation if their preconditions are satisfied in that situation.

Effect axioms describe how actions affect values of fluents. The basic general form of an effect axiom can be stated as:

if action *a* is performed in situation *s*
and condition *c* is true
then fluent *F* becomes true (or false) in the successor situation *do(a,s)*.

For example, the effect axiom:

fragile(x,s) → broken(x,do(drop(r,x),s)

means that if robot *r* drops object *x* in situation *s* and *x* is a fragile object in *s* then, in the successor situation *do(drop(r,x),s)* object *x* will be broken.

These axioms characterize all conditions under which actions can cause *F* to become true (or false) in the successor situation. Effect axioms can be seen as cause-effect rules, event subapproach, as described in our characterization. There may be several axioms with the same action or fluent. The successor state axiom for fluent *F*, defining the facts that hold at a situation, can be obtained from the effect axioms. If the successor state axioms are defined directly, then the approach can be considered temporal fact-centered.

3.4.6 Event Calculus

The event calculus was formulated by Kowalski and Mergot [19] in 1986 as a logic programming treatment of time and change. Events mark the occurrence of change, and initiate and terminate periods of time for which facts hold. Sadri and Kowalski [20] describe several variants of the event calculus. We will focus here on the simplified event calculus (SEC), which can be used when complete information about events is available. SEC consists of a core, general axiom, and any number of auxiliary domain dependent definitions. The core axiom is:

$$\begin{aligned} holdsAt(P,T) \leftarrow\ & happens(E1,T1) \wedge initiates(E1,P) \wedge T1 < T \wedge \\ & \neg\exists E2,T2\ (happens(E2,T2) \wedge terminates(E2,P) \wedge T1 < T2 \wedge T2 < T) \end{aligned}$$

This states that a property *P* (fact) which is initiated by an event *E1* that happens at time *T1*, persists until it is terminated by a subsequent event *E2*. The conjunction *happens(E1,T1)* $\wedge$ *initiates(E1,P)* corresponds to an occurrence of an insertion event of *P* at *T1*. Similarly, *happens(E2,T2)* $\wedge$ *terminates(E2,P)* corresponds to a deletion event of *P* at *T2*.

Domain dependent axioms define predicates *initiates* and *terminates*. Their general form is:

initiates(E,P) if event E occurs (and condition holds)
terminates(E,P) if event E occurs (and conditions holds)

which, abstracting from syntactical details, can be seen as cause-effect rules, event subapproach. For example, the axioms:

initiates(E,has(X,Y)) $\leftarrow$ act(E,give(Z,Y,X))
initiates(E,has(X,Y)) $\leftarrow$ act(E(steal(X,Y,Z))
terminates(E,has(Z,Y)) $\leftarrow$ act(E,give(Z,Y,X))
terminates(E,has(Z,Y)) $\leftarrow$ act(E,steal(X,Y,Z))

state that the property of possession is initiated and terminated by giving and stealing events.

3.4.7 TELOS

Mylopoulos, Borgida, Jarke and Koubarakis [21] present Telos, one of the few knowledge representation languages that deal with both valid and belief times. Propositions (facts), deductive rules and constraints in Telos have a valid and a belief time component. The primitive time unit of Telos is the interval. Propositions, rules and constraints hold and are believed during intervals. Thus, for example, an entity is instance of a type during a time period *t1*, and this is believed by the system for a time period *t2*.

Valid times may be determinate or indeterminate. For example, a system may know that a person lived in a city for some time before May 1987. Belief times are determinate intervals.

Telos allows the definition of deductive (derivation) rules. However, Telos does not include a complete behavioral model. Changes in base fact types must be given explicitly by executing TELL, UNTELL or RETELL operations. For example, the operation

TELL TOKEN martian IN Paper (at 1986/10 ..*)
author : Stanley (at 1986/10 .. *)

executed on November 25, 1986 would declare that *martian* is instance of *Paper*, and that Stanley is one of its authors, both valid for an interval starting October 1986, and believed for an interval starting on November 25, 1986 until explicitly revised by an UNTELL or RETELL operation.

3.4.8 OMT

OMT was a widely used software development methodology. It was one of the roots of the recent UML [22]. The main reference is the book by Rumbaugh, Blaha, Premerlani, Eddy and Loresen [23]. OMT takes an object-centered approach to behavioral modeling.

The OMT dynamic (behavioral) model consists of a set of state diagrams, one for each class of objects. State diagrams are based on Harel's statecharts. A state diagram relates events and states. Not all classes need a state diagram; in such cases, it may be considered that the state diagram consists of a single state. An event is a one-way transmission of information from one object to another. Every event must have a sender object and a receiver object. An event can be directed at a set of objects or at a single object. An event flow diagram summarizes events between classes. Object interaction is explicit in OMT.

When an object receives an event, its next state depends on the current state as well as the event (transition). A transition consists of a source state, an event, a condition, an action and a target state. A transition fires when its event occurs and its condition is true. A transition may be seen as a cause-effect rule. An action is an instantaneous operation that executes when the transition fires. An action can set new values for the object attributes or generate other events.

3.4.9 Syntropy

Cook and Daniels [24] describe Syntropy, one of the few methods using the object-centered/implicit interaction approach. In Syntropy, the behavior of each object type is described in a statechart, based in the Harel's notation. The important elements of a statechart are a state machine and a textual part containing, among other things, a list of events of interest to objects of the type. In principle, all events are simultaneously available to all objects in the system. The list of event types of a statechart defines the events relevant to a particular object type. A filter expression selects which objects of the type are interested in an event. Filters allow each object to decide whether it is interested in any specific instance of the event type.

A state machine describes the way an object of a particular type moves between a finite set of states. Transitions are triggered by occurrences of events. The same event might trigger transitions in many objects of the same or distinct type. A transition can be seen as a cause-effect rule. Postconditions may be attached to transitions to specify the consequences of events. An object can specify only those postconditions that apply to its own properties and associations. A transition may generate events.

3.4.10 The frame problem in the postcondition subapproaches

We have seen that the event, cause-effect rule and object-centered approaches have a postcondition subapproach. Borgida, Mylopulos and Reiter [25] describe and exemplify the problems with this subapproach: specifications become lengthy, cumbersome and difficult to change, particularly in the case of the object type-centered approach, with inheritance. The problem, known as the frame problem, is caused by the necessity of stating explicitly that some parts of the IB remain unchanged.

For example, a pre/postcondition specification of event *enroll* student *st* in course *crs* could be:

pre: size(crs) < classLimit(crs) ∧ ¬EnrolledIn(st,crs)

post: size'(crs) = size(crs) + 1 ∧ EnrolledIn'(st,crs)

The postcondition states that in the new (primed) state of the IB the size of course *crs* must be incremented by one and that *st* must be enrolled in *crs*. However, it says nothing about the size of other courses, the enrollment of other students, the professors, and so on. The intuitive meaning is the "nothing else changes", but this is hard to state explicitly.

3.4.11 TROLL

Jungclaus, Saake, Hartmann and Sernadas [26] present TROLL, a conceptual modeling language. TROLL uses an object-centered/explicit interaction approach to behavioral modeling. The structural and behavioral properties of object instances are described by templates. Among other things, a template specifies the allowed events and the valuation rules. A valuation rule describes the effects of event occurrences on attributes. Its general form is:

If *event* and *condition* then *attribute* has *value*.

The postcondition (attribute has value) of a valuation rule is restricted to a single attribute. An implicit frame rule says that attributes for which no effects of events are specified do not change their values after occurrences of such events. A valuation rule can be seen as a cause-effect rule.

Communication among objects is described by interactions. An interaction defines a calling relationship *e1* >> *e2* between events *e1* and *e2*. The meaning is that when event *e1* occurs, then *e2* must occur too. For example:

ATM(atm).check_card(n,p) >> Bank.verify_card(n.p,atm)

means that when *ATM atm* receives event *check_card* then object *Bank* must receive an event *verify_card* requesting it to verify the inserted card.

3.4.12 Catalysis

Catalysis is a recent software development methodology that uses notations based on UML. The book by D'Souza and Wills [27] is the main reference. We focus here on the system behavior specification. In this specification, the whole system is seen as if it were a single object. At any time, the system is in some state. The state changes due to the execution of actions (external events).

The main approach to behavioral modeling used by Catalysis is event-centered, postcondition subapproach. The specification of each action includes its parameters, the precondition and the postcondition. The specifications are written in the OCL language (with extensions). A postcondition can refer to another action by naming it within brackets: [[action(..)]] means that the effect specified for that action is part of this postcondition; and [[→action(..)]] means that the action must be invoked as part of the postcondition.

Catalysis allows also specifying state charts for object types. When they are used, the approach taken is that of object-centered, with explicit interaction. The secondary structuring concept in this case is the cause-effect rule. State charts and state transitions can be translated into normal action specifications.

4 Research issues

Behavioral modeling of information systems is an active area of research. The analysis presented in this paper suggests several research issues within that area. We enumerate below six of them.

We have seen that there are four basic approaches to behavioral modeling. Each of them has its strengths and weaknesses. Some partial evaluations have been done (see, for example, [28, 29]), but we still lack a systematic and comprehensive evaluation of all approaches. This evaluation should be done with respect to the desirable properties of conceptual schemas and, probably, taking into account different domain characteristics.

The possibility of using two or more approaches in different parts of the same schema could also be studied in detail. If this were possible, the designer could use the most appropriate in each case.

When behavioral modeling is done in a formal language it should be possible to automatically transform a schema in an approach into another one. Such transformation might be very useful for validation and design purposes.

Belief time has not been considered in conceptual modeling to the same extent as valid time. Given that some applications require the use of belief times, it seems worthwhile to study in greater detail its impact in behavioral modeling. In particular, it should be analyzed the effect of corrections to the past on previously accepted events and on integrity constraints. Existing work on temporal databases is relevant here [30].

Similar to the above, indeterminate valid time has not been studied sufficiently in conceptual modeling. Given that some applications require the use of indeterminate times in events and facts, it seems worthwhile to study how to accommodate it in behavioral modeling.

Most work in behavioral modeling has been concerned with changes in the IB. These are not the only possible changes, however. The conceptual schema may also change. Schema evolution is the problem of defining possible changes to a conceptual schema and their effects on the schema and the IB. Schema evolution is a difficult problem, but with a great practical interest. Some work has been done already, but we are still far from a complete solution. Existing work on database schema evolution is relevant here, but it needs to be extended to cope with events, their effects, constraints and rules.

Acknowledgments

I would like to thank Alex Gago and Juan Ramón López for their comments and suggestions. This work has been partly supported by CICYT project TIC99-1048-C02-1.

References

1. Allen, J.F. , Hayes, P.J. A Common-Sense Theory of Time. In: Proc. IJCAI'85, pp. 528-531.
2. Bubenko, J.A. jr. The temporal dimension of information modeling. In: Nijssen, G.M. (ed.) Architecture and Models in Data Base Management Systems. North-Holland, 1977, pp. 93-118.
3. Bolour, A., Anderson, T.L., Dekeyser, L.J.,Wong, H.K.T. The Role of Time in Information Processing: A Survey. ACM SIGMOD Record, 12(3), 1982, pp. 27-50.
4. Theodoulidis, C.I., Loucopoulos, P. The time dimension in conceptual modelling. Information Systems, 16(3), 1991, pp. 273-300.
5. Gregersen, H., Jensen, C.S. Temporal Entity-Relationship Models - A Survey. IEEE Trans. on Knowledge and Data Eng., 11(3), 1999, pp. 464-497.
6. Vila, L. A Survey on Temporal Reasoning in Artificial Intelligence. AICOM, 7(1), March 1994, pp. 4-28.
7. Jensen, C.S., Dyreson, C.E.,Böhlen, M.H. et al. The Consensus Glossary of Temporal Database Concepts - February 1998 Version. Temporal Databases, Dagstuhl, 1997, pp. 367-405.
8. Bettini, C., Dyreson,C.E., Evans, W.S., Snodgrass, R.T., Wang, X.S. A Glossary of Time Granularity Concepts. Temporal Databases, Dagstuhl, 1997, pp. 406-413.
9. van Benthem, J.F.A.K. The Logic of Time. D. Reidel Pub. Co., 1983.
10. Chomicki, J. Temporal Query Languages: a Survey. In: Proc. First Intl. Conf. on Temporal Logic, LNCS 827, Springer-Verlag, 1994, pp. 506-534.
11. Bubenko, J. A. jr. Information modeling in the context of system development. In: Proc. IFIP'80, pp. 395-411.
12. Gustaffsson, M.R.; Karlsson, T.; Bubenko, J.A. jr. A Declarative Approach to Conceptual Information Modeling. In: Olle, T.W.; Sol, H.G.; Verrijn-Stuart, A.A. (eds.) Information systems design methodologies: A Comparative Review. North-Holland, 1982, pp. 93-142.
13. Rolland, C.; Richard, C. The Remora Methodology for Information Systems Design and Management. In: Olle, T.W.; Sol, H.G.; Verrijn-Stuart, A.A. (eds.) Information systems design methodologies: A Comparative Review. North-Holland, 1982, pp. 369-426.
14. Lingat, J.Y.; Novecourt, P.; Rolland, C. Behaviour Management in Database Applications. In: Proc. 13th. VLDB, 1987, pp. 185-196.
15. Veloso, P.A.S.; Furtado, A.L. Towards Simpler yet Complete Formal Specifications. In: Sernadas, A.; Bubenko, J. and Olivé, A. (Eds.) Information Systems: Theoretical and Formal Aspects. North-Holland, 1985, pp. 175-190.

16. Lee, R.M.; Coelho, H.; Cotta, J.C. Temporal Inferencing on Administrative Databases. Information Systems, 10(2), 1985, pp. 197-206.
17. Reiter, R. Formalizing database evolution in the situation calculus. In: Proc. Int. Conf. on Fifth Generation Computer Systems, June 1-5, 1992, Tokyo, Japan, pp. 600-609.
18. Reiter, R. Knowledge in Action: Logical Foundations for Describing and Implementing Dynamical Systems, 1999, draft.
19. Kowalski, R.; Sergot, M. A Logic-based Calculus of Events. New Generation Computing. 4(1986), pp. 67-95.
20. Sadri, F.; Kowalski, R. Variants of the Event Calculus. In: Proc. of ICLP, MIT Press, 1995, pp. 67-82.
21. Mylopoulos, J.; Borgida, A.; Jarke, M.; Koubarakis, M. Telos: Representing Knowledge About Information Systems. ACM TODS 8(4), 1990, pp. 325-362.
22. Rumbaugh, J.; Jacobson, I.; Booch, G. The Unified Modeling Language Reference Manual. Addison-Wesley, 1999.
23. Rumbaugh, J.; Blaha, M.; Premerlani, W.; Eddy, F. and Lorenses, W. Object-Oriented Modeling and Design. Prentice Hall, 1991.
24. Cook,S.; Daniels,J. Designing Object Systems. Object-Oriented Modeling with Syntropy. Prentice Hall, 1994.
25. Borgida, A.; Mylopoulos, J.; Reiter, R. On the Frame Problem in Procedure Specifications. IEEE Trans. on SE, 21(10), 1995, pp. 785-798.
26. Jungclaus, R.; Saake, G.; Hartmann, T.; Sernadas, C. TROLL – A Language for Object-Oriented Specification of Information Systems. ACM TOIS, 14(2), 1996, pp. 175-211.
27. D'Souza, D.F.; Wills, A.C. Objects, Components and Frameworks with UML. Addison-Wesley, 1999.
28. Bubenko, J.A.; Olivé, A. Dynamic or temporal modeling?. An illustrative comparison. SYSLAB Working Paper No. 117, University of Stockholm, 1986.
29. Hagelstein, J.; Roelants, D. Reconciling Operational and Declarative Specifications. In: Proc. CAiSE'92, Manchester, LNCS 593, Springer-Verlag, pp. 221-238.
30. Tansel, A.U.; Clifford, J.; Gadia, S.; Jajodia, S.; Segev, A.; Snodgrass, R. Temporal Databases. Theory, design, and implementation. The Benjamin/Cummings Pub. Co. 1993.

Co-operative Concept Modeling

Arne Sølvberg
Department of Computer and Information Science
The Norwegian University of Science and Technology
Trondheim, Norway

Abstract

Concept models represent agreed world-views. They reflect common classifications of worldly phenomena. The appropriate people must consequently understand the concept models. As the world evolves new concepts are created and introduced to the wider community. People will have to co-operate to find the "right" concepts, and they have to be educated in the meaning of relevant concepts. For this we need appropriate model languages, and appropriate tools for supporting the collaborative efforts which are associated with creation and learning of concepts. The paper describes the architecture of a tool, and an example of using a visual language for concept modeling.

1. Introduction

Decreasing costs of computers and telecommunication have led to a deep penetration of information technology into our society, both in the workplace and in the private sphere. Information can be made available for little cost to everybody by everybody. Information systems are established by integration of commercially available software components. Computers and people interact closely and they influence on each other's behaviour. Organisations are restructured into networks of co-operating autonomous units. Competition is global.

The complexity of the information technological infrastructure is awesome. So is the complexity and volume of the information systems needed in companies. It is beyond the capacity of a single person to understand the behaviour of every information system in a company. The knowledge is distributed in many heads inside and outside of the company. Information systems engineering is transformed into a multidisciplinary design discipline, which increasingly will be based on formal system models and common engineering approaches.

The application of computers is today so widespread that it is beyond the capacity of a single company to make their information systems from scratch. Systems must therefore be established by the integration of software components available in the market. It is also beyond the capacity of most companies to maintain a framework for integration on their own. Such frameworks also have to be established from what is available in the marketplace. Companies have changed their IT development from programming to integration of available software components.

Central to this is the issue of re-uses of system components, be they software objects, model fragments, or databases. Included are tools and approaches to information systems design through integration of re-usable components, constructivity in system modelling, verification and validation. Central to these problems is the issue of modelling. We can re-use a software component only with considerable risk of failure unless we understand what the component does, that is, we must be able to describe which information it uses and produces, and forecast its dynamic behaviour with respect to its environment. We can not predict the correctness of a planned information system unless we can derive its properties through formal reasoning about its specification. For engineering methods to replace handcraft approaches depends to a large degree on the availability of cost-effective modelling approaches.

Research in information systems engineering has been a central issue in the Nordic countries since the start of organised education and research in informatics at universities in the late 1960s. Modelling during the early phases of information systems engineering, and associated tool support, have in particular been investigated. Modelling during the early phases is known as conceptual modelling, and is also called requirements engineering. Langefors [LAN66] founded this line of research. Some of the first research results of 'the Scandinavian school' were reported in [BUB70A]. The early research projects CADIS [BUB70B] and CASCADE [SOL70] were reported on here. Both of those research projects developed computer assisted tool support for conceptual modelling in information systems engineering. They were early CASE tools.

Teichroewe's work on computer assisted systems engineering tools [TEI74] [TEI76] took place during the same time period in USA. Several of the US-developed CASE-tools of the 1980's were based on Teichroewe's pioneering work. While hopes were high when the commercial CASE-tools were launched in the early 80's they did not survive for long in practical use. There may be several reasons for that. Most of the CASE-tools were single-user tools and did not support co-operative systems development. Those that did, most notably the so-called IPSEs, were mostly project support frameworks and did not come with modelling tools of acceptable quality. Almost no tools supported effectively the transformation of requirements models into executable code. Few tools provided an 'escape route' for their users, that is, a way to preserve the specifications and models beyond the life of the tool. So it became too risky for prospective buyers to base their systems development on tools and modelling languages that were comparatively untested in practical use.

This paper shortly describes some of the relevant ongoing research in my research group in Trondheim. The selected activities are concerned with developing a framework and a support tool for collaborative model development, and with finding a set of useful modelling concepts that relate well to modelling concepts used in other fields of technology and science. Only static modelling concepts, for concept modelling, are discussed. The realm of conceptual modelling of systems dynamics is not considered in this short paper.

2. Model development as a co-operative effort

Information system engineering comprises design of business processes, organisation, patterns for human-computer interaction, as well as of data system design. A characteristic feature of information systems is their abstract nature. Classical engineering is concerned with natural phenomena, which may be touched, felt and viewed. The models of classical engineering are closely related to "touchables" and may often be visualised in easily recognisable forms. Information systems are different. They comprise human phenomena, e.g., human communication, as well as economic phenomena and physical phenomena, e.g., finding out whether it will rain. Information systems engineering is more concerned than other engineering disciplines with phenomena which do not lend themselves easily to formal modelling. Information systems engineers must be trained to work with abstractions in a more profound way than their colleagues in other branches of technology, because they will be forced to make formal models on their own of phenomena that have not been modelled before.

The building of information system models is more the exception than the rule in contemporary information system design. Occasionally one may see an entity-relationship model of data, but mostly there are no models. In the world of classical engineering modelling is commonplace. Engineers work in a world of models. Engineers co-operate in making a model of a product prior to its production. Co-operation requires that every participant have access to the shared product model. In the IT age this is comparatively simple. Product models can be represented in computers. Support for change must be provided, e.g., versioning of model fragments, as the different product parts are modified and fitted to new requirements.

The engineer's habit of thinking in models will have to be widely assimilated by people who want to influence on the building of information systems. Most information systems are based on models of the world that are so widely agreed that one hardly thinks of them as being models. Even a trivial application like a payroll system is based on a model of the relationship between the employer and the employees, which is so widely accepted by all parties involved that nobody thinks of it as a model.

Co-operation is necessary in order to achieve synergy when several persons are active within the same activity area. A pre-requisite for achieving co-operation is to have effective means for the exchange of information among the actors. Central application areas are information resource management and CSCW - Computer Supported Co-operative Work.

A pre-requisite for obtaining co-operation is that the participants share a language by which they can communicate. They must have a common view with respect to what they communicate about. They must agree on the meaning of the words that they use. A shared model of what the actors co-operate about must exist.

Specification languages are mostly informal and support system-sketches rather than firm specifications, e.g., data flow diagrams in structured analysis. Most of the development approaches that address the whole life-cycle rest on a combination of modelling approaches, one model fragment for each development phase, e.g., as seen in UML [ALH98, BRJ99]. The different model fragments are usually only loosely related. Different modelling constructs are applied at different stages of systems development, and support different concerns. Specifications developed during early phases of development are mostly used only in an informal manner as a basis for the work in the later phases. Model fragmentation inhibits a safe development of a sequence of models, from the early phases to the programming stage, because specifications are not integrated over the phase boundaries.

There is a need for a comprehensive information system specification language that captures all relevant concerns during the whole development process. The modelling constructs should support the fuzzy thinking in the early phases, as well as the specification of details at the level of programming. The system model must be formal so that specifications may be automatically translated from one set of modelling constructs to another, and so that detailed specifications may be automatically evaluated with respect to previous statements of system properties. That is, the information system model should be a constructive model.

It must become possible to express models of information systems in such a way that people who are not computer specialists can understand them. Everybody is deeply affected by computers. So everybody should be able to understand what the information systems are doing. Unfortunately different audiences have different knowledge about the universe of discourses, and different conceptual knowledge as well. Transformation of detailed system specifications to more abstract levels becomes necessary in order to achieve model comprehension. This may also be viewed as a form of explanation generation [GUL96], whereby explanations of operation details in an information system are generated and fitted to the conceptual abilities of an intended audience.

An information system specification, which serves as the basis for communication and understanding, is also the basis for the coding of software. A specification must consequently contain massive amounts of formal detail and at the same time be understandable for the participants in the system development process. Two major issues are the model fragmentation problem and the complexity explosion problems.

Complexity explosion occurs when approaching the implementation level. The number of details in the specification must be enough to permit automatic generation of software. Even if some of the syntactical details of programming languages may be omitted in the specification, all of the logical details have to be stated. The system model consequently becomes difficult to comprehend for those who have no time to learn every detail. There is a need for mechanisms to "filter" away details that are not needed for the reader of the model.

It is necessary that specifications that are understandable are also executable. Otherwise software will have to be expressed in a language that is different from the specification language, and the software may become difficult to understand. There is no solution to this problem so far. It is still in the research realm. The solution seems to be in a language that comprises both concerns, in a language that permits the transformation from one form to the other, so that unwanted details can be removed when specifications are to be presented to those who know little about software.

3. Processes and tools for collaborative modelling

Concurrent engineering is commonplace. Teams of developers work in parallell on different parts of a common product model. The work is geographically distributed, often spanning several time zones, and going on around the clock. The modellers add intellectual content to the shared product model, either as revisions to previous designs, or as new model fragments. A general architecture of processes and tools is shown below.

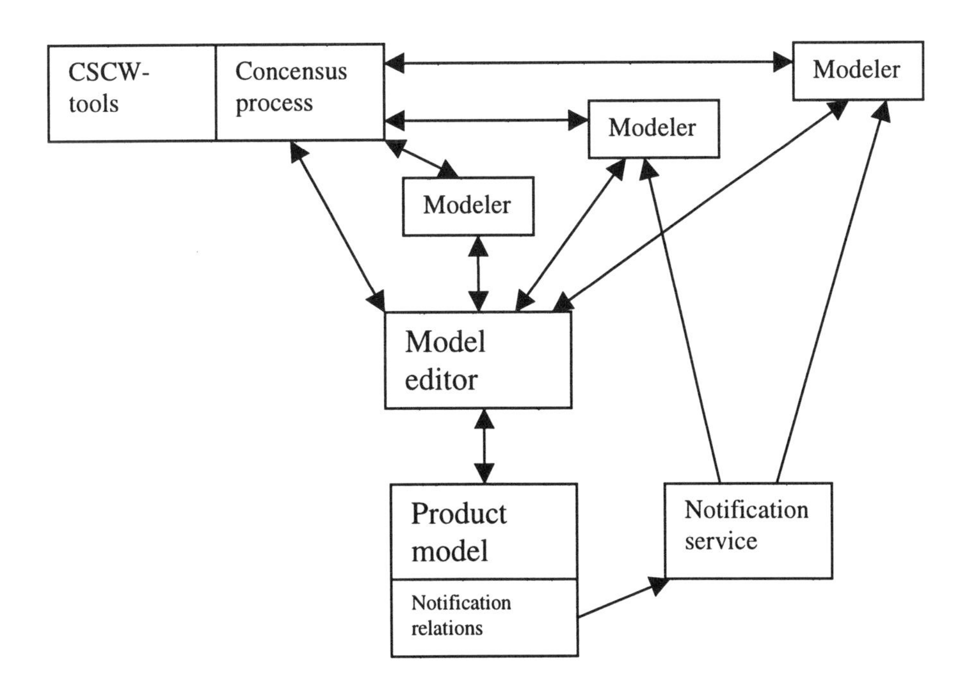

Figure 1Co-operative product model development

Modellers provide product specifications via a model editor. The product model, e.g., the information system model, is enhanced with notification data, which defines who to inform when changes are introduced to the product model. The appropriate persons are alerted to relevant changes by a notification service, and may take action as deemed proper. When changes to the product model are introduced which influence on others than the individual modeller, those that have stakes in the change have to resolve possible conflicts. They have to engage in a consensus process, which more often than not spans different geographic locations, and consequently must be supported by CSCW-tools. A proposal for a tool to support collaborative information systems development has been proposed in [FAR00]. The tool architecture is shown below.

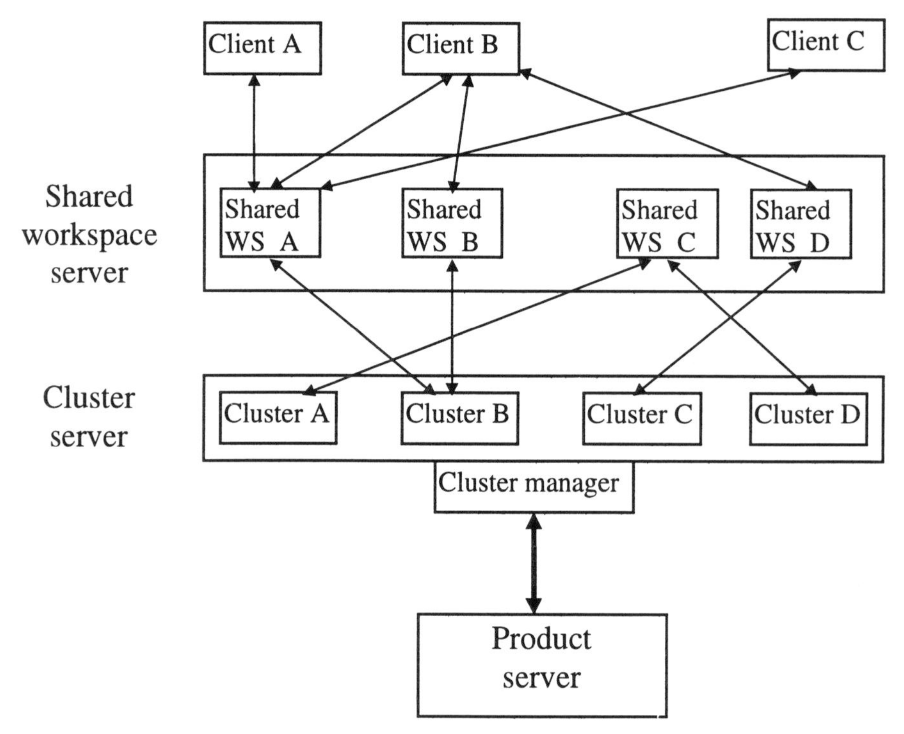

Figure 2 Tool system for co-operative product model development

The tool is intended to support collaboration in the large by supporting a shared product model, and provide change awareness to all connected shared workspaces. The framework is called IGLOO. It is based on the idea that collaboration in the large requires several concurrent workspaces. Each workspace may access several clusters, where each cluster provides a separate view of the total product model. The aggregation of local editing tasks in each workspace constitutes the total work result. Typical services are access control, version control, change management, and consistency preservation. The tool is currently being developed and a pilot version will be available mid-2000.

4. Basic concepts for information systems modelling

The way that we comprehend the basic relationships among data, machines and humans are by and large unchanged by the computer revolution. The prevailing view is that computers deal in data, and that people must interpret the data in order to convey information. Data is interpreted relative to how the human interpreter perceives the phenomena that the data refers to. More often than not there is no explicit model of the world for the human interpreter to lean to. The usual situation is that relevant parts of the appropriate world models are carried by the names of the various symbol classes that appear in the computers. The world models are thus implicit to the interpreters.

With the advent of WWW this practise is probably too lax. Everybody compete about the attention of everybody else, on a global scale. Intellectual energy is in short supply. Unless the information provider can communicate the intended message effectively the audience will be lost. The knowledge provider will almost never be in direct contact with the knowledge recipient. We generally resort to modelling when we want to express knowledge intended to be communicated to others and to be subjected to examination independently of the knowledge originator. The conversion of personal knowledge into public knowledge is accompanied by the representation of knowledge by signs of some language. The statements of the language must be such that a receiver understands them as intended by the sender.

In order to achieve this we use concepts and terms. The concept is the unit of thought. The term is the unit of language. Theories, e.g., Newton's theory of motion, are structures of thoughts, and concepts like force, mass and acceleration is the units of these thoughts. Conceptual knowledge comes wrapped in signs, e.g., mathematical notation, words or diagrams, which are the linguistic expressions of knowledge. In order to get access to the ideas of other people we thus have to understand the conceptual structures that are employed, and we have to understand the relationship between the signs and the ideas that they stand for. To contribute to this end is an important aspect of conceptual modelling of information systems.

An information system is a body of signs, and the associated processes for storing and transforming the signs, and for exchanging signs with the exterior of the information system. Each sign reflects some property of the Universe of Discourse (UoD), which is the domain of individuals referred to by the information system, or a sign may represent some property of the information system itself.

In human discourse we distinguish among the physical aspect, the conceptual aspect and the linguistic aspect of the UoD. Consider the sentence: John is a person. Each word in this sentence is a term (enclosed in simple quotes), namely 'John', 'is a' (a syntactic composite of two terms), and 'person'. The concepts (enclosed in double quotes) are "John", "class membership" (designated by the term 'is a'), and "person". The term 'John' designates the concept "John" which in turn represents the living individual John, which is called the referent. The term 'person' designates the concept "person", which represents all known (and unknown) persons, which is the so-called extension E("person").

Because the knowledge providers are in need of reaching many people they have to resort to modelling concepts which are widely shared and therefor easier to grasp than more unfamiliar concepts. For information systems modelling there is the particular need of integrating information system concepts with concepts of the various application domains. The deep penetration of computers into all kinds of technical-administrative systems means that hybrid specification languages and systems models will probably emerge on a broad scale. The hybrid languages will comprise software concepts, information system concepts, organisation concepts, and "application area" concepts, e.g., automobile-concepts, banking concepts, civil engineering concepts. The integration of these different fields requires that general laws that reflect on the static and dynamic properties of the various parts of a system must be expressed in a unified way so as to provide a complete theory for the system as a whole. E.g., the laws that reflect the general properties of information systems must be expressed in a way similar to how the laws that express the particularities of engineering models of automobiles or of merchant banks.

So it becomes desirable that concept models for information systems relate better to classifications employed in other fields of human interest. One such classification is found in the American National Standard's guidelines for thesauri construction [THES93], which recommends that distinction is made among the following kinds of concepts (non-exhaustive list):

- things and their physical parts, e.g. bird, car, mountain
- materials, e.g., water, steel, oxygen
- activities or processes, e.g., painting, golf
- events or occurrences, e.g., birthday, war, revolution
- properties or states of persons, things, materials or actions, e.g., elasticity, speed
- disciplines or subject fields, e.g., theology, informatics
- units of measurement, e.g., hertz, volt, meter

Another concept classification is found in science, where distinctions are made among individual concepts, class concepts, relation concepts, and magnitudes (quantitative concepts)[BUN98]. Distinctions are also made between specific (definite) concepts and generic (indefinite) concepts, e.g., "Einstein" is a specific concept, but "x" is a generic concept and denotes an arbitrary referent. This concept classification reflects the properties of mathematical set theory and is thus of considerable generality. It is independent of the concept classification made for thesauri purposes, and may be used for concept modelling within each group as well as for building models to relate concepts that belong to different groups in the thesaurus classification.

Class concepts and relation concepts are well known. Magnitudes (qualitative concepts) usually appear in the form of attributes. Quantitative concepts apply to properties that reflect magnitudes which are associated with individuals and/or sets, e.g., the temperature of a body, the number of elements of a set. Quantitative concepts do not represent distinct objects, that is, no distinct objects are associated with a quantitative concept, e.g., temperature, weight, mass, heat, acceleration. Functions are the structure of quantitative concepts. For example, temperature is a function T that maps the set of bodies (each of which has a temperature) into the set of real numbers. Let "b" be a generic individual concept which represents some physical body (the object variable), and let "t" be a generic individual concept which represents a numerical value (the numerical variable), then "T (b)=t" reflects the functional nature of T, and is to be read 'the temperature of b equals t'. The numerical variable t occurring in the temperature function is equal to the number of temperature units on a given scale, e.g., Celsius or Fahrenheit's. If scale and unit system is not fixed by the context we need to indicate it by a special symbol, say 's'. In short, "T(b,s)=t" is to be read 'the temperature of b equals t measured in the scale s'.

Quantitative concepts are general in the sense that they apply to large numbers of referents, e.g., every physical body has a weight and a temperature. A generalization of the definition above provides us with the definition of a quantitative concept as a function $q{:}UoD \times S \rightarrow V$, where UoD is the set of referents, S is the set of scales, V is the set of linguistic units (the values), and consequently the set of quantitative concepts is $Q=\{q \mid q{:}UoD \times S \rightarrow V\}$.

In practice we need to be able to define limited sets of properties for each concept. E.g., for the concept of person it may be appropriate to define the property of weight and of height as part of the properties that make up the concept's intension, while temperature may be irrelevant unless we are interested in sick persons. In order to delimit the number of properties to those that are relevant for a concept c we use the concept of attribute $A \subseteq C \times Q \times S$, where C is the set of concepts, and where $A=\{(c,q,s) \mid q{:}E(c) \times \{s\} \rightarrow V\}$. We see that $E(c) \subset UoD$ defines the appropriate part of the Universe of Discourse.

The knowledge, which is represented in an information system, is entirely conceptual. The data, which are stored and processed by an information system, are linguistic units, which denote concepts and referents in the Universe of Discourse. We thus have to distinguish between concepts in general and linguistic concepts, between individual concepts and class concepts, between relation concepts and those that are not, and we distinguish attributes - the relations between non-linguistic and linguistic concepts - as a separate concept class akin to quantitative concepts.

Information systems are concerned with signs that represent facts in a Universe of Discourse. A fact is what is known -or assumed - to belong to reality. One usually distinguishes the following kinds of facts: *state*, *event*, *process*, *phenomenon*, and *concrete system* [BUN98]. Ideas are formally expressed as concepts, formulas (e.g., statements) and theories, which are systems of formulae. *Models* are the formal expression of ideas.

The *state* of a thing at a given instance is the properties of the thing at that time. An *event* is any change of state over a time interval. A *process* is a time-course of events. A *phenomenon* is an event or a process such as it appears to some human subject. A *concrete system* is a physical thing or some other physical entity; e.g., a magnetic field may be considered an entity but not a thing. So an electrical field is a concrete system, but a theory of electrical fields is an idea, it is a conceptual system.

Finally, we have modelling constructs for expressing behaviour of systems and things. These are separate concept classes and represent the evolution over time of properties of concrete systems. They are intended for modelling of concrete facts, and express ideas of systems that belong to physical reality. An overview is given below.

1 Individual Concepts	(“Newton”, “x”)
2 Class Concepts	(“copper”, “living”, “person”)
3 Relation Concepts	/ Non-comparative (“spouse“, “between”) ----Comparative (“≤“, “better adapted than”) \ Operations (“∧“, “+”)
4 Linguistic concepts	/ Terms (‘John’, ‘is a’) --Datatypes (“name”, “integer”) \ Datatype-extensions (“personnel records”, “project file”)
5 Attribute concepts	/ Identifiers (“social-security-number”) ----Descriptive concepts (“first name”) \ Quantitative Concepts (“temperature”, “length”)
6 Behaviour concepts	/ State (‘waiting for input’) ----Event (‘input arrived’) \ Process (‘transform input to output’)

5. Concept modelling and term modelling

Contemporary search engines are based on searching for text patterns. They are searching for terms, not for concepts. Neither precision nor recall of web-searches is impressive with contemporary text-oriented approaches. In order to improve the situation one should search for concepts instead of terms. The current interest for so-called "ontologies" reflect on the interest for trying out an approach where terms and concepts are viewed as different and related modelling concepts.
A proposal for a solution is shown below [BRA99]. Documents are analysed relative to a conceptual model of the UoD, which the document is about. The different terms of the document text are analysed and matched to the designated concepts. The

concept model is modified and expanded as new terms give rise to new concepts. The documents are classified with respect to the concepts that they refer to, and are stored in, e.g., the WWW with this concept characterisation. Document search will be performed starting from concept models, where concepts of interest are selected, and documents are located which conform to the appropriate concept characterisation.

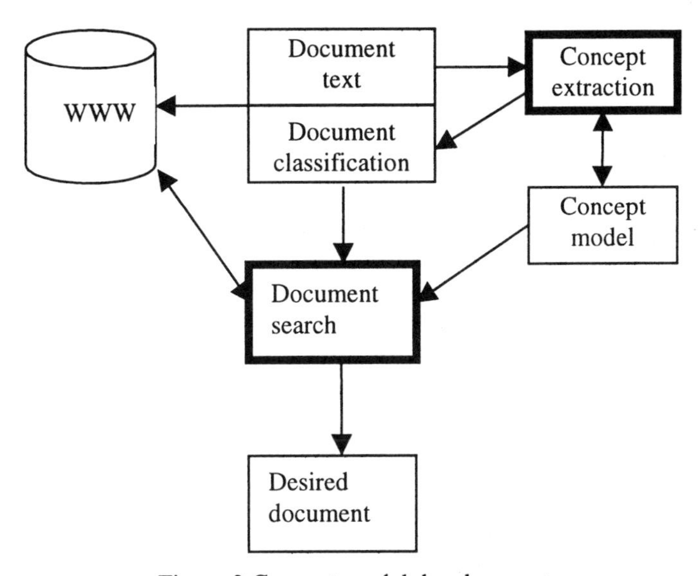

Figure 3 Concept model development

An example of a concept model is shown in the sequel. The model is adapted from [BRA99] and shows part of a concept model for health administration. A similar (and larger) model is developed to serve as a centerpiece for the classification of documents in the Norwegian health administration. The concept model is expressed in a visual language. It is developed as a collaborative effort. Selected documents are analysed and concept model is developed which reflects on the contents of the documents. The previously described tool architecture for co-operative product model development supports the modelling process.

It is beyond the scope of this paper to describe the features of the visual concept modelling language that has been used. Earlier versions of the language have been described in [SOL99A], [SOL99B]. A modelling editor with user manual can be found at http://www.idi.ntnu.no/~ppp/ . In order to help the reader to make sense of the concept model without having to invest too much time in understanding the modelling language, it should be sufficient to point out that rectangles depict class concepts, that arrows depict relation concepts, and that triangles depict operations and comparative relations.

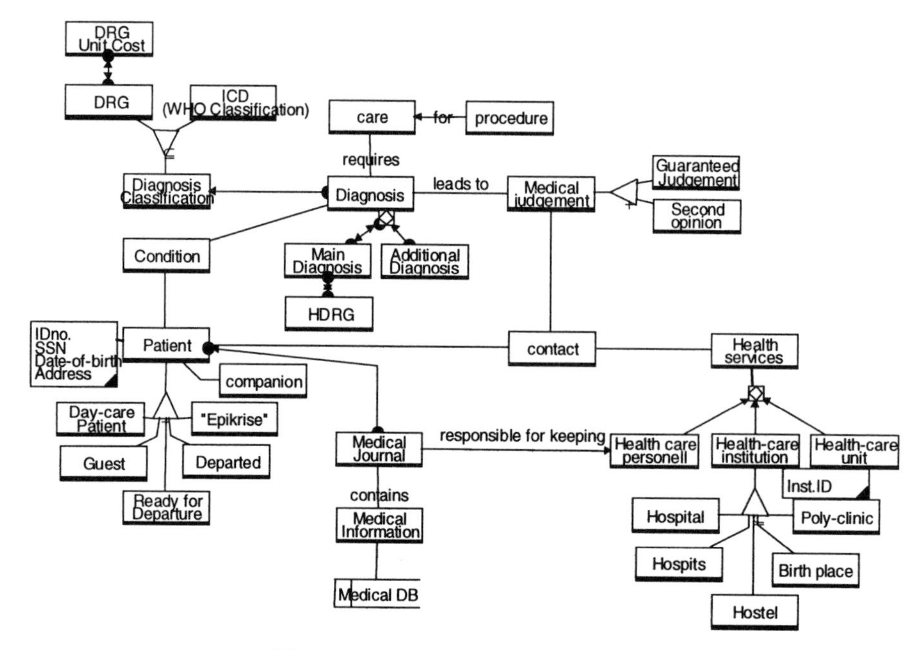

Figure 4 Example of concept model

Visual specification languages are usually less expressive and less general than the corresponding mathematical formalism, but (hopefully) comprise enough of the needed modelling constructs to make them useful. Most visual languages are tools for sketching half-baked ideas, at early stages of conceptualisation of problems. They lack mathematical underpinnings and as a consequence the models developed in these languages lack desirable rigour. Most visual languages are therefore not as useful as they could have been.

In the modelling language, which has been used here, every construct is grounded in mathematical theory; the language of mathematics is available and may be used. Because of the visual form the formality of the constructs is hidden, and the language may consequently be accessible and comprehensible for people without advanced mathematical training.

6. Summing up

We have proposed an architecture for tools to support co-operative product modelling. Furthermore, we have proposed a conceptual framework for concept modelling for information systems engineering. The framework is solidly based on elementary mathematics, and on widespread practices in science and technology. An example of using such a language has been briefly sketched. Judging from the apparent lack of success of previous efforts to implement such frameworks it will not be an easy and straightforward job to put these ideas to work. We will nevertheless continue our efforts.

References

[ALH98] Alhir S.S:UML IN A NUTSHELL, O'Reilly,1998, ISBN 1-56592-448-7

[BRA99] Brasethvik T. and Gulla J.A.: "Semantically accessing documents using conceptual model descriptions." in Chen P.P. et al. (eds.) "Advances in Conceptual Modeling", Springer LNCS #1727

[BRJ99] Booch,G.,Rumbaugh,J.,Jacobson,I.: "The Unified Modelling Language User Guide", Addison-Wesley, 1999, ISBN 0-201-57168-4

[BUB70A] Bubenko, Kællhammar, Langefors, Lundeberg, Sølvberg: Systemering 70, Studentlitteratur, Lund, Sweden, 1970

[BUB70B] Bubenko J., Kællhammar O.: Computer Aided Design of Information Systems, in [BUB70A]

[BUN98] Bunge M.: The Philosophy of Science, Transaction Publishers, 1998, ISBN 0-7658-0415-8

[FAR00] Farshchian B.A: "IGLOO: A framework for developing product--oriented shared workspace applications", Proc. COOP'2000

[GUL96] Gulla, J. A.: "A General Explanation Component for Conceptual Modeling in CASE Environments",ACM Transactions on Information Systems, 14(3), pp. 297-329, July 1996.

[LAN66]Langefors,B.: Theoretical Analysis of Information Systems, Studentlitteratur, Lund, Sweden,1966

[SOL70] Sølvberg A.: Experiments with new methods in systems work (in Norwegian), in [BUB70A]

[SOL99A] Sølvberg A.: "Data and what they refer to", in P.P.Chen et al.(eds.): Conceptual Modeling, pp.211-226, Lecture Notes in Computer Science,

[SOL99B] Sølvberg A.: "Conceptual Modeling in a World of Models" in R.Haschek (ed.): Entwicklungsmethoden fuer Informationssysteme und deren Anwendung, pp.63-77,B.G.Teubner, 1999

[TEI74] Teichroew D.: Improvements in the System Cycle. IFIP Congress 1974: pp. 972-978

[TEI76] Teichroew D., Hershey E.A.: PSL/PSA: A Computer-Aided Technique for Structured Documentation and Analysis of Information Processing Systems (Abstract). ICSE 1976: 2

[THES93] ANSI/NISO Z39.19-1993: Guidelines for the Construction, Format, and Management of Monolingual Thesauri , ISBN 1-880124-04-1

Bibliography

Ahlsén, M. and J. A. Bubenko, Jr (1991). Interoperability in Federated Information Systems (FIS). Second International Workshop on Intelligent and Cooperative Information Systems: Core Technology For Next Generation Information Systems, Ed. Brodie, M. L. and S. Ceri, Villa Olmo, Como, Italy, Dipartimento di Ellettronica, Politecnico di Milano.

Andersen, R., J. A. Bubenko, jr and A. Solvberg, Eds. (1991). Conference on Advanced Information Systems Engineering - CAiSE'91. Lecture Notes in Computer Science, Springer Verlag.

Bergenheim, A., A. Persson, D. Brash, J. A. J. Bubenko, P. Burman, C. Nellborn and J. Stirna (1998), CAROLUS - System Design Specification for Vattenfall, Vattenfall AB, ESPRIT Project No. 22927, ELEKTRA, Deliverable Document, 31 March, 1998.

Bergenheim, A., A. Persson, D. Brash, J. A. J. Bubenko, P. Burman, C. Nellborn and J. Stirna (1998), FREJA - Evaluated Design For Vattenfall, Vattenfall AB, ESPRIT Project No. 22927, ELEKTRA, Deliverable Document, Sept., 1998.

Bergenheim, A., K. Wedin, M. Waltré, J. A. Bubenko, Jr, D. Brash and J. Stirna (1997), BALDER - Initial Requirements Model for Vattenfall, Vattenfall AB, ESPRIT Project No. 22927, ELEKTRA, Document, 20 August, 1997.

Bergsten, P., J. A. Bubenko, Jr, R. Dahl, M. Gustafsson and L.-Å. Johansson (1989), RAMATIC - A CASE Shell for Implementation of Specific CASE Tools, Swedish Institute for Systems Development (SISU).

Berztiss, A. and J. A. Bubenko, jr (1995). A Software Process Model for Business Reengineering. Information Systems Development for Decentralized Organizations (ISDO95), Ed. Sölvberg, A., 21-23 August, Trondheim, Norway, Chapman & Hall.

Berztiss, A. and J. A. Bubenko (1996). Domains, patterns, reuse, and the software process. Domain Knowledge for Interactive System Design, Ed. Sutcliffe, A. G., F. v. Assche and D. Benyon, Trondheim, Norway, Chapman & Hall.

Boman, M., J. Bubenko, P. Johannesson and B. Wangler (1993). Models, Concepts, and Information: An Introduction to Conceptual Modelling for Information Systems Development, Department of Computer and Systems Sciences, Stockholm University, Stockholm.

Boman, M., J. A. Bubenko, jr, P. Johannesson and B. Wangler (1992), Studies in Federated Information Systems, Dept. of Computer and Systems Science, Royal Inst. of Technology and Stockholm University, ELECTRUM 230, S-164 40 Kista, Sweden.

Brash, D., J. Stirna and J. A. Bubenko, Jr (1998), Generic Patterns: The Silver Bullet?, Department of Computer and Systems Science, KTH/SU, Electrum 212, S-16440, Kista, Sweden., Report, ESPRIT Project No. 22927, ELEKTRA, 25 May, 1998.

Bubenko, J. A., Jr (1960), MIKE - ett interpretativt program för matrisräkning (MIKE - an interpretative program for matrix calculus (in Swedish)), Kungl Armetygförvaltningen, Report.

Bubenko, J. A., Jr (1961), Large Deflections of Rectangular Plates Solved by Finite Differences, Dept of Structural Mechanics, Chalmers Univ. of Technology, Licentiate thesis, No. 61:7.

Bubenko, J. A., Jr (1963). "Multiprogrammering - en orientering (Multiprogramming - an orientation (in Swedish))" BIT 3(1): 1-26.

Bubenko, J. A., Jr (1966). Some Aspects of the Design of Engineering Information Systems. Proceedings of the Second Conference on Power Systems Computation, Stockholm, KTH.

Bubenko, J. A., Jr (1967). Databehandlingsteknik, Studentlitteratur, Lund.

Bubenko, J. A., Jr (1968). Simulating a Management Game with Programmed Decisions. Digest of the Second Conference on Applications of Simulation, New York, SHARE/ACM/IEEE/SCi.

Bubenko, J. A., Jr (1970), Composition and Intensional Definition of Binary Relations - a Possible Extention to CADIS System 1, Inst för Informationsbehandling ADB, KTH/SU, Stockholm, CADIS Report, No. 43, March, 1970.

Bubenko, J. A., Jr (1970). Computer Aided Design of Information Systems - Objectives for Research in Information System Design. Systemering 70. Eds. Bubenko, J. A., Jr, O. Källhammar, B. Langefors, M. Lundeberg and A. Sölvberg, Studentlitteratur. Lund.

Bubenko, J. A., Jr (1970), Notes on processes in an e-system which are triggered by changes in the data base, Inst. för Informationsbehanldling ADB, KTH/SU, CADIS Report, No. 11, 11 Jan. 1970.

Bubenko, J. A., Jr (1970), Some Aspects of Crude Level Analysis of Information Systems, Inst. för Informationsbehandling -ADB, KTH/SU, CADIS Report, No. 28, May, 1970.

Bubenko, J. A., Jr (1970). Some Methods for Interactive Data Processing System Analysis. Systemering 70. Eds. Bubenko, J. A., Jr, O. Källhammar, B. Langefors, M. Lundeberg and A. Sölvberg, Studentlitteratur. Lund.

Bubenko, J. A., Jr (1970). SRTS -1 (Special Real-Time Simulator) - an Introductory Description. Systemering 70. Eds. Bubenko, J. A., Jr, O. Källhammar, B. Langefors, M. Lundeberg and A. Sölvberg, Studentlitteratur. Lund, .

Bubenko, J. A., Jr (1973). Computer Based Tools for System Documentation: Objectives and Problems. NordDATA 1973, Copenhagen, Denmark.

Bubenko, J. A., Jr (1973), Contributions to Formal Description, Analysis and Design of Data Processing Systems, Inst. för Informationsbehandling -ADB, PhD Thesis, TRITA-IBADB Nr. 1010, Sept. 1973.

Bubenko, J. A., Jr (1973), Finite Input Population Multiserver Systems with Load-dependent Service Times, Inst. för Informationsbehandling -ADB, KTH/SU, TRITA-IBADB, No. 3071, May, 1973.

Bubenko, J. A., Jr (1974). System Design in the Future. Tietojenkäsittelypäivä, OTADATA, Otaniemi, Helsinki Inst. of Technology.

Bubenko, J. A., Jr (1975). Some Theoretical and Practical Observations in a Database Design Case. Systemering 75. Eds. Lundeberg, M. and J. A. Bubenko, Jr, Studentlitteratur. Lund, .

Bubenko, J. A., Jr (1976). Information Analysis and Database Design. ONLINE, London, U.K.

Bubenko, J. A., Jr (1977), IAM - Inferential Abstract Modeling - an Approach to Design of Information Models for Large Shared Data Bases, IBM Thomas J. Watson Research Center, Research Report, RC 6343, Jan. 1977.

Bubenko, J. A., Jr (1977). IAM: an Inferential Abstract Modeling Approach to Design of Conceptual Schema. SIGMOD International Conference on Management of Data, Toronto, Canada, ACM.

Bubenko, J. A., Jr (1977). Information Modeling and Reliability. INFOTECH State of the Art Conference on "Software Reliability", London, INFOTECH Ltd.

Bubenko, J. A., jr (1977). The Temporal Dimension in Information Modelling. IFIP WG 2.6 Working Conference on Architecture and Models in Data Base Management Systems, Ed. Nijssen, G. M., Nice, France, North Holland.

Bubenko, J. A., Jr (1977). Validity and Verification Aspects of Information Modeling. 3rd International Conference on Very Large Data Bases, Tokyo, Morgan - Kaufmann.

Bubenko, J. ,A., Jr (1978). “Some Fundamentals and Problems of Information Analysis” DATA, Copenhagen, Denmark 1978(3).

Bubenko, J. A., Jr (1979). Data Models and Their Semantics. INFOTECH State of the Art Conference on Data Design, London, INFOTECH Ltd.

Bubenko, J. A., Jr (1979). On the Role of Understanding Models in Conceptual Schema Design. 5th International Conference on Very Large Data Bases (VLDB), Rio de Janeiro, Brazil, Morgan Kaufmann.

Bubenko, J. A., Jr. (1980). Information Modeling in the Context of Systems Development. Information Processing 80, Ed. Lavington, S. H., Tokyo, Japan and Melbourne, Australia, North Holland Publishing Company.

Bubenko, J. A., Jr, Ed. (1983). Information Modelling, Studentlitteratur and Chartwell-Bratt, Lund.

Bubenko, J. A., Jr. (1985), Knowledge for Schema Restructuring and Integration Tools, SYSLAB, DSV, Internal Working Note, SYSLAB IWN No.1, Dec. 1985.

Bubenko, J. A., Jr. (1986). Information System Methodologies - a Research View. Information Systems Desigm Methodologies: Improving the Practice. Eds. Olle, T. W., H. G. Sol and A. A. Verrijn-Stuart, North Holland. Amsterdam, : 289-318.

Bubenko, J. A., Jr (1987). Information Analysis and Conceptual Modeling. Databases. Ed. Paredaens, J., Academic Press. London, : 141-190.

Bubenko, J. A., Jr. (1988), Problems and Unclear Issues with Hierarchical Business Activity and Data Flow Modelling, The Royal Institute of Technology and the University of Stockholm, SYSLAB Working Paper no. 134.

Bubenko, J. A., Jr. (1988), Selecting a Strategy for Computer-Aided Software Engineering (CASE), Swedish Institute for Systems Development (SISU), Box 1250, S-164 28 Kista, Sweden, and SYSLAB, University of Stockholm and the Royal Institute of Technology, S-106 09, Stockholm, Sweden, Report, SYSLAB Report No. 59, June, 1988.

Bubenko, J. A., Jr. (1989), Ramatic - a CASE Shell, SISU, Box 1250, S-164 28 Kista, E2469/SISU/T6.1/1, May 1989.

Bubenko, J. A., Jr (1991). Next Generation Information Systems: an Organizational Perspective. International workshop on "Intelligent Information Systems: Development Tools, Environments, and Methodologies", Eds. Balzer, B. and J. Mylopoulos, Niagara-on-the-lake, Ontario, Canada.

Bubenko, J. A., Jr (1991). Towards a Corporate Knowledge Repository. Informatiesystemen in Beweging. Eds. Bots, P. W. G., H. G. Sol and I. G. Sprinkhuizen-Kuyper, Kluwer Bedrijfswetenshappen. Deventer, The Netherlands, .

Bubenko, J. A., jr (1992). On the Evolution of Information Systems Modelling - a Scandinavian Perspective (Invited paper). 25th Anniversary Conference, Jyväskylä University, Oct. 9, 1992. Ed. Lyytinen, K. Jyväskylä, Finland, .

Bubenko, J. A., Jr (1993), Basic Concepts of the F3 EM Configuration Model, Swedish Institute for Systems Development (SISU), P.O. Box 1250, S-16426, Kista, Sweden, f3@sisu.se, Fax +46-8-7526800., Tecnical Note, ESPRIT Project P6612, F3, P6612.SISU.TN.007.draft, March, 1993.

Bubenko, J. A., Jr (1993). Extending the Scope of Information Modelling. Fourth International Workshop on the Deductive Approach to Information Systems and Databases, Ed. Olivé, A., Costa Brava, Catelonia, Departament de Llenguatges i Sistemes Informatics of the Universitat Politecnica de Catalunya, Barcelona.

Bubenko, J. A., Jr (1994). “Enterprise Modelling” Ingénierie de Systemes d'Information 2(6): 657-678.

Bubenko, J. A., Jr (1995). Challenges in Requrements Engineering (invited talk). RE´95 - Second IEEE International Symposium on Requirements Engineering, Ed. , 27-29 March, York, U.K., IEEE Computer Society Press.

Bubenko, J. A., Jr. (1996). Business, Requirements, and IS Engineering - Relationships and Research Issues. Second International Baltic Workshop on Databases and Information Systems, Ed. Haav, H.-M. and B. Thalheim, Tallinn, Estonia, Institute of Cybernetics.

Bubenko, J. A., Jr. (1996). Enterprise Knowledge Management and Organisational Learning (position paper). Workshop I of CAiSE'96: Requirements Engineering in a Changing World, Crete, Greece.

Bubenko, J. A., Jr (1996), Some Observations of Technology Transfer in "Method Development" Projects, Swedish Institute for Systems Development - SISU, ELECTRUM 212, S-16440, Kista, Sweden, Paper presented at European Commission`s, Directorate General JRC Workshop "Requirements Engineering Projects and Industrial Uptake" 12-13 September, 1996, Brussels, Belgium, 1996-09-12.

Bubenko, J. A., Jr (1998). Challenges in Information System Engineering. 3rd Baltic Workshop on Data Bases and Information Systems, Ed. Barzdins, J., Riga, Latvia, Institute for Mathematics and Computer Science, University of Latvia.

Bubenko, J. A., Jr (1998). Participative Development: The Need For New Roles and Specialties. 8th Annual Workshop on Information Technologies and Systems (WITS), Ed. March, S. T. and J. A. Bubenko, Jr, Helsinki, Finland.

Bubenko, J. A., Jr and S. Berild (1974). CADIS System 4: a Tool for Incremental Description of and Analysis of Systems. International Symposium "Ansätze zur Organisationstheorie Rechnergestutzter Informationssysteme", Ed. , St Augustin, Germany, GMD and BIFOA, Köln.

Bubenko, J. A., Jr, S. Berild, E. Lindencrona-Ohlin and S. Nachmens (1975), Information Analysis and Design of Data Base Schemata, Inst. för Informationsbehandling, ADB, KTH/SU, TRITA - IBADB, No. 3091, June 1975.

Bubenko, J. A., Jr, S. Berild, E. Lindencrona-Ohlin and S. Nachmens (1976). From Information Requirements to DBTG Data Structures. ACM Sigplan/SIGMOD Conference on Data Abstraction, March 22-24, Salt Lake City, Utah, ACM.

Bubenko, J. A., jr, M. Boman, P. Johannesson and B. Wangler (1992), Information Systems Engineering Research at SYSLAB 1985-92, SYSLAB, Department of Systems and Computer Science, Royal Inst. of Technology and Stockholm University, ELECTRUM 230, S-164 40, Kista, Sweden, Report No. 92-022-DSV, October, 1992.

Bubenko, J. A., jr, D. Brash and J. Stirna (1998), EKD User Guide, Dept. of Computer and Systems Science, KTH and Stockholm University, Sweden, ESPRIT Project No. 22927, ELEKTRA, Document, Feb. 5, 1998.

Bubenko, J. A., Jr., A. Caplinskas, J. Grundspenkis, H.-M. Haav and A. Solvberg, Eds. (1994). Proceedings of the Baltic Workshop on National Infrastructure Databases: Problems, methods and Experiences. Vilnius, Lithuania, Institute of Mathematics and Informatics, Akademijos 4, 2600 Vilnius, Lithuania. Technical editor: S. Maskeliunas.

Bubenko, J. A., Jr and O. Dopping (1971). Databehandlingens XYZ, Studentlitteratur, Lund.

Bubenko, J. A., Jr, M. R. Gustafsson, C. Nellborn and W. Song (1992), Computer Support for Enterprise Modelling and Requirements Acquisition, SISU, Box 1250, S-16428, Kista, Sweden, Deliverable P6612/SISU/3-1-3-R1.B, P6612.SISU.RP.002.1, Oct. 1992.

Bubenko, J. A., jr, M. R. Gustafsson, C. Nellborn and W. W. Song (1993), An Initial Draft of the Enterprise Model, Swedish Institute for Systems Development (SISU), P.O. Box 1250, S-16426, Kista, Sweden, f3@sisu.se, Fax +46-8-7526800., Draft of deliverable 3-1-3-R2, ESPRIT project P6612, F3., P6612.SISU.RP.004.draft, March, 1993.

Bubenko, J. A., Jr and Å. Jacobsson (1961), Stödkrafter och Moment i Balk på Flera Stöd, Speciellt Tillämpbart på Propelleraxelledningar, Stiftelsen för Skeppsbyggnadsteknisk Forskning, Report, Program Nr 7.

Bubenko, J. A., jr and M. Kirikova (1994). "Worlds" in Requirements Acquisition and Modelling. 4th European - Japanese Seminar on Information Modelling and

Knowledge Bases, Ed. Kangassalo, H. and B. Wangler, Kista, Sweden, IOS, The Netherlands.

Bubenko, J. A., jr and M. Kirikova (1997). Improving the Quality of Requirements Specifications by Enterprise Modelling. Perspectives on Business Modelling. Ed. Nilsson, A. G., C. Tolis and C. Nellborn, (submitted for publication).

Bubenko, J. A., Jr and O. Källhammar (1971). CADIS: Computer-Aided Design of Information Systems. Computer-Aided Information Systems Analysis and Design. Ed. Bubenko, J. A., Jr, B. Langefors and A. Sölvberg, Studentlitteratur, Lund, Sweden.

Bubenko, J. A., Jr, O. Källhammar and S. Berild (1970), PP/I - Ett Hjälpmedel för Manipulering av Associativa Datastrukturer, Inst. för Informationsbehandling - ADB, KTH/SU, Stockholm, CADIS Report, No. 34, Sept. 1970.

Bubenko, J. A., jr, O. Källhammar, B. Langefors, M. Lundeberg and A. Sölvberg, Eds. (1970). Systemering -70. Lund, Studentlitteratur.

Bubenko, J. A., jr, B. Langefors and A. Solvberg, Eds. (1971). Computer-Aided Information System Analysis and Design. Lund, Studentlitteratur.

Bubenko, J. A., Jr. and E. Lindencrona (1984). Konceptuell modellering - Informationsanalys, Studentlitteratur, Lund.

Bubenko, J. A., jr and T. Ohlin (1971). Introduktion till Operativsystem, Studentlitteratur, Lund.

Bubenko, J. A., jr and A. Olive (1987), Dynamic or Temporal Modelling - an Illustrative Example, SYSLAB, Department of Computer and Systems Science, Royal Inst. of Technology, Stockholm, Sweden, SYSLAB Working paper 117.

Bubenko, J. A., Jr., M. P. Papazoglou and M. Norrie (1988). INTENT: An Integrated Environment for Distributed Heterogenous Databases. Workshop on Object Based Concurrent Programming, San Diego, CA.

Bubenko, J. A., jr, C. Rolland, P. Loucopoulos and V. DeAntonellis (1994). Facilitating "Fuzzy to Formal" Requirements Modelling. IEEE International Conference on Requirements Engineering, Ed. , Colorado Springs, Colorado, USA and Taipei, Taiwan, ROC, IEEE.

Bubenko, J. A., Jr and A. Sölvberg, Eds. (1975). Data Base Schema Design and Evaluation. Röros, Norway, Nordic Workshop sponsored by NORDFORSK.

Bubenko, J. A., Jr and B. Wangler (1992). Research Directions in Conceptual Specification Development. Conceptual Modeling, Databases and CASE: An Integrated View of Information Systems Development. Eds. Loucopoulos, P. and R. Zicari, John Wiley & Sons, Ltd. London.

Bubenko, J. A., Jr and B. Wangler (1993). Objectives Driven Capture of Business Rules and of Information System Requirements. IEEE Systems Man and Cybernetics '93 Conference, Le Touquet, France.

Gustafsson, M. R., T. Karlsson and J. A. Bubenko, Jr. (1982). A Declarative Approach to Conceptual Information Modeling. Information Systems Design Methodologies: A Comparative Review, Eds. Olle, T. W., H. G. Sol and A. A. Verrijn-Stewart, Noordwijkerhout, The Netherlands, North-Holland.

Gustas, R., J. A. Bubenko, Jr and B. Wangler (1995). Goal Driven Enterprise Modelling: Bridging Pragmatic and Semantic Descriptions of Information Systems. 5th European - Japanese Seminar on Information Modelling and Knowledge Bases, May 30-June 3, Sapporo, Japan.

Gustas, R., J. A. Bubenko, Jr and B. Wangler (1995). Goal Driven Enterprise Modelling: The Basis for Description Semantic and Pragmatic Dependencies of Information Systems. IFIP 8.1 Working Conference on Information System Concepts: Towards a consolidation of Views, Marburg, Germany.

Jarke, M., J. Bubenko Jr, C. Rolland, A. Sutcliffe and Y. Vassiliou (1993). Theories Underlying Requirements Engineering: An Overview of NATURE at Genesis. IEEE International Symposium on Requirements Engineering, San Diego, CA.

Jarke, M., J. A. Bubenko, Jr and K. Jeffery, Eds. (1994). Advances in Database Technology - EDBT'94, 4th International Conference on Extending Database Technology, Cambridge, U.K., March 1994. Lecture Notes on Computer Science (779). Berlin, Springer Verlag.

Jarke, M., J. A. Bubenko, jr, C. Rolland, A. Sutcliffe and Y. Vassiliou (1993). Theories Underlying Requirements Engineering: An Overview of NATURE at Genesis. IEEE Symposium on Requirements Engineering, RE'93, San Diego, CA, Jan. 4-6, 1993.

Jarke, M., K. Pohl, S. Jacobs, J. A. Bubenko, jr, P. Assenova, P. Holm, B. Wangler, C. Rolland, V. Plihon, J. Schmitt, A. G. Sutcliffe, S. Jones, N. A. M. Maiden, D. Till, Y. Vassiliou, et al. (1993). Requirements Engineering: An Integrated View of Representation. 4th European Software Engineering Conference, Eds. Sommerville, I. and P. Manfred, Garmisch-Partenkirchen, Springer-Verlag.

Johannesson, P., M. Boman, J. Bubenko and B. Wangler (1997). Conceptual Modelling, Prentice Hall.

Kirikova, M. and J. A. Bubenko, jr (1994). Enterprise Modelling: Improving the Quality of Requirements Specification. Information systems Research seminar In Scandinavia, IRIS-17, Ed. , Oulu, Finland.

Kirikova, M. and J. A. Bubenko, jr (1994). Software Requirements Acquisition through Enterprise Modelling. Software Engineering and Knowledge Engineering - SEKE'94, Jurmala, Latvia.

Lindencrona-Ohlin, E. and J. A. Bubenko, jr. (1983), Towards a Formal Syntax for a Data Modeling Language - DMOL, Department of Computer and Systems Sciences, Stockholm University and the Royal Institute of Technology, Working Paper, SYSLAB WP No. 63 version 2, December, 1983.

Lundeberg, M. and J. Bubenko, Eds. (1975). Systemeering 1975. Lund, Studentlitteratur.

Nellborn, C., M. R. Gustafsson and J. A. Bubenko, jr (1992), Enterprise Modelling - an Approach to Capture Requirements, SISU, Box 1250, S-16428 Kista, Sweden, Deliverable P6612/SISU/3-1-3-R1A, P6612.SISU.RP.001.1, Oct. 1992.

Rolland, C., G. Grosz, J. A. Bubenko, jr, A. Persson, J. Stirna, P. Loucopoulos and N. Prekas (1999), Newton: Validated ESI Knowledge Base, University of Paris - 1, Sorbonne, ESPRIT Project No. 22927, ELEKTRA, Deliverable Document., Oct. 1999.

Sernadas, A., J. Bubenko and A. Olive, Eds. (1985). Information Systems: Theoretical and Formal Aspects - TFAIS. Amsterdam, North Holland.

Song, W., P. Johannesson and J. Bubenko (1996). "Semantic Similarity Relations and Computations in Schema Integration" Journal of Data and Knowledge Engineering 19(1): 65-67.

Song, W. W., M. R. Gustafsson, C. Nellborn and J. A. Bubenko, Jr (1993), Requirements on an Enterprise Modelling Tool, Swedish Institute for Systems Development (SISU), P.O. Box 1250, S-16426, Kista, Sweden, f3@sisu.se, Fax +46-8-7526800., Technical Note, ESPRIT project P6612, F3, P6612.SISU.TN.006.draft, Spring, 1993.

Song, W. W., P. Johannesson and J. A. Bubenko, Jr (1992). Semantic Similarity Relations in Schema Integration. 11th International Conference on the Entity-Relationship Approach, Ed. Tjoa, A. M., Karlsruhe, Germany.

Theodoulidis, C., B. Wangler, J. A. Bubenko and P. Loucopoulos (1990), A Conceptual Model for Temporal Database Applications, Dept. of Computer and Systems Sciences, Stockholm University, report, SYSLAB Report No. 71, March 1990.

van_Tongeren, H. and J. A. Bubenko, Jr (1966). Administrativ Rationalisering - ADB Systemarbete, Studentlitteratur, Lund,

The Authors

Carlo Batini
Member of the Executive Board, Italian Authority for IT in the Public Administration, Rome, Italy

Carlo Batini obtained the graduation in Engineering at University of Roma in 1972, and the post graduation in Computer Science in 1973. Since 1986 he is full professor at University of Roma “La Sapienza”. Since 1993 he is on leave from University, being a member of the executive board of the Italian Authority for IT in the Public Administration. His research activity has concerned methodologies for conceptual data base design, automatic layout of diagrams and visual interfaces. His teaching activity is in the field of Information systems.

Address: Carlo Batini, Autorità per l’Informatica nella Pubblica Amministrazione (AIPA), Roma, Italy, batini@aipa.it

Alfs Berztiss
Professor Emeritus, Department of Computer Science, University of Pittsburgh, Pittsburgh, USA, and SYSLAB, University of Stockholm, Sweden

Until his retirement in 1999 Alfs Berztiss belonged to the faculty of the Department of Computer Science of the University of Pittsburgh. Since 1984 he has spent several months each year at the University of Stockholm. He is the author of three books and numerous papers. He belongs to ACM, IEEE, and IFIP Working Groups 8.1, 8.3, and 8.5. His current research interests are domain modeling, transactional computing, and management of uncertain information.

Address: Alfs Berztiss, Department of Computer Science, University of Pittsburgh, Pittsburgh, PA 15260, USA, alpha@cs.pitt.edu

Sjaak Brinkkemper
Chief Architect, Baan Company, Barneveld, the Netherlands

Dr. Sjaak Brinkkemper is a Chief Architect in the Research and Development department of Baan Company (www.baan.com). He is responsible for the overall software process improvement initiatives of the areas of Requirements Management and Design. Before joining Baan he held academic positions at the University of Twente in the Netherlands, the University of Texas at Austin (USA) and the University of Nijmegen, also in the Netherlands. He is a member of IFIP Working Group 8.1 on the "Design and Evaluation of Information Systems", of the ACM, of the Computer Society of the IEEE, and of the Netherlands Society for Informatics. He has published more than 100 papers and four books on his research interests: information systems methodology, meta-modelling, method engineering, and software product development.

Address: Sjaak Brinkkemper, Baan Company, P.O. Box 143, 3770 AC Barneveld, the Netherlands, SBrinkkemper@Baan.nl

Michael L. Brodie

Chief Scientist, GTE Laboratories Incorporated, Waltham, USA

Dr. Michael L. Brodie is Sr. Technologist, GTE Laboratories, Waltham, MA and Chief Scientist (SAP Program), GTE Corporation. He works on large-scale strategic Information Technology (IT) challenges for GTE Corporation's senior executives. His industrial and research focus is on large-scale information systems - their total life cycle, business and technical contexts, core technologies, and "integration" within in a large scale, operational telecommunications environment. For three years, his primary focus was GTE's SAP implementation and the related 200 Enterprise Planning and Management (EP&M) systems. He co-led the EP&M Information Technology Team for the upcoming GTE-Bell Atlantic merger which was responsible for over 300 EP&M systems required to support the merged entity. He is currently the chairman of the Global SAP Telecommunications Special Interest Group (TSIG); an international organization of 200 Telecom organizations focused on the evolution of Enterprise Resource Planning for Telecommunications. His current focus is on the interaction of economics, business, and technology in transforming our world for the better.

Address: Michael L. Brodie, GTE Laboratories Incorporated, 40 Sylvan Road, Waltham, Massachusetts, MA 02451-1128 USA, brodie@gte.com

Alfonso F. Cardenas

Professor, Computer Science Department, University of California, Los Angeles

Alfonso Cárdenas is a Professor of the Computer Science Department, School of Engineering and Applied Science, UCLA, and a consultant in computer science and management. He obtained the B.S. degree from San Diego State University and the M.S. and Ph.D. degrees in Computer Science, at the University of California, Los Angeles, in 1969. His major areas of interest include database management, distributed multimedia (text, image/picture, voice) systems, information systems planning and development methodologies, software engineering, and legal and intellectual property issues. He has been a consultant to many user and vendor organizations in several countries, and member of various review boards, conference organization and program committees, and professional societies. He is past-president (1984-1989) and member of the Board of Trustees of the Very Large Data Base Endowment. He is the author of many journal and conference publications, and co-editor with Dennis McLeod of the book Research Foundations in Object-Oriented and Semantic Database Systems (Prentice Hall, 1990), author of Data Base Management Systems (W. C. Brown Publishers, 1979 and 1984), co-editor and co-author of Computer Science (John Wiley & Sons, 1972).

Address: Alfonso F. Cardenas, Department of Computer Sciences, University of California at Los Angeles, Los Angeles, CA., U.S.A, cardenas@cs.ucla.edu

Jaelson Castro
Associate Professor, Centro de Informatica, Universidade Federal de Pernambuco, Brazil

Jaelson Castro received his Ph.D. degree from Imperial College, University of London in 1990. He previously received the B.Sc. and M.Sc. degrees from the Universidade Federal de Pernambuco, Brazil, where he is currently Associate Professor. His research interests include software engineering, requirements engineering and object oriented development.

Address: Jaelson Castro, Universidade Federal de Pernambuco, Centro de Informática, Recife, Brazil, jbc@di.ufpe.br

Angelo E. M. Ciarlini
Departamento de Informática, Pontifícia Universidade Católica do Rio de Janeiro, Brazil.

Address: Angelo E. M. Ciarlini, Departamento de Informática, Pontifícia Universidade Católica do Rio de Janeiro, 22.453-900 Rio de Janeiro, Brazil, angelo.ciarlini@uol.com.br

Antonio Furtado
Professor, Department of Computer Science, Pontifical Catholic University of Rio de Janeiro, Brazil

Antonio L. Furtado (Ph.D., University of Toronto) is Professor in the Department of Computer Science of the Pontifical Catholic University of Rio de Janeiro, Brazil. His areas of research include conceptual modelling of information systems, textual databases, and the application of methods related to logic programming to databases and knowledge bases.

Address: Antonio L. Furtado, Departamento de Informática, Pontifícia Universidade Católica do Rio de Janeiro, 22.453-900 Rio de Janeiro, Brazil, furtado@inf.puc-rio.br

Matthias Jarke
Professor, Information Systems Group, RWTH Aachen, Germany and Executive Director, GMD-FIT Institute, Birlinghoven, Germany

Matthias Jarke is Professor of Information Systems and Chairman of the Informatics Department at Aachen University of Technology (RWTH Aachen), Germany. Prior to joining Aachen in 1991, he held faculty positions at New York University and the University of Passau. His research area is information systems support for cooperative work processes in engineering, business, and culture. His publications include about 150 refereed journal and conference publications and several books, most recently "Fundamentals of Data Warehouses". He has served as coordinator of three ESPRIT projects in Information Systems Engineering, and is Chief Editor of the journal, Information Systems. In January, 2000, he was appointed Executive Director of GMD-FIT, the Institute of Applied Information Technology at the GMD National Computer Science Research Labs in Birlinghoven near Bonn, and elected Vice President of GI, the German Computer Society.

Address: Matthias Jarke, RWTH Aachen, Informatik V, and GMD-FIT, Ahornstr. 55, 52074 Aachen, Germany , jarke@informatik.rwth-aachen.de

P. Jayaweera
Department of Computer and Systems Science, Stockholm University/KTH Sweden

Address: P. Jayaweera, Department of Computer and Systems Science, Stockholm University/KTH, Stockholm, Sweden, prasad@dsv.su.se

Keith Jeffery
Director, CLRC-Rutherford Appleton Laboratory, Chilton, UK.

Keith Jeffery is currently Director, IT at the CLRC-Rutherford Appleton Laboratory in UK. His department provides national services and undertakes research and development projects funded by the UK Research Councils, government departments, the European Commission and commerce and industry internationally. Keith is a Fellow of both the Geological Society of London and the British Computer Society. He is an Honorary Fellow of the Irish Computer Society and holds three honorary visiting professorships. He is a trustee (past secretary and currently vice-president) of the Endowment Board of the VLDB (Very Large Database) Conference, and is a member of the boards controlling the EDBT (Extending Database Technology) conference, CAiSE (Conference on Advanced Systems Engineering) conference and OOIS (Object-Oriented Information Systems) conference. He has numerous publications in refereed journals, books and conference proceedings.

Address: Keith G Jeffery, CLRC-Rutherford Appleton Laboratory, Chilton, Oxfordshire, OX11 0QX UK, K.G.Jeffery@rl.ac.uk

Paul Johannesson
Associate Professor, Department of Computer and Systems Sciences, Stockholm University and the Royal Institute of Technology, Sweden

Paul Johannesson received his BSc in Mathematics, and his PhD in Computer Science from Stockholm University in 1983 and 1993, respectively. He became docent in 1995. He holds a position as associate professor at Stockholm University, where he lectures in information systems and databases. Johannesson has published work on federated information systems, translation between data models, languages for conceptual modelling, schema integration, the use of linguistic instruments in information systems, and analysis patterns in systems design. He is the co-author of a text book on conceptual modelling, published by Prentice Hall in 1997. Johannesson has participated in the ESPRIT projects KIWIS and NATURE. He is currently a member of a national project on process brokers. Johannesson has been a member of several international program committees; among these are the ER conference and the CAiSE conference.

Address: Paul Johannesson, Department of Computer and Systems Science, Stockholm University/KTH, Stockholm, Sweden, pajo@dsv.su.se

Hannu Kangassalo

Professor, Department of Computer and Information Sciences, University of Tampere, Finland

Hannu Kangassalo studied at the University of Tampere, in Finland. Started teaching there 1970. Acting professor 1984-90. Visiting professor at the University of Tokio 1990. Professor 1998. Main research interests are on information systems design methodologies, especially conceptual modelling, different conceptual modelling approaches, conceptual schema languages, and ontologies. Developed COMIC System for conceptual modelling for information systems and data bases. Chairman of IFIP TC 8 WG 8.1 as of 1998.

Address: Hannu Kangassalo, University of Tampere, Department of Computer and Information Sciences, P.O.Box 607, FIN-33014 University of Tampere, Finland, hk@cs.uta.fi

Eva Lindencrona

Director, Swedish Institute of Technology, Stockholm, Sweden

Address: Eva Lindencrona, Swedish Institute for Systems Development, Electrum 230, SE-164 40 Kista, Sweden, eva@sisu.se

Paul Lindgreen

Professor Emeritus, Copenhagen Business School, Danmark.

Paul Lindgreen is retired as associate professor in systems analysis at Copenhagen Business School, and more than 35 years activities in the area of systems utilizing electronic data technology. He held an early affiliation with the Danish company "Regnecentralen" for 17 years mainly occupied with the development of compilers and other basic system software and later on with administrative data processing systems, the interest has gradually turned into the more theoretical areas of formal analysis of organisations, systems, communication, semiotics, language and information. Are presently he is working on a doctoral thesis on feasible concepts and terminology in the area of informatics.

Address: Paul Lindgreen, Pildamsvej 20, DK-2610 Rdo. Copenhagen, Denmark, P_lindgreen@hotmail.com

Peter C. Lockemann
Professor of Informatics, Fakultät für Informatik, Universität Karlsruhe, Germany

Peter Lockemann has been Professor of Informatics at the University of Karlsruhe since 1972 and a director of the Computer Science Research Center at Karlsruhe since 1985. His current research interests are in the area of distributed and collaborative data-intensive computer applications. Applications are in areas such as engineering databases, traffic databases, electronic commerce, digital libraries, environmental information systems. Technological emphasis is on data mining, constraint evaluation and enforcement, active databases, interoperable databases, and the integration of telecommunications and database technologies. He is the author of four textbooks and well over 100 research papers in journals, conference proceedings and books, and an editor of six books. An additional 350 papers have been published by his young researchers and Ph.D. students. He received numerous research grants from public institutions and private industry, and devotes considerable effort to issues of technology transfer to high-technology companies.

Address: Peter C. Lockemann, Fakultät für Informatik, Universität Karlsruhe, Postfach 6980, 76128 Karlsruhe. lockeman@ira.uka.de

Pericles Loucopoulos
Professor, Department of Computation, UMIST, Manchester, UK

Peri Loucopoulos is a professor of Information Systems in the Department of Computation at UMIST where he has worked since 1984 following a period of six years in industry. His research interests are in the areas of: requirements engineering, conceptual modelling and system development methods. He is the editor-in chief of the Journal of Requirements Engineering and serves on the editorial board of numerous journals. He is the author of 4 books and of over 100 journal, invited and refereed conference papers.

Address: Pericles Loucopoulos, Department of Computation, UMIST, PO Box 88, Manchester, M60 1QD, UK, pl@co.umist.ac.uk

Gergely Lukács
Doctoral Student, Fakultät für Informatik, Universität Karlsruhe, Germany.

Gergely Lukács is a doctoral student at the Budapest University of Technology and Economics, currently doing research at the University of Karlsruhe. His interests include management of imperfect data, decision support and information integration in distributed systems.

Address: Gergely Lukács, Fakultät für Informatik, Universität Karlsruhe, Postfach 6980, 76128 Karlsruhe.

Kalle Lyytinen
Professor, Department of Computer Science and Information Systems, University of Jyväskylä, Finland

Kalle Lyytinen is a professor in Information Systems at the University of Jyväskylä, Finland and currently the Dean of the Faculty of Information Technology. He serves on the editorial boards of several leading IS journals including Information Systems Research, Accounting Management and Information Technologies, Requirements Engineering Journal, and Information Systems Journal. He is currently also the Senior Editor of MISQ. He has published over 70 articles and edited or written six books. He has been the principal leader of the MetaPHOR group which as implemented two metaCASE environments (MetaEdit, MetaEdit+) and published over 40 artiocles in areas like CASE, metaCASE, method engineering, and repository management. His research interests include information system theories, system design, CASE and metaCASE system failures and risk assessment, computer supported cooperative work, and diffusion of complex technologies.

Address: Kalle Lyytinen, Department of Computer Science and Information Systems, University of Jyväskylä, PL 35, FIN-40351 Jyväskylä, Finland, kalle@jytko.jyu.fi

Massimo Mecella
Università "La Sapienza" Dipartimento di Informatica e Sistemistica, Roma, Italy

Address: Massimo Mecella, Università "La Sapienza" Dipartimento di Informatica e Sistemistica, Roma, Italy, mecella@iol.it

John Mylopoulos
Professor, Department of Computer Science, University of Toronto, Canada

John Mylopoulos received his Ph.D. degree from Princeton University in 1970 and is currently professor of Computer Science at the University of Toronto. His research interests include conceptual modelling, requirements engineering and information systems. Mylopoulos was the recipient of the first ever Outstanding Services Award given out by the Canadian AI Society (CSCSI), a co-recipient of the most influential paper award of the 1994 International Conference on Software Engineering, a fellow of the American Association for AI (AAAI), and is currently serving as elected president of the VLDB Endowment Board. He has also served on the editorial boards of several international journals, including the ACM Transactions on Software Engineering and Methodology (TOSEM), the ACM Transactions on Information Systems (TOIS), the VLDB Journal and Computational Intelligence.

Address: John Mylopoulos, Department of Computer Science, University of Toronto, Toronto, Canada, jm@cs.toronto.edu

Björn E. Nilsson

Senior Consultant, Adera+, Stockholm, Sweden

Nilsson has an MA from Stockholm University (1974) and holds a Ph.D. in administrative information processing (1979). He has worked with methods development for Statistics Sweden (1968-83) and with IT-strategies and architectures for the Swedish Defence Rationalisation Institute (1983-87). During a couple of years he has acted as vice president of the research institute SISU, Swedish Institute for Systems Development (1987-95). Since then, he is a senior consultant within the Adera group.

Address: Bjorn E. Nilsson, Adera+, 11, rue Plaetis, L-2338 Luxembourg, bjorn.nilsson@aderagroup.com

Tomas Ohlin

Professor, Institution for Economic Information Systems, Linkoping University, Sweden

Tomas Ohlin was an early developer of information services, and was doing experiments with Sweden´s first public e-mail system in the middle 70s, long before Videotex and Internet. He published a text on electronic democracy in the beginning of the 70s. He then spent a number of years with research administration in public research boards. Later, he worked in public committees on information society, defining infrastructures of different types. In the beginning of the 90s, he became Secretary General at Sweden´s first IT Commission. He has published three books and numerous papers. Back in the university environment, his current interests include formats and conditions for increased user participation in social information systems.

Address: Tomas Ohlin, Institution for Economic Information Systems, Linköping University, Sweden, tomas.ohlin@telo.se

Antoni Olivé

Professor, Universitat Politecnica Catalunya, Barcelona, Spain

Antoni Olive is currently professor of information systems. He has worked in this field during over 20 years, mainly in the university environment. His main interests have been, and are, conceptual modeling, requirements engineering, information systems design and databases. He has taught extensively on these topics. He has also conducted research on these topics, which has been published in international journals and conferences. He is a member of IFIP WG8.1, (Design and evaluation of information systems) where he served as chairman during 1989-1994.

Address: Antoni Olivé, Departament de Llenguatges i Sistemes Informàtics, Universitat Politècnica de Catalunya, Barcelona (Catalonia), olive@lsi.upc.es

S.J. Paheerathan
Department of Computer and Systems Sciences, Stockholm University, Sweden

Address: S.J Paheerathan, Department of Computer and Systems Sciences, SU/KTH, Electrum 230, SE-164 40 Kista, Sweden

Barbara Pernici
Professor, Politecnico di Milano, Italy

Barbara Pernici is full professor of Computer Engineering at Politecnico di Milano. She has a doctor in engineering degree from Politecnico di Milano and a MS in Computer Science from Stanford University. Her research interests include office and information systems conceptual modeling, computer based design support tools, reuse of conceptual components, temporal deductive databases, applications of database technology.

Address: Barbara Pernici, Politecnico di Milano, Milano, Italy,
pernici@elet.polimi.it

Colette Rolland
Professor, Université de Paris1 - Panthéon Sorbonne, Paris, France

Colette Rolland is currently Professor of Computer Science in the department of Mathematics and Informatics at the University of PARIS-1 Panthéon Sorbonne where she has worked since 1979. Her research interests lie in the areas of information modelling, databases, temporal data modelling, object-oriented analysis and design, requirement engineering, design methodologies and CASE and CAME tools, change management and enterprise knowledge development. She has supervised 62 PhD thesis and has an extensive experience in participating in national and European projects under the ESPRIT program and conducting co-operative projects with the industry. She is the representative in IFIP TC8 on Information Systems and has been chairperson of the IFIP WG8.1.

Address: Colette Rolland, Université de Paris 1 Panthéon Sorbonne, PMF, CRI, 90 rue de Tolbiac, 75013 Paris, France, rolland@univ-paris1.fr

Arne Sølvberg
Professor, Department of Computer and Information Science, The Norwegian University of Science and Technology, Trondheim, Norway

Arne Sølvberg is Professor of Computer Science at The Norwegian University of Science and Technology, Trondheim, Norway, since 1974. He received a siv.ing. (M.Sc.) degree in Applied Physics in 1963, and a dr.ing. (Ph.D.) degree in Computer Science in 1971, both from The Norwegian Institute of Technology (now incorporated in The Norwegian University of Science and Technology) . His main fields of competence are information systems design methodology, database design, information modelling, CASE tools and software engineering environments. He has been active in international organizations for research cooperation. He was Norwegian national representative to IFIP General Assembly in 1979-82. He has been chairman of IFIP WG8.1 for Information Systems Design in 1982-88. He was a trustee in the VLDB Endowment until 1994. He was a co-founder of the CAiSE conference series. He has been a Visiting Scientist with IBM San Jose Research Labs, with The University of Florida, The Naval Postgraduate School, and most recently with the University of California at Santa Barbara.

Address: Arne Sølvberg, Department of Computer and Information Science, Norwegian University of Science and Technology, Trondheim, Norway, Arne.Soelvberg@idi.ntnu.no

Yannis Vassiliou
Professor, Department of Electrical and Computer Engineering, National Technical University of Athens, Greece

Professor Vassiliou has a Ph.D and M.SC. in Computer Science from the University of Toronto. Since July 1993, he is a Professor at the National Technical University of Athens, Department of Electrical and Computer Engineering, Computer Science Division. From September 1984 until November 1993, he was the Director of the Institute of Computer Science of F.O.R.T.H (Foundation for Research and Technology - Hellas). He has been a Professor at the University of Crete, Computer Science Department and an Associate Professor at New York University, Graduate School of Business Administration. Since September 1987, a Tenured Member of the Scientific Council of the Greek Parliament. Scientific and Management Consultant for a number of research centres, companies, banks, and government agencies. Consultant and Expert Reviewer for the Commission of the European Union. His current research interests are is the areas of Data Warehouses, Data Base Management Systems: applications in Interactive Software Development, Transactional Systems, Conceptual and Business Modeling and Information Systems. Prof. Vassiliou has over 120 publications in international journals, conferences and books.

Address: Yannis Vassiliou, Department of Computer and Electrical Engineering, National Technical University of Athens, Athens, Greece, yv@cs.ntua.gr

Benkt Wangler
Professor, Department of Computer Science, University of Skövde, Sweden

Benkt Wangler is a professor at University of Skövde, Sweden. He holds a PhD from KTH (Royal Institute of Technology) in Stockholm. His areas of interest include information systems and enterprise modelling as well as the role of IT in organisations in general. Wangler was previously working as a systems analyst and database expert for Sperry Corporation in Sweden and became later a research leader at SISU (Swedish Institute of Technology), where he, initiated, participated in and led the Swedish participation in several European projects concerning databases and information systems and enterprise development issues.

Address: Benkt Wangler, Department of Computer Science, University of Skövde, Skövde, Sweden, benkt@ida.his.se

Anthony I. Wasserman
Principal, Software Methods and Tools, San Francisco, USA

Anthony I. (Tony) Wasserman is Principal of Software Methods and Tools (www.methods-tools.com), which provides technical and executive consulting services on a wide variety of software and web development topics. He was founder and CEO (1983-1993) of Interactive Development Environments, Inc. (IDE), developer of the innovative Software through Pictures modeling environment. From 1972 through 1986, Tony was a University of California professor, where he developed the User Software Engineering methodology and supporting environment. He was selected as a fellow of both the Association for Computing (ACM) and the IEEE Computer Society for his contributions to software engineering.

Address: Anthony I. Wasserman, Software Methods and Tools, 176 Gold Mine Drive, San Francisco, CA 94131 USA, tonyw@methods-tools.com

Zheying Zhang
Doctoral Student, Department of Computer Science and Information Systems, University of Jyväskylä, Finland

Zheying Zhang received her Master's degree from the University of Jyväskylä in 1997. Currently she is a doctoral student in the Jyväskylä Graduate School in Computing and Mathematical Sciences (COMAS). She has been working in the RAMSES project in metaPHOR group that studies the feasibility of reuse strategies in a MetaCASE environment. Her research interests include software reuse, method engineering, and method component classification and retrieval in MetaCASE.

Address: Zheying Zhang, Department of Computer Science and Information Systems, University of Jyväskylä, PL 35, FIN-40351 Jyväskylä, Finland, zhezhan@cc.jyu.fi

For all inquiries on this book you can contact Sjaak Brinkkemper at the address above.

Author Index